METHODS IN MOLECULAR BIOLOGY

Series Editor
John M. Walker
School of Life and Medical Sciences
University of Hertfordshire
Hatfield, Hertfordshire, AL10 9AB, UK

For further volumes:
http://www.springer.com/series/7651

Kidney Research

Experimental Protocols

Second Edition

Edited by

Tim D. Hewitson

Department of Nephrology, Royal Melbourne Hospital, Melbourne, VIC, Australia

Edward R. Smith

Department of Nephrology, Royal Melbourne Hospital, Melbourne, VIC, Australia

Stephen G. Holt

Department of Nephrology, Royal Melbourne Hospital, Melbourne, VIC, Australia

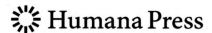

Humana Press

Editors
Tim D. Hewitson
Department of Nephrology
Royal Melbourne Hospital
Melbourne, VIC, Australia

Edward R. Smith
Department of Nephrology
Royal Melbourne Hospital
Melbourne, VIC, Australia

Stephen G. Holt
Department of Nephrology
Royal Melbourne Hospital
Melbourne, VIC, Australia

ISSN 1064-3745 ISSN 1940-6029 (electronic)
Methods in Molecular Biology
ISBN 978-1-4939-3351-8 ISBN 978-1-4939-3353-2 (eBook)
DOI 10.1007/978-1-4939-3353-2

Library of Congress Control Number: 2015957191

Springer New York Heidelberg Dordrecht London

Printed on acid-free paper

Humana Press is a brand of Springer
Springer Science+Business Media LLC New York is part of Springer Science+Business Media (www.springer.com)

Preface

Acute and chronic forms of kidney disease are major causes of morbidity and mortality worldwide. In compiling this volume, we have attempted to collate both established and novel experimental methods that can be used to study this debilitating condition.

The first edition of this book gathered together a collection of techniques that focused on solving the inherent problems that exist in renal research, most notably the complexity of renal anatomy and physiology and the associated plethora of different cell types found in the kidney. It is hoped that the new volume complements the first edition by emphasizing recent innovations that are relevant to the questions currently being asked in basic renal research. In addition the present volume takes a more holistic approach by considering associated disorders such as complications seen with peritoneal dialysis, the cardiorenal syndrome, and vascular calcification.

Part I of the second edition provides a number of in vitro, in vivo, and ex vivo models of kidney disease and associated complications. The first two chapters describe the isolation and propagation of podocytes (Chapter 1) and fibroblasts (Chapter 2), while Chapters 3 and 4 address the pathology of peritoneal dialysis complications. An important technique for studying the poorly understood sympathetic nervous system link between renal and cardiac disease is provided in Chapter 5. In a number of organs, decellularized extracellular matrix is being used to provide both a scaffold for organ regeneration studies and a more accurate ex vivo representation of the in vivo microenvironment. Jin et al. provide a protocol for the kidney in Chapter 6.

Part II looks at recent advances in imaging techniques. Estimation of glomerular number has long been a tedious and time-consuming process, poorly suited to comparative studies with even small numbers of animals. The exciting developments described in Chapter 7 provide a means of measuring glomerular endowment through magnetic resonance imaging. Likewise in Chapter 8 Ashraf et al. offer us the tantalizing opportunity to image living tissue in a time-critical manner that yields novel insights into intact cell functions.

The next two parts provide protocols to study two important clinical complications experimentally. Part III covers recent advances in studying metabolism in renal ischemia and reperfusion via telemetry (Chapter 9), magnetic resonance imaging (Chapters 10 and 11), and multiphoton microscopy (Chapter 12). Part IV addresses the study and measurement of vascular calcification, an important manifestation of uremia-induced disturbances in bone metabolism and mineral handling. This includes a state-of-the-art review on the topic (Chapter 13), in addition to protocols of an in vitro model of mineralization (Chapter 14), the isolation and quantitation of novel mediators of extraosseous mineral stress (Chapter 15), and fluorescent imaging of vascular calcification (Chapter 16).

Finally, the last three chapters (Part V) provide analytical techniques that are both topical and of widespread relevance to the study of experimental renal disease. These are improved methodologies for studying protein phosphorylation in signal transduction (Chapter 17), laser capture microdissection (Chapter 18), and in situ insights into the epigenome (Chapter 19).

Nephrology is a discipline that combines both basic and clinical sciences. In editing this book, we have attempted to provide a collection of protocols useful to those with some laboratory experience in Nnephrology and those new to the field. Protocols are written in the standard *Methods in Molecular Biology* format which includes a general introduction, an extensive list of materials required, and stepwise instructions to follow. As is characteristic of this series, a complementary notes section with key tips for troubleshooting accompanies each protocol. In many cases, the authors provide case studies as well as detailed protocols to illustrate their application.

We are indebted to all the authors for their generosity in sharing their expertise and experience for this book and for the ongoing support of the editor-in-chief, Professor John Walker. We trust that those using this new edition will find it a valuable aid in studying the mechanisms of kidney disease.

Melbourne, Australia
May 2015

T.D. Hewitson
E.R. Smith
S.G. Holt

Contents

Contributors

FELIX AIGNER • *Daniel Swarovski Research Laboratory, Department of Visceral, Transplant and Thoracic Surgery, Center of Operative Medicine, Medical University Innsbruck, Innsbruck, Austria*

KAREN ARAKELYAN • *Institute of Physiology, Charité—Universitätsmedizin Berlin, Campus Mitte, Berlin, Germany; Center for Cardiovascular Research, Berlin, Germany*

MUHAMMAD IMTIAZ ASHRAF • *Daniel Swarovski Research Laboratory, Department of Visceral, Transplant and Thoracic Surgery, Center of Operative Medicine, Medical University Innsbruck, Innsbruck, Austria*

EDWIN J. BALDELOMAR • *Department of Physics, College of Natural Sciences, University of Hawaii at Manoa, Honolulu, HI, USA*

SCOTT C. BEEMAN • *Department of Radiology, Washington University School of Medicine, Saint Louis, MO, USA*

KEVIN M. BENNETT • *Department of Biology, College of Natural Sciences, University of Hawaii at Manoa, Honolulu, HI, USA*

JOHN F. BERTRAM • *Department of Anatomy and Developmental Biology, Monash University, Melbourne, VIC, Australia*

KATHLEEN CANTOW • *Institute of Physiology, Charité—Universitätsmedizin Berlin, Campus Mitte, Berlin, Germany; Center for Cardiovascular Research, Berlin, Germany*

JENNIFER R. CHARLTON • *Division of Nephrology, Department of Pediatrics, University of Virginia Medical Center, Charlottesville, VA, USA*

MICHAEL CHRISTIE • *Department of Anatomical Pathology, The Royal Melbourne Hospital, Melbourne, VIC, Australia*

MATTHIAS DERWALL • *Department of Anesthesiology, University Hospital RWTH Aachen, Aachen, Germany*

MASAHIRO ERIGUCHI • *Department of Medicine and Clinical Science, Graduate School of Medical Sciences, Kyushu University, Fukuoka, Japan*

ROGER G. EVANS • *Department of Physiology, Monash University, Melbourne, VIC, Australia*

DIETMAR FRIES • *Department of General and Surgical Critical Care Medicine, Medical University Innsbruck, Innsbruck, Austria*

GLENDA C. GOBE • *Centre for Kidney Disease Research, School of Medicine, Translational Research Institute, The University of Queensland, Brisbane, QLD, Australia*

BRADLEY D. HANN • *College of Tropical Agriculture and Human Resources, University of Hawaii at Manoa, Honolulu, HI, USA*

PAUL HENGSTER • *Daniel Swarovski Research Laboratory, Department of Visceral, Transplant and Thoracic Surgery, Center of Operative Medicine, Medical University Innsbruck, Innsbruck, Austria*

MARTIN HERMANN • *Department of General and Surgical Critical Care Medicine, Medical University Innsbruck, Innsbruck, Austria; Department of Anesthesiology and Critical Care Medicine, Medical University Innsbruck, Innsbruck, Austria*

TIM D. HEWITSON • *Department of Nephrology, The Royal Melbourne Hospital, Melbourne, VIC, Australia; Department of Medicine, University of Melbourne, Melbourne, VIC, Australia*

Tsunehito Higashi • *Department of Cellular Pharmacology, Hokkaido University Graduate School of Medicine, Hokkaido, Japan*
Stephen G. Holt • *Department of Nephrology, The Royal Melbourne Hospital, Melbourne, VIC, Australia; Department of Medicine, University of Melbourne, Melbourne, VIC, Australia*
Takahiro Horinouchi • *Department of Cellular Pharmacology, Hokkaido University Graduate School of Medicine, Hokkaido, Japan*
Yasuhiko Ito • *Renal Replacement Therapy, Division of Nephrology, Nagoya University Graduate School of Medicine, Nagoya, Japan*
Zhang Jianse • *Anatomy Department, Wenzhou Medical University, Wenzhou, China; Institute of Bioscaffold Transplantation and Immunology, Wenzhou Medical University, Wenzhou, China*
Mei Jin • *Anatomy Department, Wenzhou Medical University, Wenzhou, China; Institute of Bioscaffold Transplantation and Immunology, Wenzhou Medical University, Wenzhou, China*
Nazanin Kabgani • *Department of Internal Medicine II, Nephrology and Clinical Immunology, RWTH Aachen University Hospital, Aachen, Germany*
Tom C. Karagiannis • *Epigenomic Medicine, Baker IDI Heart & Diabetes Institute, Melbourne, VIC, Australia; Department of Pathology, University of Melbourne, Melbourne, VIC, Australia*
Kristen J. Kelynack • *Department of Nephrology, The Royal Melbourne Hospital, Melbourne, VIC, Australia*
Maarten P. Koeners • *School of Physiology and Pharmacology, University of Bristol, Bristol, UK; Department of Physiology, Auckland Bioengineering Institute, University of Auckland, Auckland, New Zealand; Department of Nephrology, University Medical Centre Utrecht, Utrecht, The Netherlands*
Simon C. Malpas • *Department of Physiology, Auckland Bioengineering Institute, University of Auckland, Auckland, New Zealand; Millar Ltd, Auckland, New Zealand*
Selly Marzully • *Epigenomic Medicine, Baker IDI Heart & Diabetes Institute, Melbourne, VIC, Australia*
Soichi Miwa • *Department of Cellular Pharmacology, Hokkaido University Graduate School of Medicine, Hokkaido, Japan*
Masashi Mizuno • *Renal Replacement Therapy, Division of Nephrology, Nagoya University Graduate School of Medicine, Nagoya, Japan*
Marcus J. Moeller • *Department of Internal Medicine II, Nephrology and Clinical Immunology, RWTH Aachen University Hospital, Aachen, Germany*
Thoralf Niendorf • *Berlin Ultrahigh Field Facility (B.U.F.F.), Max Delbrück Center for Molecular Medicine (MDC), Berlin, Germany*
Connie P.C. Ow • *Department of Physiology, Monash University, Melbourne, VIC, Australia*
Andreas Pohlmann • *Berlin Ultrahigh Field Facility (B.U.F.F.), Max Delbrück Center for Molecular Medicine (MDC), Berlin, Germany*
David M. Russell • *Department of Physiology, Auckland Bioengineering Institute, University of Auckland, Auckland, New Zealand; Millar Ltd, Auckland, New Zealand*
Chrishan S. Samuel • *Department of Pharmacology, Monash University, Melbourne, VIC, Australia*
Washington Y. Sanchez • *Therapeutics Research Centre, School of Medicine, Translational Research Institute, The University of Queensland, Brisbane, QLD, Australia*

ERDMANN SEELIGER • *Institute of Physiology, Charité—Universitätsmedizin Berlin, Campus Mitte, Berlin, Germany; Center for Cardiovascular Research, Berlin, Germany*

DAVID M. SMALL • *Centre for Kidney Disease Research, School of Medicine, Translational Research Institute, The University of Queensland, Brisbane, QLD, Australia*

BART SMEETS • *Department of Internal Medicine II, Nephrology and Clinical Immunology, RWTH Aachen University Hospital, Aachen, Germany; Department of Pathology, Radboud University Medical Center, Nijmegen, The Netherlands*

EDWARD R. SMITH • *Department of Nephrology, The Royal Melbourne Hospital, Melbourne, VIC, Australia*

WERNER STREIF • *Department of Pediatrics I, Medical University Innsbruck, Innsbruck, Austria*

SVEN-JEAN TAN • *Department of Nephrology, The Royal Melbourne Hospital, Melbourne, VIC, Australia; Department of Medicine, University of Melbourne, Melbourne, VIC, Australia*

KOJI TERADA • *Department of Cellular Pharmacology, Hokkaido University Graduate School of Medicine, Hokkaido, Japan*

JAKOB TROPPMAIR • *Daniel Swarovski Research Laboratory, Department of Visceral, Transplant and Thoracic Surgery, Center of Operative Medicine, Medical University Innsbruck, Innsbruck, Austria*

KAZUHIKO TSURUYA • *Department of Medicine and Clinical Science, Graduate School of Medical Sciences, Kyushu University, Fukuoka, Japan; Department of Integrated Therapy for Chronic Kidney Disease, Graduate School of Medical Sciences, Kyushu University, Fukuoka, Japan*

KATHERINE VERVERIS • *Epigenomic Medicine, Baker IDI Heart & Diabetes Institute, Melbourne, VIC, Australia; Department of Pathology, University of Melbourne, Melbourne, VIC, Australia*

TERESA WU • *School of Computing, Informatics, and Decision Systems Engineering, Arizona State University, Tempe, AZ, USA*

YU YALING • *Anatomy Department, Wenzhou Medical University, Wenzhou, China; Institute of Bioscaffold Transplantation and Immunology, Wenzhou Medical University, Wenzhou, China*

XUEQING YU • *Institute of Nephrology, The First Affiliated Hospital Sun Yat-Sen University, Guangzhou, China*

MIN ZHANG • *School of Computing, Informatics, and Decision Systems Engineering, Arizona State University, Tempe, AZ, USA*

WANG ZHIBIN • *Anatomy Department, Wenzhou Medical University, Wenzhou, China; Institute of Bioscaffold Transplantation and Immunology, Wenzhou Medical University, Wenzhou, China*

QIN ZHOU • *Institute of Nephrology, The First Affiliated Hospital Sun Yat-Sen University, Guangzhou, China*

Part I

In Vitro, In Vivo and Ex Vivo Models

Chapter 1

Isolation and Primary Culture of Murine Podocytes with Proven Origin

Bart Smeets, Nazanin Kabgani, and Marcus J. Moeller

Abstract

Genetic studies on hereditary kidney diseases and in vivo experimental model studies have revealed a critical role for the podocyte in glomerular function and disease. Primary podocyte cultures as well as immortalized podocyte cell lines have been used extensively to study podocyte function. Although, primary cells often more closely resemble the in vivo cells, they may have only a finite replicative life span before they reach senescence. Therefore, the success of studies using primary cell cultures depends on standardized isolation and culture protocols that allow reproducible generation of stable primary cultures.

This chapter describes the isolation of primary podocytes with a proven origin using the novel technology of cell-specific genetic tagging. Podocytes are isolated from glomeruli from a podocyte-specific transgenic reporter mouse. The podocyte-specific reporter gene beta-galactosidase is used to identify and specifically isolate the labeled podocytes from other glomerular cells by FACS.

Key words Podocytes, Cell culture, Glomerulus, Primary cell culture, Cell-specific tagging, Pod-rtTA/LC1/R26R mice

1 Introduction

Within the glomerulus, the filtration unit of the kidney, podocytes form the visceral epithelial layer on the glomerular capillaries. Podocytes are essential for the structure and function of the glomerular filtration barrier. In the last two decades, it has become apparent that podocyte injury is an early and critical event in many forms of glomerular disease [1]. Cultured podocytes are an essential and indispensable tool to study biological processes in podocytes in a defined environment, without confounding influences of renal hemodynamics and paracrine signaling by different neighboring cell types [2]. The earliest reports of primary podocyte cultures date from the late 1970s [3, 4]. Strikingly, the methods used and the challenges encountered then are still relevant today.

The main challenge is to identify and separate podocytes from other kidney cell types in culture. Generally, podocytes are isolated

Tim D. Hewitson et al. (eds.), *Kidney Research: Experimental Protocols*, Methods in Molecular Biology, vol. 1397, DOI 10.1007/978-1-4939-3353-2_1,

from cellular outgrowths from isolated glomeruli. The use of isolated glomeruli is necessary to enrich the fraction of podocytes, as podocytes contribute little of the total kidney cells. The glomerulus still contains four different cell types, i.e., the glomerular endothelial cells, mesangial cells, podocytes, and parietal epithelial cells, all of which may be present in the cellular glomerular outgrowth. Since podocytes have a lower cell turnover, they are easily overgrown by the other glomerular cell types. Identification of podocytes is mainly done on the basis of cell morphology of cell-specific marker expression. However, it is well known that virtually all cells undergo significant phenotypic changes as soon as they are removed from their physiological environment and placed into a culture flask. Even in the relative short culture period to grow glomerular outgrowths, rapid phenotypic changes of the outgrowing podocytes can be observed [5]. These phenotypic changes make clear-cut identification of podocytes in culture problematic.

To address these problems, at least in part, irreversible genetic tagging of podocytes in vivo is used to unequivocally determine the origin of cellular outgrowths from isolated glomeruli in primary culture (*see* Fig. 1 for details of the transgenic mouse model). This cell-specific genetic label is used to isolate the primary podocytes from glomerular outgrowths using fluorescence-activated cell sorting (FACS) (Fig. 2) [5]. This approach also avoids several cloning

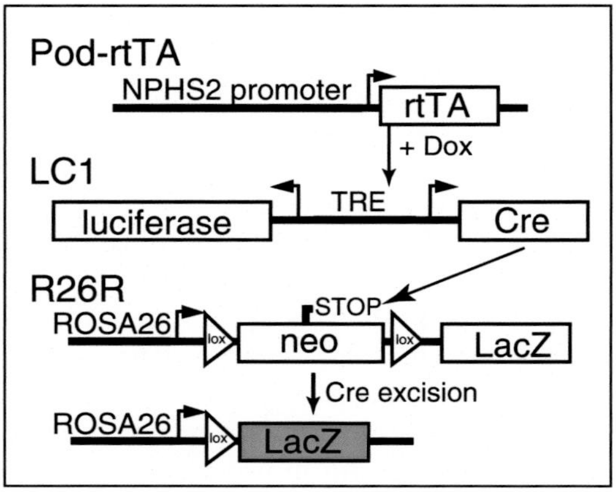

Fig. 1 Genetic map of the transgenic Pod-rtTA/LC1/R26R mice. Pod-rtTA: The podocyte promoter (derived from the human podocin gene, NPHS2) drives expression of the tetracycline-inducible reverse transactivator (rtTA2-M2) in podocytes. LC1: Expression of Cre recombinase and a reporter gene (luciferase, not relevant in this context) are expressed under the control of a "tetracycline-inducible" promoter containing Tet-responsive elements (TRE). Expression is activated reversibly after administration of doxycycline (Dox) in podocytes. R26R: A stop signal (a neocassette) is irreversibly excised by Cre recombination, so that expression of beta-galactosidase (LacZ) is activated irreversibly under the control of the ubiquitously active R26R locus

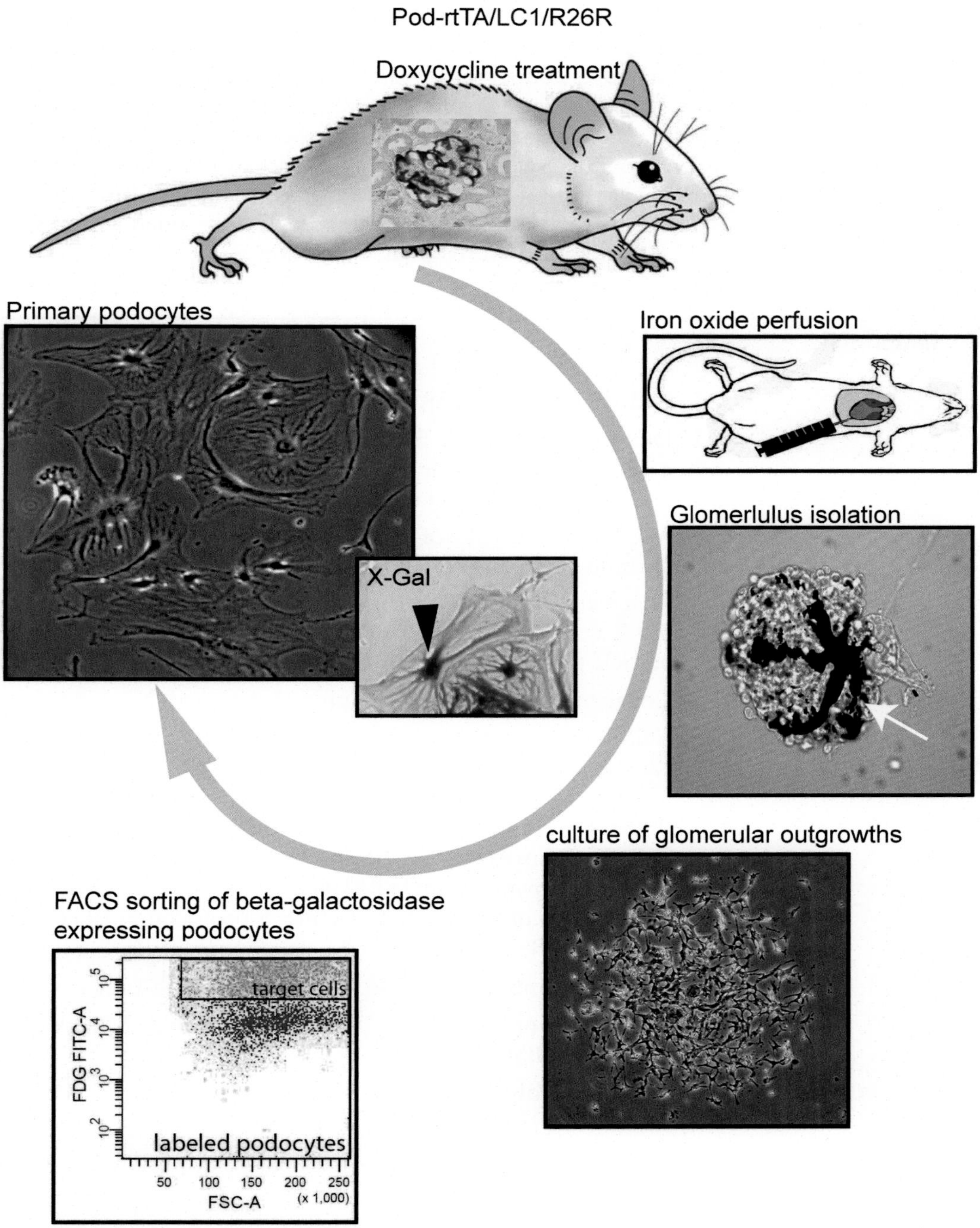

Fig. 2 Flow scheme for the isolation of genetically tagged podocytes. Podocytes are specifically and irreversibly labeled by administration of doxycycline in transgenic Pod-rtTA/LC1/R26R mice. *Inset* in mouse illustrates X-Gal staining of labeled podocytes (*blue staining*). The mice are perfused with magnetic iron oxide which accumulates in the capillaries of the glomeruli (*white arrow*). Glomeruli are isolated and cultured. Single-cell suspensions of the glomerular outgrowths are treated with FDG, which is hydrolyzed by β-gal into an insoluble fluorescent product, and are subjected to FACS. After FACS, X-Gal staining can be performed to assess the purity of the podocyte culture (*blue staining, black arrowhead*). The primary podocytes form large arborized cell bodies with several intracytoplasmic extensions, i.e., thickenings

steps by limited dilution including multiple cell passages and population doublings, which could affect the phenotype of the cells.

In short, podocytes are irreversibly genetically tagged in vivo in the inducible podocyte reporter mouse (Pod-rtTa/LC1/R26R) [6–8]. The glomeruli containing the labeled podocytes, i.e., expressing beta-galactosidase (β-gal), are isolated and cultured. The β-gal-expressing podocytes are directly isolated from the cellular glomerular outgrowths by FACS. The obtained cultures are >97 % pure, based on the β-gal expression.

2 Materials

1. Pod-rtTA/LC1/R26R mice.
2. Doxycycline solution: 5 % sucrose (w/v) and 1 mg/mL doxycycline hydrochloride (Fargon GmbH&Co, Barsbüttel, Germany), in normal tap water.
3. Anesthetic: mixture of 2 % xylazine (20 mg/mL) and 10 % ketamine (100 mg/mL).
4. Iron oxide solution: normal saline (0.9 % NaCl) containing 0.9 % w/v iron (II,III) oxide (Fe_3O_4, 98 % purity, 20–30 nm particle powder; Alfa Aesar GmbH, Karlsruhe, Germany).
5. Collagenase Type 4 (Worthington, Lakewood, NJ, USA).
6. EGM medium: Endothelial Growth Media (EGM™) Bullet Kit (Lonza, Walkersville, MD USA). To formulate EGM medium, basal medium EBM™ is supplemented with the Bullet Kit components (i.e., human epidermal growth factor [hEGF], hydrocortisone, bovine brain extract [BBE], ascorbic acid, fetal bovine serum [FBS], and gentamicin/amphotericin-B [GA]), according to manufacturer's protocol. Final concentration of FBS is 20 % v/v.
7. RPMI medium: RPMI 1640 supplemented with 10 % v/v FBS.
8. HBSS-Tween: 0.05 % v/v Tween®20 in Hank's balanced salt solution (HBSS).
9. Fluorescein di-β-galactopyranoside (FDG) (Molecular Probes, Leiden, the Netherlands).
10. Cell culture antibiotics: penicillin-streptomycin solution (10,000 U/mL).
11. Trypsin-EDTA cell culture formulation to remove cells from cell culture vessel.
12. 50 mL centrifuge tubes.
13. Six-well cell culture plates.
14. 75 cm^2 cell culture flask.
15. Automatic FACS cell sorter (e.g., BD FACSAria cell sorter, BD Biosciences, San Jose, CA, USA).

16. Centrifuge (e.g., Multifuge 3L-R, Thermo Fischer Scientific Inc., Waltham, MA USA).
17. Cell strainer, 70 μm (Becton Dickinson, Bradford, MA, USA).
18. Glutaraldehyde fixative: Prepare a 2 % glutaraldehyde solution in PBS (v/v) from a 25 % glutaraldehyde commercial stock.
19. Triton™ X-100.
20. X-Gal staining solution: 1 mg/mL X-Gal, 5 mM potassium ferricyanide, 5 mM potassium ferrocyanide, and 2 mM $MgCl_2$ in PBS [pH 7.8].
21. Aqueous mounting media.

3 Methods

3.1 Isolation of Podocytes

All transgenic animals used to establish these methods were housed under SPF-free conditions, and the procedures were approved by the local state government authorities of LANUV Cologne. All procedures are performed using aseptic techniques.

1. To induce the genetic labeling of the podocytes, the podocyte reporter mice (Pod-rtTA/LC1/R26R) receive the doxycycline solution via drinking water for a total of 10 days. The solution needs to be protected from light and exchanged every 2 days.
2. Mice are anesthetized with ketamine (80 mg/kg bodyweight) and xylazine (4 mg/kg bodyweight) and perfused via the left heart ventricle with 20–40 mL iron oxide solution (*see* **Note 1**).
3. Kidneys are transferred into 50 mL RPMI medium containing 1 % (v/v) penicillin-streptomycin solution and cut into small fragments.
4. The kidney fragments are incubated for 30 min at 37 °C with 1 mg/mL collagenase Type 4.
5. The small kidney fragments are gently pressed and sieved through a 70 μm nylon mesh cell strainer/sieve. During this process the strainer is repeatedly rinsed with cold HBSS-Tween. The filtrate is collected in 50 mL centrifuge tubes.
6. HBSS-Tween is added to the filtrate to a final volume of 40 mL per 50 mL falcon tube and incubated on ice for 20 min. During this period the relatively large glomeruli form a sediment, whereas cell debris and smaller structures will remain largely in the supernatant. The supernatant is discarded and the sediment is resuspended in 40 mL of HBSS-Tween and incubated on ice for another 20 min.
7. The sediment is resuspended in 1 mL of HBSS-Tween and placed in a magnetic particle concentrator. The "magnetic" glomeruli containing the iron oxide particles will be attracted to the magnet. Other cells/structures will stay in the solution and can be discarded carefully.

8. The glomeruli still bound to the magnet are washed twice with 1 mL HBSS-Tween.
9. The washed glomeruli are resuspended in EGM medium with 1 % penicillin-streptomycin solution and incubated for 7 to 14 days at 37 °C in a humidified atmosphere containing 95 % air 5 % CO_2. During this culture period, the glomeruli will adhere to the culture dish and glomerular cell outgrowths are formed (*see* **Note 2**).
10. The cellular outgrowths are trypsinized, and the cells are resuspended in RPMI medium containing 1 % penicillin-streptomycin solution and cultured in a 75 cm^2 flask in a humidified atmosphere containing 95 % air 5 % CO_2.
11. When the cells have grown to 80 % confluence, the cells are trypsinized and resuspended in RPMI medium, centrifuged at $600 \times g$, and resuspended in 1 mL RPMI medium containing 1 % penicillin-streptomycin solution.
12. The cell suspension is placed in a magnetic particle concentrator. Free iron oxide particles and remaining glomerular structures containing iron oxide will be attracted to the magnet. The oxide-free cell suspension can be transferred to a clean 1.5 mL test tube.
13. Fluorescein di-β-galactopyranoside (FDG) is added to the cell suspension to a final concentration of 2 mM and incubated for 1 min at 37 °C (*see* **Note 3**).
14. Subsequently, the samples are diluted tenfold with RPMI medium and incubated on ice for 60 min. The fluorescent-labeled podocytes are sorted from non-labeled cells at 514 nm using FACS (BD FACSAria II cell sorter).
15. Primary podocytes are cultured in RPMI medium containing 1 % penicillin-streptomycin solution at 37 °C in a humidified atmosphere containing 95 % air 5 % CO_2 (*see* **Note 4**).

3.2 Characterization of Isolated Podocytes

Primary podocytes have a distinct phenotype. When the culture is below 80 % confluence, the cells show a large cell body with several arborized intracytoplasmic extensions (Fig. 2) (*see* **Note 4**).

The podocyte origin and purity of the culture can be verified by enzymatic X-Gal staining. The enzymatic X-Gal staining is used to detect the constitutive beta-galactosidase expression in the labeled podocytes. For the enzymatic X-Gal staining:

1. Cells are washed with HBSS and fixed with 2 % glutaraldehyde for 5 min.
2. Cells are washed with HBSS and treated with 1 % Triton® X-100 for 5 min.
3. The cells are washed again with HBSS and incubated overnight at 37 °C in a humidified atmosphere in X-Gal staining solution.

4. On the next day, samples are counterstained with eosin, washed in tap water, and mounted using aqueous mounting media.
5. Examine microscopically.

4 Notes

1. A minimum of 20 mL Iron oxide solution is used for the perfusion. Successful perfusion of the kidney, with the black-colored iron oxide solution, will turn the color of the perfused kidney gray. When necessary, larger volumes of the iron oxide solution, up to 40 mL, can be used.
2. The isolated glomeruli in culture should not be disturbed for at least a week; any movement of the culture dish will prevent adherence of the glomeruli to the culture plate and the formation of glomerular cell outgrowths.
3. FDG is a substrate for beta-galactosidase. Nonfluorescent FDG is sequentially hydrolyzed by β-galactosidase, first to fluorescein monogalactoside (FMG) and then to highly fluorescent fluorescein.
4. Primary podocyte should be passaged when the monolayer is about 90 % confluent. The cells show restricted growth if the cell density is too low. Therefore, primary podocytes should be seeded to a confluence of 40 % or higher.

Acknowledgments

This work was supported by the German Research Foundation (Grant BO 3755/1-1 to BS), by the Genzyme Renal Innovation Program (GRIP, to BS). Additional support came from the eRARE consortium "Rare-G" (01 GM 1208A to MJM) and TP17 of SFB/Transregio 57 of the German Research Foundation (to MJM). MJM is a member of the SFB/Transregio 57 DFG consortium "mechanisms of organ fibrosis." We thank Regina Lanzmich for excellent technical support and helpful discussions, and we thank the Q1 platform of the SFB/Transregio 57 for implementing the above-described protocol into routine practice.

References

1. Wiggins RC (2007) The spectrum of podocytopathies: a unifying view of glomerular diseases. Kidney Int 71:1205–1214
2. Shankland SJ, Pippin JW, Reiser J, Mundel P (2007) Podocytes in culture: past, present, and future. Kidney Int 72:26–36
3. Kreisberg JI, Hoover RL, Karnovsky MJ (1978) Isolation and characterization of rat glomerular epithelial cells in vitro. Kidney Int 14:21–30
4. Striker GE, Killen PD, Farin FM (1980) Human glomerular cells in vitro: isolation and characterization. Transplant Proc 12:88–99

5. Kabgani N, Grigoleit T, Schulte K et al (2012) Primary cultures of glomerular parietal epithelial cells or podocytes with proven origin. PLoS One 7:e34907
6. Shigehara T, Zaragoza C, Kitiyakara C et al (2003) Inducible podocyte-specific gene expression in transgenic mice. J Am Soc Nephrol 14:1998–2003
7. Moeller MJ, Sanden SK, Soofi A, Wiggins RC, Holzman LB (2003) Podocyte-specific expression of cre recombinase in transgenic mice. Genesis 35:39–42
8. Soriano P (1999) Generalized lacZ expression with the ROSA26 Cre reporter strain. Nat Genet 21:70–71

Chapter 2

Propagation and Culture of Human Renal Fibroblasts

Sven-Jean Tan and Tim D. Hewitson

Abstract

The renal fibroblast and phenotypically related myofibroblast are universally present in all forms of progressive kidney disease. The in vitro study of the fibroblast, its behavior, and factors affecting its activity is therefore key to understanding both its role and significance. In this protocol, we describe a reproducible method for selective propagation and culture of primary human renal fibroblasts from the human kidney cortex. Techniques for their isolation, subculture, characterization, and cryogenic storage and retrieval are described in detail.

Key words Culture, Explants, Fibroblasts, Kidney, Propagation

1 Introduction

Progression of all kidney disease inevitably follows a common pathway characterized by progressive tubulointerstitial fibrosis, which continues to provide the most consistent and accurate histological predictor of disease outcome [1]. While other cells may contribute to the pathophysiology of fibrosis and scarring, the renal interstitial fibroblast is widely recognized as the pivotal effector cell in this process. Therefore, the study of the fibroblast, its behavior, and factors affecting its activity through an in vitro cell culture system is central to understanding this cell and its characteristics.

The study of fibroblasts involves cultivation of this cell type from either immortalized cell lines or via establishing primary cell cultures. Immortalized cell lines are widely accessible and while their use is able to provide reliability and consistency, viral mutation to exploit their proliferation characteristics poses a problem as these cells can differ in its behavior and phenotype compared to endogenous non-transformed cells. On the contrary, primary cell cultures provide an environment to study cells in conditions as close to normal as possible.

Tim D. Hewitson et al. (eds.), *Kidney Research: Experimental Protocols*, Methods in Molecular Biology, vol. 1397,
DOI 10.1007/978-1-4939-3353-2_2, © Springer Science+Business Media New York 2016

Primary human renal fibroblasts are therefore an attractive option for the study of the fibroblast as this provides a more realistic system, comparable to what occurs in human kidney disease. However, accessibility to human tissue can be problematic and the culture of fibroblasts has typically been derived from animal cells. A number of methods and techniques have been described to isolate renal interstitial fibroblasts from various segments of kidney that include differential sieving, centrifugation, collagenase digestion, magnetic bead separation, and outgrowth culture [2–5].

In this chapter, we describe a simple, effective, and reproducible method used in our laboratory to selectively culture primary human fibroblast from human kidney tissue [6, 7]. The method is based on the outgrowth of cells from explanted kidney tissue that has been established using animal models for propagation of animal renal fibroblasts [2]. This method involves mincing and scratching cortical human kidney tissue into a gelatin-coated petri dish surface. The tissue is incubated with enriched medium for 14 days to encourage cellular outgrowth until the cell monolayer has reached 75 % confluency. These cells may demonstrate initially a cobblestone appearance representative of tubuloepithelial cells before being replaced by the classic "fingerprint" appearance that is characteristic of a fibroblast cell monolayer (Fig. 1). These cells are passaged and subcultured until 100 % confluence is achieved, before characterization using immunocytochemistry techniques, also described in this chapter. These fibroblasts are characterized by positive staining for α-smooth muscle actin, vimentin, and collagen I together with negative staining for endothelial and epithelial

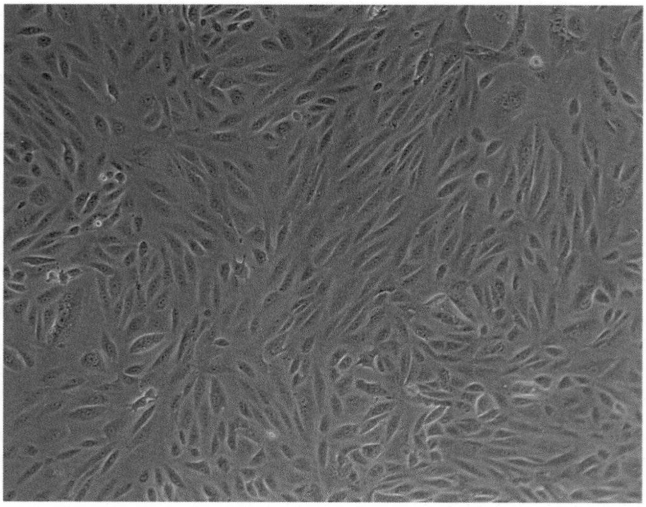

Fig. 1 Phase contrast microscopic appearance of confluent human renal fibroblasts. Cells show characteristic "fingerprint" monolayers without forming hillocks

markers. Once the human renal fibroblast culture is established, these cells can be cryogenically stored until required for experimentation. Appropriate methodology for cryogenic storage and retrieval are also described here.

2 Materials

2.1 General Sterile Cell Culture Materials

1. Glass media bottles: 100, 200, and 500 mL.
2. 60 mm^2 cell culture petri dishes.
3. 50 mL centrifuge tubes.
4. Tissue culture flasks: 25 and 75 cm^2.
5. Six-well cell culture plates.
6. Graduated 5 and 10 mL pipettes.
7. Cell scrapers.
8. Disposable 50 mL syringes.
9. Disposable 0.2 μm pore syringe filters.
10. 0.2 μm 500 mL bottle-top vacuum filter unit.
11. Sterile containers for tissue collection.
12. Disposable basic sterile dressing pack for dissection.
13. N° 22 scalpel blades.
14. Pasteur transfer pipettes.
15. Waste bottle.
16. Laboratory grade double deionized water (ddH_2O).

2.2 Specialist Cell Culture Equipment

1. Biohazard hood.
2. Inverted light microscope.
3. Vacuum pump.
4. Tissue culture incubator at 37 °C with 95 % O_2/5 % CO_2.
5. Water bath.
6. Autoclave.
7. Pipet-Aid™ (BD Biosciences, San Jose, CA, USA) or equivalent pipet controller to aspirate and change media.
8. Hemocytometer.
9. Mechanical tally counter.
10. Cryogenic dewar for storage of frozen cells.
11. –70 °C freezer.

2.3 Reagents

1. 0.01 M phosphate-buffered saline (PBS), pH 7.4: For each 1 L weigh 8.0 g NaCl, 0.2 g KCl, 2.9 g $Na_2HPO_4 \cdot H_2O$, and

0.2 g KH_2PO_4 and transfer to the bottle containing 980 mL of ddH_2O. Adjust to pH 7.4 with NaOH at 25 °C. Make up to 1 L with ddH_2O.

2. Coating solution: 2 % solution of Type B gelatin from bovine skin (Sigma-Aldrich, St Louis, MO, USA).
3. Tissue collection buffer: Combine 100 mL Hank's balanced salt solution with 2.5 μL gentamicin (50 mg/mL gentamicin sulfate). Filter through a 0.2 μm pore syringe filter using a 50 mL syringe (*see* **Note 1**).
4. Enriched medium (Dulbecco's Modified Eagle Medium [DMEM] with 20 % serum): Combine 80 mL fetal calf serum (FCS), 10 mL 1 M HEPES, 4 mL of 200 mM glutamine, and 8 mL 5000 U/mL penicillin-5000 μg/mL streptomycin, and make up to 400 mL with 1× DMEM. This solution is sterilized by passing through a sterile 0.2 μm pore cellulose acetate filter unit into a sterile 500 mL glass bottle by vacuum extraction. The bottle is capped and stored at 4 °C until use (*see* **Note 2**).
5. Maintenance medium (DMEM with 10 % serum): as above for enriched medium but containing 40 mL FCS per 400 mL medium.
6. Ethylenediaminetetraacetic acid (EDTA): 0.02 % solution.
7. EDTA/trypsin solution: 1:10 dilution of 1× trypsin in 0.02 % EDTA.
8. Freezing medium: 1 mL dimethyl sulfoxide (DMSO) added to 9 mL of maintenance medium.
9. Liquid nitrogen (N_2).

2.4 Immunoperoxidase Cytochemistry

2.4.1 General Materials

1. Established primary human renal cells.
2. Sterile glass cover slips (22 × 40 mm).
3. Sterile 0.01 M PBS, pH 7.4.
4. Borosilicate glass test tubes.
5. Filter paper.
6. Fixative, e.g., ice-cold methanol, acetone, or 4 % paraformaldehyde in 0.01 M PBS (*see* **Note 3**).
7. Glass microscope slides.
8. Hydrophobic wax pen.

2.4.2 Immunoperoxidase Cytochemistry

1. Primary antibody (Table 1).
2. Antibody diluent (Dako, Glostrup, Denmark).
3. 0.035 % hydrogen peroxide in methanol: Dilute 10 μL of 35 % hydrogen peroxide in 990 μL of 100 % methanol. Make as required.

Table 1
Immunocytochemical antiserum specifications

Antibody	Clone	Raised in	Dilution	Specificity	Manufacturer
α-SMA	1A4	Mouse	1:50	Smooth muscle isoform of actin	Dako, Glostrup, Denmark
Collagen I	–	Rabbit	1:1000	Collagen type I	Southern Biotech, Birmingham, AL, USA
Cytokeratin	MNF116	Mouse	1:100	Pan-keratin	Dako
Desmin	D33	Mouse	1:100	Desmin	Dako
E-cadherin	36/E-cadherin	Mouse	1:100	E-cadherin	BD Biosciences, San Jose, CA, USA
Vimentin	V9	Mouse	1:50	Vimentin	Dako

4. Peroxidase IgG kits containing normal serum and appropriate biotinylated secondary antibody (2°Ab) (Table 1) (Vectastain™ Kit; Vector Laboratories, Burlingame, CA, USA).
5. Avidin-biotin complex (ABC) Elite kit (Vector Laboratories).
6. Chromogen substrate (e.g., SIGMAFAST™, Sigma-Aldrich: dissolve 1× 3′3-diaminobenzidine (DAB) tablet and 1× UreaH_2O_2 tablet in 1 mL 0.01 M PBS).
7. Harris hematoxylin.
8. Aqueous mounting media.
9. Humidified chamber (e.g., plastic container lined with blotting paper and soaked with 0.01 M PBS).

3 Methods

The methodology described in Subheadings 3.1–3.3 provides techniques to establish a primary human fibroblast cell line originating from explanted human renal cortical tissue. The procedures outline (a) the preparation of cell culture reagents, (b) explanting of human kidney tissue, and (c) subculturing of the established primary cell line. Subheading 3.4 offers methods in characterizing a primary human fibroblast cell line. Subheadings 3.5 and 3.6 describe the methods for cryogenic storage and subsequent experimental use once the primary human fibroblast cell line has been established.

All cell culture work described in this protocol should be performed in a biohazard cabinet. All reagents for use with cell culture work should be prewarmed to 37 °C prior to use in order to maintain cellular viability and integrity, unless otherwise specified.

3.1 Gelatin Coating of Petri Dishes

Petri dishes are coated with 1 % gelatin solution prior to explanting of kidney tissue. This will enable minced tissue to secure to the petri dish and provide a substrate for the initial cultivation of cell populations.

1. Dilute 2 % gelatin solution with 0.01 M PBS (1:1 dilution) and filter sterilize into a centrifuge tube using a disposable syringe and a 0.2 μm pore syringe filter.
2. Coat entire surface of the petri dish with 1 mL of 1 % gelatin solution and incubate for ≥30 min at 37 °C.
3. Remove excess gelatin solution and rinse the petri dish with 0.01 M PBS prior to explanting.

3.2 Propagation of Human Renal Interstitial Cultures from Kidney Explants

Collection of and conducting research with any human tissue should be performed in strict accordance with the Human Research Ethics guidelines and the Declaration of Helsinki. Collection of human tissue is subject to approval by the appropriate Human Research Ethics Committee and requires individual human/ patient consent.

1. Retrieve the human kidney cortical tissue in tissue collection buffer as soon as possible after removal of tissue. Keep this on ice until required for explanting (*see* **Note 1**).
2. Set out a sterile field in the biohazard hood using a basic dressing pack.
3. Pour out the kidney tissue and tissue collection buffer onto one small section of a sterile tray (e.g., dressing pack tray) and use the sterile disposable forceps to gently transfer the kidney into another clean section of the tray.
4. Cover the kidney tissue with warm enriched medium to keep moist.
5. Dissect this piece of tissue with surgical forceps into ~0.25 cm^2 pieces.
6. Transfer one piece of ~0.25 cm^2 kidney tissue to a gelatin-coated petri dish (which has been rinsed).
7. Dice this piece of tissue until a minced consistency is achieved.
8. Make crosswise scratches across the surface of the petri dish using the surgical blade in order to adhere the minced tissue to the gelatin surface. Continue until 75 % of this dish is covered.
9. Cover explanted tissue with ~2 mL of enriched medium using a plastic Pasteur pipette, then label, and date.
10. Incubate overnight at 37 °C with 95 % O_2/5 % CO_2.
11. Supplement with an additional 1 mL of enriched medium 24 h after the explanting procedure.

12. Continue to incubate at 37 °C with 95 % O_2/5 % CO_2 to allow initial cell outgrowth for another 48 h (i.e. total 72 h post explanting).
13. Aspirate explanting medium into a waste bottle using a 5 mL glass pipette to remove cellular debris and nonadherent cells.
14. Replenish with 2 mL of enriched medium.
15. Change medium three times a week for ~14 days until the monolayer has covered approximately 75 % of the dish surface.

3.3 Subculture of Primary Human Cell Cultures

Cells are lifted from their growing surface by treatment with a solution that is a combination of EDTA and the enzyme trypsin (*see* Subheading 2.3, **item** 7). Tissue culture medium supplemented with FCS commonly contains a mixture of divalent cations and proteins that inhibit proteins. Therefore cells are first washed with Ca^{2+}/Mg^{2+}-free PBS to reduce the concentration of these inhibitors before the addition of this treatment solution of EDTA/trypsin. EDTA is a calcium chelator whose properties are exploited to bind any remaining divalent ions. Once the cells have detached using this EDTA/trypsin solution, adding an excess of fresh medium to prevent further digestion and cellular damage quenches the enzyme reaction.

3.3.1 Lifting Explant Monolayer Cultures

1. Ensure monolayer is subconfluent (~75 % of the dish surface is covered).
2. Remove medium from the petri dish and discard into a waste bottle.
3. Wash the cell monolayer twice with 2 mL of 0.01 M PBS to remove residual medium.
4. Remove any residual fluid.
5. Replace with 1 mL of EDTA/trypsin solution.
6. Incubate for 5–15 min at 37 °C (*see* **Note 4**).
7. Check for cell detachment under an inverted light microscope.
8. If required, gently tap the base of the petri dish to detach any remaining cells.
9. Add 2 mL of enriched medium to inhibit any further trypsin reaction.
10. If required, gently scrape the explant layer with a cell scraper to assist the detachment of cells.
11. Transfer cell suspension into a 25 cm^2 tissue culture flask.
12. Add 2–3 mL of enriched medium to this culture flask.
13. Label and date the culture flask.
14. Incubate overnight at 37 °C with 95 % O_2/5 % CO_2.
15. Change medium three times a week until cultures are 100 % confluent (*see* **Note 5**).

3.3.2 Maintenance of Primary Human Renal Fibroblast Subcultures

1. Ensure monolayer is confluent.
2. Remove medium from 25 cm^2 tissue culture flask and discard into a waste bottle.
3. Wash the cell monolayer twice with 2 mL of 0.01 M PBS to remove residual medium.
4. Remove any residual fluid.
5. Replace with 1 mL of EDTA/trypsin solution.
6. Incubate for 5–10 min at 37 °C (*see* **Note 4**).
7. Check for cell detachment under an inverted light microscope.
8. If required, gently tap the base of the petri dish to detach any remaining cells.
9. Add 2 mL of enriched medium to inhibit any further trypsin reaction.
10. Transfer cell suspension into a 75 cm^2 tissue culture flask.
11. Add 7–8 mL of enriched medium to this culture flask.
12. Incubate overnight at 37 °C with 95 % O_2/5 % CO_2.
13. Change medium three times a week until cultures are 100 % confluent (*see* **Note 6**).

3.4 Characterization of Primary Human Renal Fibroblasts

Primary cell cultures isolated using this technique yield a mixed population of cells due to the complex composition of the kidney and the heterogeneity of cell lineages, consistent with the anatomical location of the initial kidney tissue sample [8]. Therefore it is important that the kidney cortex is used to isolate these primary human cells.

The identification of fibroblasts is difficult as these cells demonstrate low immunogenicity. Thus, the characterization of fibroblasts is performed through a series of inclusion and exclusion criteria based on their morphological and biochemical properties. Morphologically, these cells display an elongated, spindle-like shape and a "fingerprint"-patterned confluent cell monolayer (Fig. 1). These cells are further defined as fibroblasts based on positive staining for vimentin (cells of mesenchymal origin), α-smooth muscle actin (smooth muscle cells and myofibroblasts), and collagen I (fibrogenic cells), together with negative staining for desmin (podocytes, smooth muscle cells), E-cadherin and cytokeratin (epithelial cells) (Table 2) (Fig. 2).

3.4.1 Growth Characteristics

The initial cell outgrowths from human cortical kidney tissue obtained through the methods described in this chapter will demonstrate a cobblestone morphology consistent with an epithelial cell origin. Prolonged culture (14 days and upward) as described here will allow the development of a more

Table 2
Cell markers and corresponding phenotypes

Markers	Cell expression
α-SMA	Myofibroblasts, vascular smooth muscle cells, some mesangial cells
Collagen I	Fibrogenic cells
Cytokeratin	Epithelial cells
Desmin	Vascular smooth muscle cells, mesenchymal cells, mesangial cells, some myofibroblasts
E-cadherin	Epithelial cells
Vimentin	Fibroblasts, mesangial cells, vascular smooth muscle cells

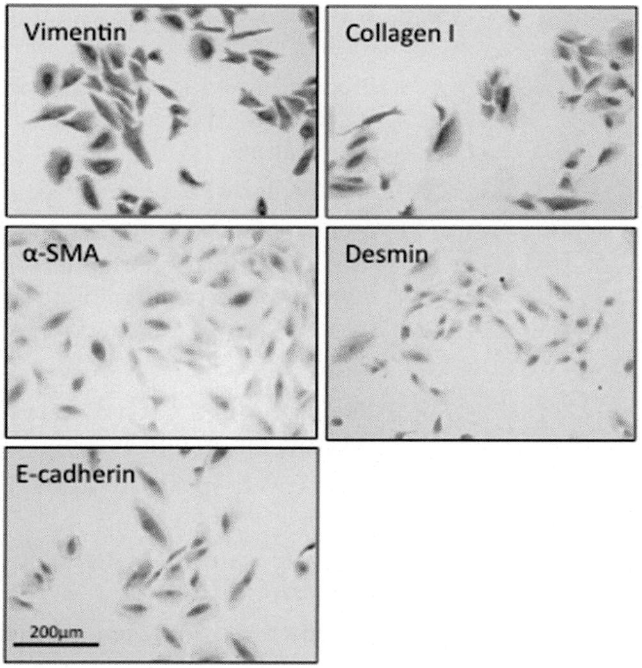

Fig. 2 Immunocytochemical characterization of human renal fibroblasts based on positive staining for vimentin, collagen I, α-smooth muscle actin (α-SMA), and negative staining for desmin and E-cadherin

homogenous population of fibroblasts, evident by their change in morphology into elongated, spindle-like shape cells. When the cell population is confluent, the cell monolayer resembles a "fingerprint" pattern without the "hillocks" typical of mesangial cells.

3.4.2 Immunocytochemistry/Immunoperoxidase Staining

1. Primary cells should be grown until ~75 % confluent on sterile cover slips in six-well plates or petri dishes with either enriched or maintenance medium.
2. Remove medium and wash the cell monolayer twice with 2 mL of 0.01 M PBS to remove residual medium.
3. Fix cells with fixative of choice for 5 min at 4 °C (*see* **Note 3**).
4. Remove fixative and air-dry.
5. Outline cover slip with a hydrophobic wax pen to conserve antibody.
6. Wash cover slips twice with 0.01 M PBS (minimum of 5 min/wash).
7. Block with 0.035 % hydrogen peroxide in methanol for 20 min at room temperature to quench endogenous peroxidase activity.
8. Wash cover slips twice with 0.01 M PBS (minimum of 5 min/wash).
9. Block nonspecific binding sites by incubating sections with the appropriate diluted normal serum for 10 min at room temperature.
10. Gently blot with a tissue paper to remove excess serum.
11. Incubate for 2 h at room temperature in a humidified chamber with the primary antibody of interest diluted with antibody diluent to the required dilution (Table 1).
12. Wash cover slips twice with 0.01 M PBS (minimum of 5 min/wash).
13. Incubate with the appropriate biotinylated secondary antibody (Vector) for 10 min at room temperature.
14. Wash cover slips twice with 0.01 M PBS (minimum of 5 min/wash).
15. Incubate with ABC Elite solution (Vector) for 15 min at room temperature.
16. Prepare the chromogen substrate solution.
17. Wash cover slips twice with 0.01 M PBS (minimum of 5 min/wash).
18. Apply chromogen substrate to sections for 2–10 min. Monitor staining on wet slides using a light microscope.
19. Wash in H_2O for 5 min to terminate reaction.
20. Counterstain with Harris hematoxylin for 1 min.
21. Wash in tap water for 5 min.
22. Mount cover slips onto glass microscope slides with an aqueous mounting medium and examine using a light microscope.

3.5 Freezing Cell Cultures for Cryogenic Storage

A confluent cell monolayer can be passaged and frozen in 10 % DMSO-supplemented medium for cryogenic storage. DMSO is added as a cryoprotectant to protect the cells. DMSO penetrates the cells and binds to water molecules, thereby, preventing the loss of water from the cytoplasm during the freezing process and preventing cellular dehydration and maintaining intracellular physiology. Furthermore, DMSO is able to slow the rate of freezing and prevents the formation of ice crystals within the cell.

1. Remove culture medium from the tissue culture flask and discard into a waste bottle.
2. Wash the cell monolayer twice with 5 mL of 0.01 M PBS to remove residual medium.
3. Remove any residual fluid.
4. Replace with 2 mL of EDTA/trypsin solution.
5. Incubate for 5–10 min at 37 °C (*see* **Note 4**).
6. Check for cell detachment under an inverted light microscope.
7. If required, gently tap the base of the petri dish to detach any remaining cells.
8. Add 5 mL of maintenance medium to inhibit any further trypsin reaction.
9. Transfer the contents into a 50 mL centrifuge tube.
10. Centrifuge the cell suspension for 5 min at 2000 × *g*.
11. Remove the supernatant from the cell pellet and discard.
12. Resuspend the cell pellet with 1–1.5 mL of freezing medium and transfer into a sterile cryogenic vial (*see* **Note 7**).
13. Label and date the vial with appropriate details using a lead pencil.
14. Store at −70 °C overnight.
15. Transfer vial to a dewar filled with liquid nitrogen for long-term cryogenic storage.

3.6 Thawing of Cryogenically Stored Cells

1. Prepare hood with all required equipment: Warm medium, sterile pipettes, centrifuge tube, and culture flask.
2. Remove the cryogenic vial from storage and quickly thaw in a 37 °C water bath.
3. Transfer the contents to a 50 mL centrifuge tube.
4. Slowly add 10 mL of warm maintenance medium in a drop-wise fashion to the cell suspension. Simultaneously agitate the tube gently.
5. Aspirate the suspension and transfer the cells into a 75 cm^2 tissue culture flask.

6. Vent the cap and incubate the cells overnight at 37 °C with 95 % O_2/5 % CO_2 to allow the cells to adhere to the flask surface.
7. Remove the culture media (containing DMSO from cryogenic storage) 24 h later together with cells that have not adhered.
8. Replace it with 10 mL of fresh maintenance medium.
9. Return the flask to the incubator.
10. Change medium three times a week and maintain these subcultures until ready for experimentation.

4 Notes

1. Prepare collection buffer immediately prior to kidney tissue retrieval. Store collection buffer at 4 °C. Transport kidney tissue cold/on ice. Retrieve human kidney cortical tissue in tissue collection buffer as soon as possible subsequent to the removal of tissue. Once kidney tissue is submerged in tissue collection buffer, keep cold/on ice until required for explanting.
2. All cell culture reagents (especially growth medium) should be prewarmed to 37 °C prior to use in order to maintain cellular viability and integrity.
3. Fixation allows preservation of tissue or cells as close to its natural state as possible. The two main types of chemical fixatives used in cytochemistry are cross-linking fixatives (e.g., paraformaldehyde) and precipitating fixatives (e.g., methanol, acetone). Fixation by precipitation may destroy fewer antibody-binding sites though does not preserve three-dimensional organization and is not recommended for study of cytoskeletal proteins.
4. When lifting the cell monolayer from the tissue explant, the cells should be left to incubate with the EDTA/trypsin solution for a minimum of 5 min and no longer than 15 min. The dissociation of cells from the surface should be checked by direct visualization using a light microscope. When lifting cell monolayers from subsequent subcultures, the cells should be left to incubate with the EDTA/trypsin solution for a minimum of 5 min and no longer than 10 min. Prolonged incubation with EDTA/trypsin solution may cause cellular damage.
5. When initial primary human renal fibroblasts are 100 % confluent, the cell monolayer can be passaged using the trypsinization method described. Transfer these cells into a larger tissue culture flask to establish subcultures.
6. When secondary passage of primary human fibroblast cultures has been achieved, cells may be characterized for experimentation or cryogenically stored as described within this chapter.

7. An exothermic reaction may occur when DMSO is added to a warm medium. Therefore, the freezing medium should be prepared with cold media prior to use and refrigerated until use. Ensure that the cell pellet is evenly resuspended in the cold freezing medium and that no clumps are visible.

Acknowledgments

The authors are supported by a Postgraduate Research Scholarship (SJT) and Project grant 1078694 (TDH) from the National Health and Medical Research Council of Australia (NHMRC).

References

1. Bohle A, Mackensen-Haen S, von Gise H (1987) Significance of tubulointerstitial changes in the renal cortex for the excretory function and concentration ability of the kidney: a morphometric contribution. Am J Nephrol 7(6):421–433
2. Grimwood L, Masterson R (2009) Propagation and culture of renal fibroblasts. In: Becker GJ, Hewitson TD (eds) Kidney research. Humana Press, New York, pp 25–37
3. Sharpe CC, Dockrell ME, Noor MI, Monia BP, Hendry BM (2000) Role of Ras isoforms in the stimulated proliferation of human renal fibroblasts in primaru culture. J Am Soc Nephrol 11(9):1600–1606
4. Sommer A, Schaller R, Funfstuck R, Bohle A, Bomer FD, Muller GA, Stein G (1999) Abnormal growth and clonal proliferation of fibroblasts in an animal model of unilateral ureteral obstruction. Nephron 82(1):39–50
5. Clayton A, Steadman R, Williams JD (1997) Cells isolated from the human cortical interstitium resemble myofibroblasts and bind neutrophils in an ICAM-1-dependent manner. J Am Soc Nephrol 8(4):604–615
6. Kelynack KJ, Hewitson TD, Nicholls KM, Darby IA, Becker GJ (2000) Human renal fibroblast contraction of collagen I lattices is an integrin-mediated process. Nephrol Dial Transplant 15(11):1766–1772
7. Hewitson TD, Martic M, Kelynack KJ, Pedagogos E, Becker GJ (2000) Pentoxifylline reduces in vitro renal myofibroblast proliferation and collagen secretion. Am J Nephrol 20(1):82–88
8. Grupp C, Muller GA (1999) Renal fibroblast culture. Exp Nephrol 7(5–6):377–385

Chapter 3

Isolation and Propagation of Rat Peritoneal Mesothelial Cells

Qin Zhou and Xueqing Yu

Abstract

With the development of peritoneal dialysis in many countries, there has been much interest in the cell biology of peritoneal mesothelial cells. In this chapter we describe a reliable and reproducible method for the culture of rat primary mesothelial cells (RPMCs). This chapter outlines how to isolate mesothelial cells from rat peritoneum. The subculture of primary peritoneal mesothelial cells and the characterization by immunofluorescence is also described in detail.

Key words Isolation, Rat primary mesothelial cells, Subculture, Characterization, Immunofluorescence

1 Introduction

Peritoneal dialysis is an effective and safe alternative treatment for patients with end stage renal disease [1]. The patient's abdominal peritoneum is used as a membrane across which fluids and dissolved substances (urea, creatinine, glucose, electrolytes, albumin, and other small molecules) are exchanged from the blood. The peritoneal mesothelium is composed of an extensive monolayer of mesothelial cells that lines the body's serous cavity and internal organs and acts as a protective surface. Mesothelial cells are critical components in maintaining the integrity and functional properties of the peritoneum. During peritoneal dialysis, the peritoneum is continuously exposed to bioincompatible dialysis fluids that cause denudation of mesothelial cells and, ultimately, tissue fibrosis and failure of ultrafiltration. With the increasing population of peritoneal dialysis patients, there has been much interest in the cell biology and molecular aspects of peritoneal mesothelial cells.

The primary function of mesothelial cells is to provide nonadhesive barrier for organs, immunological defense, transport of fluids and substance across the serosal cavities, and synthesis of cytokines, matrix proteins, and growth factors [2]. In general,

Tim D. Hewitson et al. (eds.), *Kidney Research: Experimental Protocols*, Methods in Molecular Biology, vol. 1397,
DOI 10.1007/978-1-4939-3353-2_3, © Springer Science+Business Media New York 2016

characterization of peritoneal mesothelial cells, and their distinction from endothelial cells and fibroblasts, is based on their cobblestone phenotype and immunoreactivity with monoclonal antibodies to cytokeratin 18(CK-18) and vimentin. Pure peritoneal mesothelial cells are positively stained for CK-18 and vimentin, in the absence of staining for alpha-smooth muscle actin (α-SMA) [3, 4].

To date, there are two mesothelial cell lines used in many peritoneal studies. The first is a human peritoneal mesothelial cell line (HMrSV5), which was originally established and well documented by Dr. Pierre Ronco (Department of Nephrology, Tenon Hospital, Paris, France) [5]. This cell line originates from a fully characterized primary culture of human peritoneal mesothelial cells (HPMC) after infection of a T-antigen-encoding retroviral vector. The ATCC biological resource center also makes available MeT-5A cells, a human mesothelial cell line, isolated from the pleural fluids of noncancerous patients, which has been transformed with the pRSV-T plasmid, containing the SV40 virus early region [6].

Immortalized cell lines have been specifically selected for their ability to rapidly proliferate but present the problem of being virally mutated cells, which differ in behavior to non-transformed primary cells [4]. The use of primary peritoneal mesothelial cells in culture offers researchers a convenient and valuable tool to investigate their functional parameters in the absence of systemic modulating factors. However, the primary cells are much more difficult to culture, because of their limited proliferative potential.

Normal omental specimens obtained from surgery are the source for the isolation of human peritoneal mesothelial cells (HPMC) [7]. The use of peritoneal mesothelial cells from rodents is likewise common. Rat primary peritoneal mesothelial cells (RPMCs) have several advantages over the use of HPMC. They can be maintained for several passages without transfection or transformation, and the age and sex of animal are easy to replicate.

In this chapter, we describe two methods used in our laboratory for the preparation and culture of RPMCs [8–11].

2 Materials

2.1 Animals

1. Male Sprague-Dawley rats weighing 180–200 g.
2. Anesthetic (in accordance with local guidelines).
3. Sharp ophthalmic scissors.

2.2 General Sterile Cell Culture Materials

1. Disposable 2-mL, 5-mL, and 15-mL pipettes.
2. 15-mL and 50-mL centrifuge tubes.
3. 25-cm^2 and 75-cm^2 tissue culture flasks with vent caps.
4. 2-mL cryogenic vials for freezing cells.

5. Sterile 200-μL and 1000-μL pipette tips.
6. Disposable 60-mL syringe.
7. Disposable 0.2-μm pore syringe filters.
8. 10-cm cell culture dishes.
9. Disposable 70-μm cell strainer (BD Biosciences, San Diego, CA, USA).
10. Hemocytometer.
11. Trypan blue cell stain 20 mm × 20 mm cover slips.
12. Ultrapure double-deionized water (ddH_2O).

2.3 Specialist Cell Culture Equipment

1. Autoclave.
2. Water bath.
3. Biohazard safety hood.
4. Cell culture incubator set at 37 °C with 5 % CO_2.
5. Upright microscope.
6. Inverted phase contrast microscope.
7. −30 °C freezer.
8. −70 °C freezer.
9. Liquid nitrogen tank for storage of frozen cells.
10. Centrifuge (e.g., Eppendorf™ R5804; Eppendorf, Hamburg, Germany).

2.4 Reagents

1. Sterile 0.01 M 1× Dulbecco A phosphate-buffered saline (PBS), pH 7.4.
2. 1× Hanks' balanced salt solution (HBSS).
3. Fetal bovine serum (FBS).
4. Coating solution: collagen (type I calf skin) (Sigma-Aldrich, St. Louis, MO, USA).
5. Dulbecco's Modified Eagle Medium: nutrient mixture F-12 (DMEM/F12).
6. 100× penicillin/streptomycin −10,000 U/mL.
7. 1 M HEPES.
8. 100× Insulin-Transferrin-Selenium cell culture supplement (Gibco, Grand Island, NY, USA).
9. Enriched medium: Add 15-mL FBS, 2.5 mL of 1 M HEPES, 1-mL penicillin/streptomycin, and 1 mL of 100× Insulin-Transferrin-Selenium to 70 mL of DMEM/F12 medium. Make up to 100 mL with DMEM/F12 and filter through a 0.2-μm filter using a 60-mL syringe immediately prior to use.
10. Maintenance medium: DMEM/F12 supplemented with 10 % FBS (v/v) and 1 % penicillin/streptomycin (v/v).

11. 0.25 % trypsin-EDTA (1×) solution with red phenol.
12. Enzyme dissociation solution: 0.1 % trypsin (w/v)/0.02 % EDTA (v/v) in ddH_2O. Dissolve cell culture grade trypsin powder and ethylenediaminetetraacetic acid (EDTA) powder in ddH_2O, sterile-filter through a 0.2-μm filter, and keep at 4 °C.
13. Freezing medium: Cell Culture Freezing Medium (e.g., Recovery™; Gibco) (*see* **Note 1**).

2.5 Immuno fluorescence

2.5.1 General Materials

1. Established RPMCs.
2. 10-mm cover slips (EM Sciences, Hatfield, PA, USA) and 6-well plate (Costar; Corning Inc., NY, USA).
3. 1× PBS, pH 7.4.
4. Disposable 0.2-μm pore syringe filters.
5. 60-mL syringe.
6. 4 % paraformaldehyde (PFA): Slowly dissolve 4 g PFA (Sigma-Aldrich, St. Louis, MO, USA) in 100 mL 1× PBS, and then heat at 50 °C until solution clears; add NaOH to adjust PH to 7.4. Use immediately or aliquot and store at −20 °C for up to 3 months.
7. Glass microscope slides with frosted edges.
8. Hydrophobic barrier pen (e.g., Dako Pen; Dako, Glostrup, Denmark).
9. Humidified chamber (a sealed box with paper tissue at the bottom and soaked with ddH_2O to maintain humidity).

2.5.2 Immuno fluorescence

1. Antibodies (Table 1).
2. Fixatives: methanol, 4 % PFA.
3. Permeabilizing solution: Triton X-100.
4. Blocking buffer: 1× PBS containing 3 % BSA (w/v) and 10 % goat serum (v/v).

Table 1
Cell characterization

Antibody	IgG source	Dilution	Application	Company
α-SMA	Mouse	1:200	WB/IF	Sigma
Cytokeratin 18	Mouse	1:100	WB/IF	Sigma
Vimentin	Mouse	1:100	WB/IF	Sigma
Alexa 488	Goat	1:1000	IF	Invitrogen

WB, western blotting; IF, immunofluorescence

5. Nuclei staining solution: Prepare a stock solution by dissolving 1 mg DAPI (4′, 6-diamidino-2-phenylindole) powder in 1 mL ddH_2O. Store at 4 °C in dark. To prepare working solution, dilute DAPI stock 1:1000 in ddH_2O and filter to remove undissolved material prior to use.
6. Anti-fade mounting medium.
7. Immunofluorescent or confocal microscope.

3 Methods

3.1 Preparation

3.1.1 Medium and Reagent

1. Preheat enriched medium and maintenance medium in 37 °C water bath.
2. Fill a 10-cm dish with 10-mL PBS, and place on ice before surgery.

3.1.2 Collagen Coating of Culture Dishes

1. Add 5 mL of coating solution (collagen solution) to each 25-cm^2 flask, and incubate for 30 min prior to starting.

3.2 Isolation of Mesothelial Cells from Rat Peritoneum

This procedure is performed with a modification of the technique reported by Shostak [12]. The experimental protocols were approved by the Animal Care and Use Committee of the Sun Yat-sen University.

3.2.1 Isolation of Rat Peritoneal Mesothelial Cells by Intraperitoneal Injection

1. Humanely kill rats with an overdose of anesthetic. When dead, quickly inject intraperitoneally with 10 mL/100 g bodyweight of 0.25 % trypsin in EDTA. Place the animal in a faceup position for 10 min, and then facedown 10 min.
2. To harvest RPMCs, the abdominal fluid of individual rats is collected in a 50-mL tube in a biohazard hood. Collect about 15–20 mL cell suspension for each rat.
3. Centrifuge cell suspension at 180 × *g* for 10 min at 4 °C.
4. Remove the supernatant, wash cell pellets with 25 mL D-Hanks' balanced salt solution once to remove the trypsin, and repeat using the same centrifuge speed and time as above.
5. Resuspend the cell pellets in 25 mL maintenance medium (*see* **Note 2**).
6. Centrifugate again at 180 × *g* for another 10 min, suspend in 15-mL tube with 10 mL enriched medium, and gently pipette the mixture to resuspend cells.
7. Use a hemocytometer to count viable cells, based on trypan blue exclusion.
8. Aspirate the collagen solution off the 25-cm^2 culture flasks, and add 4 mL enriched medium to each 25-cm^2 flask.

9. Seed cells at $1–2 \times 10^6$ with a final volume of 5 mL enriched medium in 25-cm² tissue culture flask. Incubate at 37 °C in a humidified 5 % CO_2 atmosphere.
10. Refresh the medium when cells adhere to the surface of the flask (*see* **Note 3**).

3.2.2 Isolation of Rat Peritoneal Mesothelial Cells by Dissection of Peritoneum

1. Humanely kill rats with an overdose of anesthetic, and collect specimens of peritoneum (mainly mesenterium). Put peritoneal tissue in a 10-cm dish with 10 mL PBS on ice.
2. Wash tissue specimens extensively with PBS three times to remove contaminating red blood cells.
3. Dissect a 2–3-cm² piece of peritoneum with sharp ophthalmic scissors (*see* **Note 4**).
4. Digest the tissue with pre-warmed enzyme dissociation solution (0.1 % trypsin/0.02 % EDTA) for 30 min at 37 °C with continuous rotation.
5. Sieve the cell suspension with 70-μm strainer and collect the sieved solution (*see* **Note 5**).
6. Spin down the cell suspension by centrifugation at $180 \times g$ for 10 min at 4 °C.
7. Wash the cell pellet twice in maintenance medium (use the same rpm and time as in the step above), and gently pipette the mixture up and down a few times.
8. Use a hemocytometer to count viable cells, based on trypan blue exclusion.
9. Seed cells at $1–2 \times 10^6$ with a final volume of 5 mL enriched medium in 25-cm² tissue culture flask. Incubate at 37 °C in a humidified 5 % CO_2 atmosphere.
10. Refresh the medium every 2–3 days.

These two methods have proven to be both efficient and reproducible for isolating rat peritoneal mesothelial cells (*see* **Note 6**).

3.3 Subculture of Primary Peritoneal Mesothelial Cells

Passage the cells to 75-cm² flasks for amplifying when cultures are over 95 % confluent in 25-cm² flasks (*see* **Note 7**).

3.3.1 Passage of Rat Primary Cell

1. UV disinfect biohazard hood for 15 min, and then lay out sterile pipettes, centrifuge tubes, and flasks. Warm medium to 37 °C in a water bath.
2. Remove medium from the flasks and discard as biological waste.
3. Replace with 1 mL of 0.25 % trypsin/EDTA solution, and incubate for 3–5 min at 37 °C.
4. Under an inverted light microscope, check that most of the cells have detached from the flask surface.

5. Gently tap the base of flasks to detach any remaining cells, and add 5 mL of pre-warmed maintenance medium.
6. Gently pipette up and down to assist the detachment of cells.
7. Transfer cell suspension into a sterile 15-mL tube.
8. Spin down cell suspension by centrifugation at 180×*g* for 5 min at 4 °C.
9. Aspirate supernatant, and resuspend the cell pellets by adding 5 mL enriched medium. Pipette up and down gently.
10. Transfer cells into a 75-cm^2 flask with enriched medium to a final volume of 20 mL.
11. Return to the 37 °C, 5 % CO_2 incubator.
12. Change medium twice weekly until cultures are over 95 % confluent.

3.3.2 Freezing Cell Cultures for Cryogenic Storage

Once the monolayer is over 95 % confluent in a 75-cm^2 flask, total cell number will reach 0.5–1 × 10^7 cells. The cells can be passage and frozen in freezing medium for cryogenic storage. Recovery™ is an optimized fully supplemented formulation which avoids the messy mixing of DMSO. DMSO protects cells by slowing the rate of freezing and the formation of ice crystals within the cell, but has cytotoxicity at culture temperature.

1. UV disinfect biohazard hood, and then prepare sterile pipettes, centrifuge tubes, and flasks, and warm up the medium at 37 °C water bath.
2. Remove medium from monolayer and discard as biological waste.
3. Dislodge cells with 2 mL of dissociation solution (0.25 % trypsin/EDTA), and incubate flask at 37 °C for 3–5 min until cells detach from the dish surface.
4. Gently tap the base of the flask to separate the cells.
5. Add 15 mL of warm maintenance medium to inhibit the trypsin reaction.
6. Gently aspirate cell suspension and transfer contents into a 50-mL centrifuge tube.
7. Centrifuge the cell suspension for 5 min at 180×*g*, 4 °C.
8. Remove the supernatant from the cell pellet and discard.
9. Resuspend the cell pellet in 1.8 mL of cold freezing medium and transfer into a sterile cryogenic vial.
10. Store at −70 °C overnight but no more than 1 week.
11. Transfer vial to liquid nitrogen for long-term cryogenic storage.

3.3.3 Thawing of Cryogenically Stored Cells

1. UV disinfect biohazard hood, prepare sterile pipettes, centrifuge tubes, and flasks, and warm up the culture medium.
2. Remove the cryogenic vial from liquid nitrogen and thaw it in a 37 °C water bath as quickly as possible.
3. Aspirate the cell suspension to a 25-cm^2 cell culture flask, slowly add 5 mL of warm enriched medium, and mix the cells by pipetting up and down to separate cells.
4. Return the flask to an incubator for 24 h, until the cells adhere to flask surface.
5. Remove the DMSO-containing medium from cell layer, slowly replace with 5 mL of fresh and pre-warmed enriched medium, and return the flask to the incubator.
6. Transfer cells to 75-cm^2 cell culture flask if the monolayer is over 95 % confluent in the 25-cm^2 flask.
7. Change medium every 2–3 days and maintain until ready for experiments.

3.4 Characteristics of Primary Cells

3.4.1 Morphologic Characteristics

RPMCs are characterized by the typical cobblestone appearance at confluence over 90 % when observed by inverted optical microscopy (Fig. 1).

3.4.2 Immuno fluorescence

Cells are identified as mesothelial based on their cobblestone appearance, as well as by immunofluorescence staining performed with monoclonal antibodies to cytokeratin 18(CK-18), vimentin, and alpha-smooth muscle actin (α-SMA). Mesothelial cells are

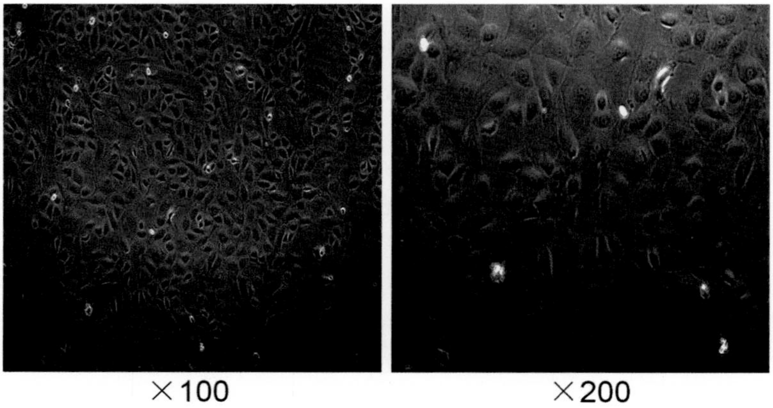

Fig. 1 Phenotype of rat peritoneal mesothelial cells (RPMCs) in culture. Confluent monolayers are obtained after 3–5 days of culture; cells display an obvious cobblestone morphology at confluence. Original magnification ×100 (*left*), ×200 (*right*)

positively stained for CK-18 and vimentin and negative for α-SMA by immunofluorescence. To examine by immunofluorescence:

1. Culture RPMCs on 10-mm cover slips for 24 h or until 80 % confluence.
2. Fix cells in PBS containing 4 % paraformaldehyde (w/v) with 0.1 % Triton X-100 (v/v) for 10 min at room temperature and then in −20 °C methanol for 10 min [13].
3. After washing with PBS, block nonspecific antibody binding sites on cells by incubating with blocking buffer (PBS containing 3 % bovine serum albumin and 10 % goat serum) for 30 min at 37 °C.
4. Incubate cells on cover slips overnight at 4 °C, with the relevant primary antibodies (Table 1) diluted in the blocking solution.
5. After washing with PBS, incubate with Alexa 488 secondary antibody diluted 1:1000 in blocking solution for 1 h at 37 °C.
6. Stain nuclei with a working solution of DAPI for 5 min.
7. Mount samples in anti-fade mounting medium, and put them in a dark place until dry.
8. Analyze and collect images with an immunofluorescent or confocal imaging system.
9. Staining characteristics of cells are shown in Fig. 2. Cells are mostly positive for cytokeratin and vimentin and negative for α-SMA.

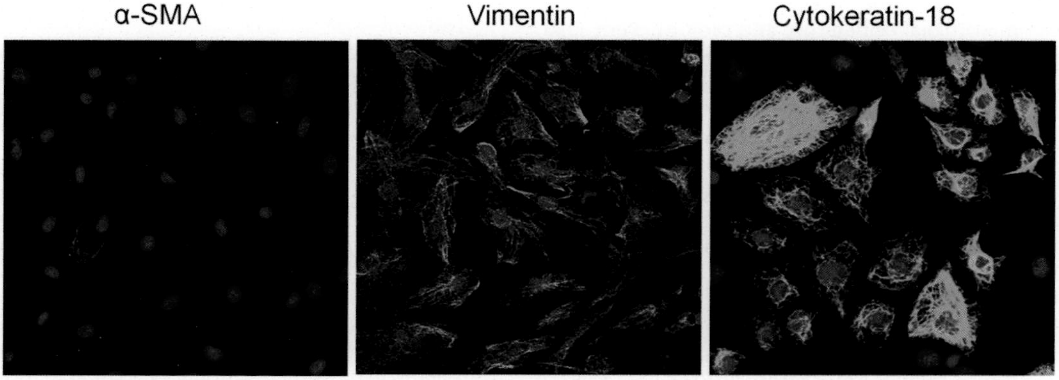

Fig. 2 Immunofluorescent staining of rat peritoneal mesothelial cells. Over 90 % confluent mesothelial cells are positive for cytokeratin and vimentin and negative for α-SMA. Original magnification ×400. Images collected using a Zeiss LSM 510 confocal imaging system

4 Notes

1. Alternatively mix dimethyl sulfoxide (DMSO) in medium 10 % v/v (1-mL DMSO in 9 mL of DMEM/F12 medium supplemented with either 10 % or 20 % FBS).
2. Residual trypsin activity is easily neutralized by the addition of medium containing FBS.
3. Viable cells grow rapidly after adherence; thus, refresh medium every 2–3 days.
4. Keep the culture dish on ice in this step.
5. Remember to wet each sieve first with HBSS otherwise the cell suspension will not pass through.
6. Subheading 3.2.1 has a higher yield of mesothelial cells than Subheading 3.2.2, but there is a greater risk of contamination with endothelial cells from abdominal microvessels with an intraperitoneal injection of trypsin solution into the rat abdomen.
7. RPMCs between passages 1 and 3 are suitable for experiments.

References

1. Jiang Z, Yu X (2011) Advancing the use and quality of peritoneal dialysis by developing a peritoneal dialysis satellite center program. Perit Dial Int 31:121–126
2. Yung S, Chan TM (2007) Mesothelial cells. Perit Dial Int 27(Suppl 2):110–115
3. Yanez-Mo M, Lara-Pezzi E, Selgas R et al (2003) Peritoneal dialysis and epithelial-to-mesenchymal transition of mesothelial cells. N Engl J Med 348:403–413
4. Yung S, Li FK, Chan TM (2006) Peritoneal mesothelial cell culture and biology. Perit Dial Int 26:162–173
5. Rougier JP, Moullier P, Piedagnel R, Ronco PM (1997) Hyperosmolality suppresses but TGF beta 1 increases MMP9 in human peritoneal mesothelial cells. Kidney Int 51:337–347
6. Rampino T, Cancarini G, Gregorini M et al (2001) Hepatocyte growth factor/scatter factor released during peritonitis is active on mesothelial cells. Am J Pathol 159:1275–1285
7. Li S, Zhou Y, Fan J et al (2011) Heat shock protein 72 enhances autophagy as a protective mechanism in lipopolysaccharide-induced peritonitis in rats. Am J Pathol 179:2822–2834
8. Liu Q, Zhang Y, Mao H, Chen W, Luo N, Zhou Q, Yu X (2012) A crosstalk between the Smad and JNK signaling in the TGF-beta-induced epithelial-mesenchymal transition in rat peritoneal mesothelial cells. PLoS One 7, e32009
9. Nie J, Dou X, Hao W et al (2007) Smad7 gene transfer inhibits peritoneal fibrosis. Kidney Int 72:1336–1344
10. Zhou Q, Yang M, Lan H, Yu X (2013) miR-30a negatively regulates TGF-beta1-induced epithelial-mesenchymal transition and peritoneal fibrosis by targeting Snai1. Am J Pathol 183:808–819
11. Margetts PJ, Kolb M, Galt T, Hoff CM, Shockley TR, Gauldie J (2001) Gene transfer of transforming growth factor-beta1 to the rat peritoneum: effects on membrane function. J Am Soc Nephrol 12:2029–2039
12. Shostak A, Pivnik E, Gotloib L (1996) Cultured rat mesothelial cells generate hydrogen peroxide: a new player in peritoneal defense? J Am Soc Nephrol 7: 2371–2378
13. Zhou Q, Fan J, Ding X, Peng W, Yu X, Chen Y, Nie J (2010) TGF-beta -induced MiR-491-5p expression promotes Par-3 degradation in rat proximal tubular epithelial cells. J Biol Chem 285:40019–40027

Chapter 4

Rat Models of Acute and/or Chronic Peritoneal Injuries Including Peritoneal Fibrosis and Peritoneal Dialysis Complications

Masashi Mizuno and Yasuhiko Ito

Abstract

Peritoneal injury is a major cause of discontinuation from long-term peritoneal dialysis. However, the precise mechanisms underlying such injury remain unclear. Suitable animal models of peritoneal injury may be useful to analyze pathogenic mechanisms and facilitate the development of therapeutic approaches. We describe herein two rat models of peritoneal injury that we have recently proposed.

Key words Peritoneal dialysis, Peritoneal injuries, Fibrosis, Zymosan

1 Introduction

In patients with end-stage renal disease, peritoneal dialysis (PD) represents a key option for renal replacement therapy. PD offers patients a number of advantages when compared to hemodialysis. However, continuing PD therapy for the long term is sometimes difficult. Peritoneal injury represents a major reason for discontinuation of PD therapy. PD-related physical stress and peritonitis have been associated with induction of peritoneal injuries and might play roles in the development of encapsulating peritoneal sclerosis (EPS), a lethal complication in PD patients [1]. Fibrosis, infiltration of macrophages, neoangiogenesis, lymphangiogenesis, and dysregulation of the complement system have been reported as factors related to peritoneal injuries [2–7]. However, the detailed mechanisms behind the development and progression of peritoneal injuries remain unclear, as does the final pathway to the EPS that arises in some PD patients. To analyze these mechanisms and develop specific therapies for peritoneal injuries and EPS, in vivo experiments are required, because the information that can be obtained from in vitro experiments is limited.

Tim D. Hewitson et al. (eds.), *Kidney Research: Experimental Protocols*, Methods in Molecular Biology, vol. 1397, DOI 10.1007/978-1-4939-3353-2_4, © Springer Science+Business Media New York 2016

Several animal models have been proposed for the analysis of peritoneal injuries, the most common of which are the chlorhexidine gluconate-induced model [8] and recurrent exposure of peritoneal dialysate. While these are well-known animal models of PD-related peritonitis [9, 10], they have only gone some way in clarifying the mechanisms underlying the development and progression of peritoneal injuries [6].

We describe herein two novel animal models for the analysis of PD-related peritoneal injures.

1.1 Acute Peritoneal Injury Induced by Mechanical Scraping of the Peritoneum in Rats

This model induces acute, temporary inflammation in the peritoneum by mechanical scraping [2]. The condition mimics peritoneal injuries caused by PD-related physical stress. Pathological changes with the accumulation of inflammatory cells are found from day 3. On day 18 after the induction of peritonitis, accumulation of inflammatory cells almost disappears and mild fibrosis can be found along the subperitoneum (Fig. 1).

1.2 A Rat Model of Peritonitis Induced with Injection of a Zymosan Suspension after Mechanical Scraping of the Peritoneum

This model is a modification of the above peritonitis model caused by mechanical scraping, adding five daily administrations of zymosan [6]. Zymosan is a cellular component of yeast as a fungus and activates the complement system through the alternative pathway [11]. In PD patients, fungal peritonitis is known as a severe complication with poor prognosis. A single episode of fungal infection has been reported as sufficient to cause the development of EPS [1]. This model is ideally suited to studying the pathogenesis of fungal peritonitis in PD patients. In this model, a large accumulation of inflammatory cells in the peritoneum is clearly observed on day 3 with thickening of the subperitoneum. Interestingly, in most animals, an accumulation of inflammatory cells is still apparent on day 36 (Fig. 1).

2 Materials

2.1 Animals

We use 7-week-old Sprague-Dawley rats (Japan SLC, Hamamatsu, Japan), 210–230 g initial body weight. Animals are acclimatized for a few days before any treatment and maintained on a 12-h light/dark cycle under conventional laboratory conditions, with free access to water and food.

2.2 Acute Peritoneal Injury Model

1. Disinfectant for surgical wounds: Ethanol (70–80 % v/v) to clean the animal skin before the surgical procedure (*see* **Note 1**).
2. Anesthetic: Pentobarbital sodium (*see* **Note 2**).
3. Surgical instruments: Operating scissors (straight sharp/blunt) and straight iris scissors, thumb forceps, blunt-nose thumb forceps, Kocher forceps, mosquito forceps, and a needle holder. Preoperatively, surgical instruments should be sterilized by dry-heat sterilization (2 h at 160 °C). Surgical procedures are performed under aseptic conditions.

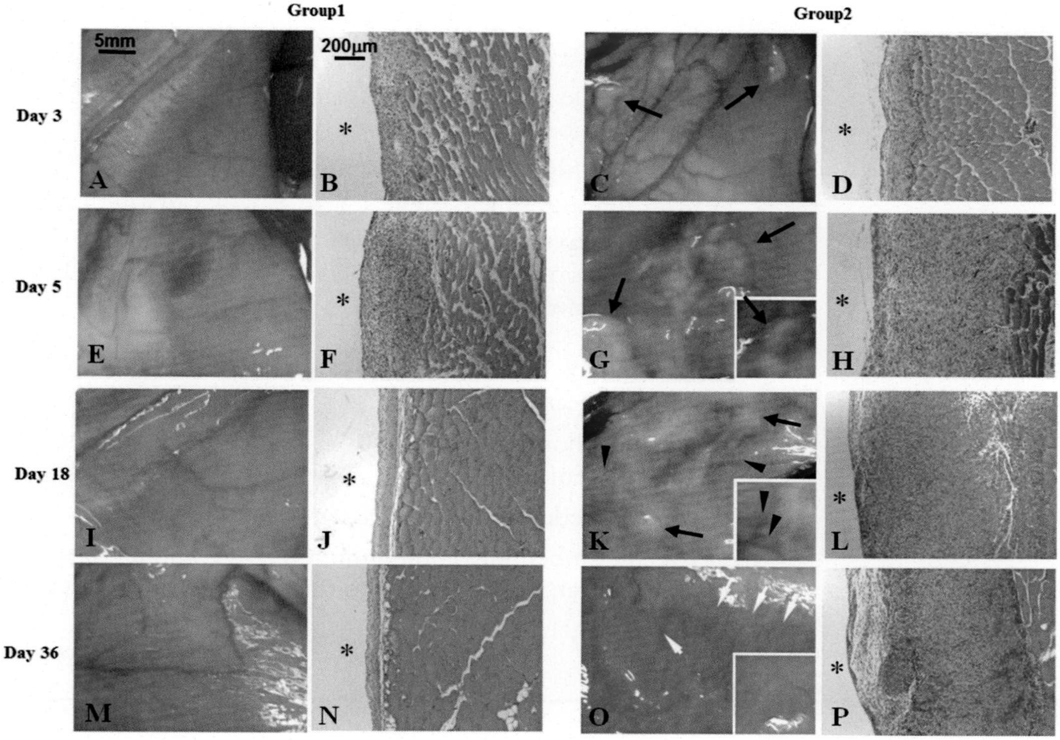

Fig. 1 Histopathological changes in peritonitis induced by mechanical scraping or in peritonitis induced by mechanical scraping following repeated zymosan injections. Frames **a**, **b**, **e**, **f**, **i**, **j**, **m**, and **n** are peritoneum with mechanical scraping in rats, and frames **c**, **d**, **g**, **h**, **k**, **l**, **o**, and **p** are peritoneum with mechanical scraping and repeated zymosan injections. *Left frame* of each set shows the macroscopic appearance of parietal peritoneum. *Right side* of each set shows the microscopic appearance stained with H&E. The original magnifications were 100. The timing of sacrifice is displayed on the left side for each set. *Arrows* in frame **c** indicate small plaques. On day 5, fusion of the plaques was observed, *arrowed* around *red swelling* (magnified and shown with *arrow* in *right bottom corner*). On day 18, the surface of peritoneum was covered with white fibrous tissue and subperitoneal vessels were obscured (*arrowheads* in frame **k**). Fibrous tissue formation was accompanied with neoangiogenesis (*arrowheads* in *right bottom corner* in frame **k**). *White arrowhead* shows vessels obscured under thickened peritoneum on day 36 (frame **o**). External face of peritoneum. Scale bars are in the *upper left corner* of frame **a** for macroscopic appearance and **b** for microscopic appearance. Reproduced from ref. [6], with permission of the American Association of Immunologists (*Copyright 2009. The American Association of Immunologists, Inc.*)

4. Sterilized cotton gauze.
5. Ethanol.
6. Sterile isotonic saline (0.9 % NaCl).
7. Pre-sterilized or disposable sutures (3-0 nylon) and surgical needles (reverse cutting edge and 1/2 circle by choice).
8. 15-mL sterile polypropylene tubes (BD Falcon Conical Centrifuge tubes; BD Biosciences, Bedford, MA, USA).

2.3 Peritonitis Model

The following additional materials are required for the zymosan A model:

1. 50-mL sterile polypropylene tubes (BD Falcon Conical Centrifuge tubes).
2. Zymosan A (Sigma-Aldrich, St. Louis, MO, USA).
3. Dianeal PD-4 4.25 % used as a 4.25 % acidic PD fluid (PDF) (pH ~5.0; Baxter, Tokyo, Japan) [6, 7] or Dianeal-N PD-4 2.5 % used as a 1.5 % neutral PDF (pH ~6.4; Baxter [12] to prepare a zymosan suspension.

2.4 Tissue Processing

1. Fixative: 10 % buffered formalin to fix peritoneal tissues for light microscopy (LM) study.
2. Graded alcohols (50, 75, 100 % v/v) for dehydration and clearing solvent.
3. Paraffin embedding wax and cassettes.
4. Cryogenic embedding compound: OCT™ compound (Sakura Fine Technical, Tokyo, Japan) for immunofluorescence (IF).
5. Liquid nitrogen.

3 Methods

The following animal studies have been approved by the Animal Experimentation Committee of Nagoya University Graduate School of Medicine and are performed according to the Animal Experimentation Guidelines of Nagoya University Graduate School of Medicine (Nagoya, Japan).

3.1 Acute Peritoneal Injury Model [2]

1. At the beginning of the surgery process (operation), the rat is intraperitoneally injected with 50 mg/kg of pentobarbital sodium (*see* **Note 2**).
2. Under anesthetic, fix the rat on a corkboard in a supine operating position.
3. Before starting the operation, shave abdominal hair using hair clippers. The bare abdominal skin is then disinfected with ethanol (*see* **Note 1**).
4. The skin is incised at the abdominal midline with surgical scissors or a scalpel. A midline incision is advantageous because it encounters fewer blood vessels and therefore less bleeding (Fig. 2).
5. After the skin incision, the peritoneum and abdominal muscles are cut together using operating scissors (straight sharp/blunt) (Fig. 2). When opening the abdominal cavity, the peritoneal muscle with the peritoneum should be suspended to separate from the intestines. To prevent injuries to the liver, spleen, and

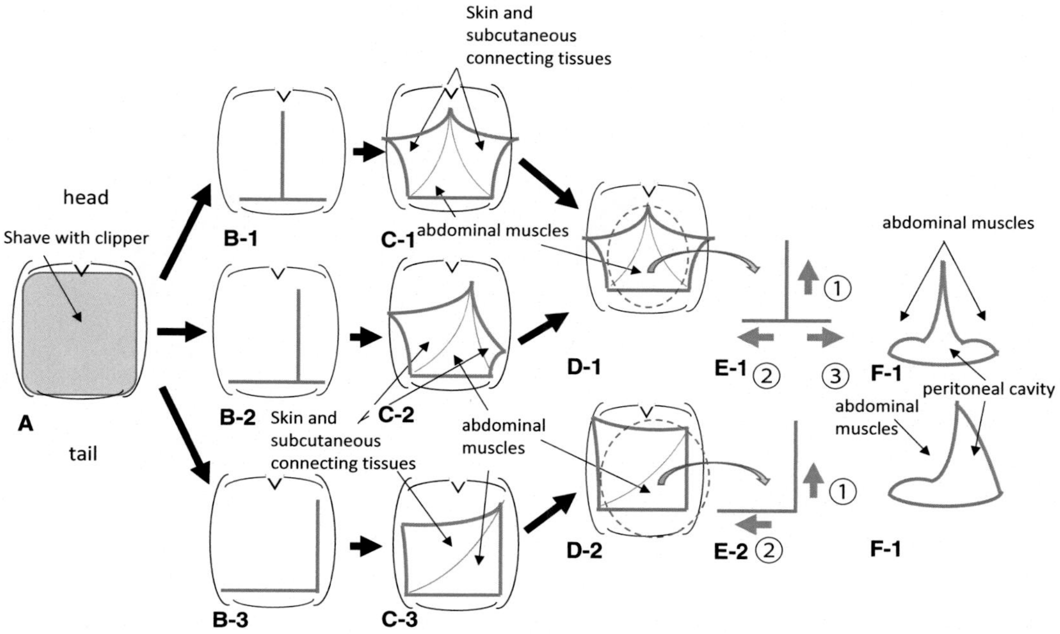

Fig. 2 Incision of abdominal muscle and peritoneum. After incising the skin, the peritoneum is opened using two approaches (frames *D-1*, *E-1*, and *F-1* or frames *D-2*, *E-2*, and *F-2*). The approach shown in *D-1*, *E-1*, and *F-1* is suitable for the skin incision approach of frames *B-1* and *B-2*. Another approach in frames *D-2*, *E-2*, and *F-2* is suitable for the skin incision approach of frames *B-3*. The *red circle* indicates the position to incise abdominal muscle and open the peritoneum. *Black arrows* show the order in which to approach incision processes in rat skin

intestines, the blunt side of the scissors should be inside the peritoneal cavity. If bleeding arises along the edges of incised abdominal muscles, clamp with mosquito forceps to stop bleeding.

6. To scrape the surface of the rat peritoneum, the edge of the opened peritoneum is tightly pinched with blunt-nose thumb forceps to prevent movement of the peritoneum during the scraping process (Fig. 3). The right parietal peritoneum needs to be mechanically scraped twice per second for 60 s using the top of the rim of a sterile 15-ml polypropylene tube. The direction of scraping should be changed from vertically to horizontally or vice versa every 5 s (*see* **Note 3**).
7. If planning to measure peritoneal function at the end of the experiment (e.g., peritoneal equivalent test [2, 13]), it is essential to scrape both right and left sides of the peritoneum.
8. After the scraping of the peritoneum, the abdominal incision is sutured with 3-0 nylon to close the incision. Firstly, the peritoneum and abdominal muscles are sutured together. After closing the peritoneum completely, the skin is sutured separately (*see* **Note 4**).

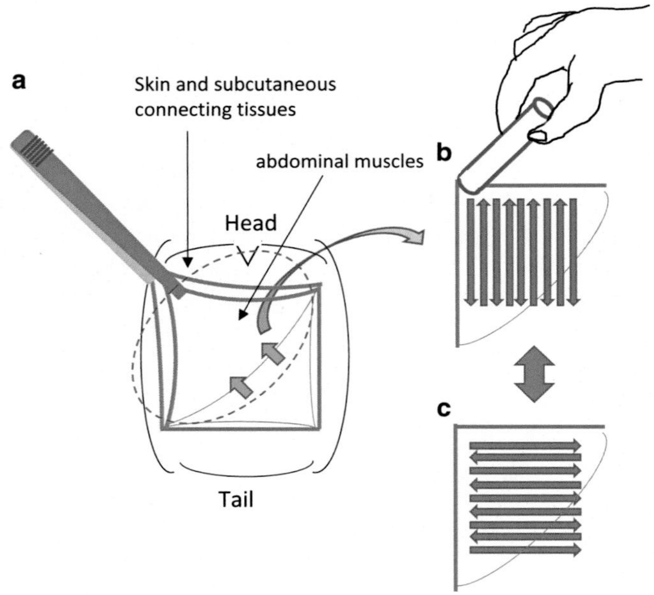

Fig. 3 Scraping the parietal peritoneum. The *red circle* is the area in which to scrape the parietal peritoneum in *A*. *B* and *C* show closeup views of the *red circle* in *A*. *Light blue-colored arrows* show the direction of scraping in the parietal peritoneum. Directions of peritoneal scraping are performed between *B* and *C* in turn for 60 s (*see* **Note 2**)

9. Postoperatively care for rats by keeping in a cage on a warm plate (37 °C) until awake, to prevent loss of body temperatures. Rats can then be permitted ad libitum access to tap water. A subcutaneous injection of 0.9 % isotonic saline is useful for fluid support.

3.2 Peritonitis Model [6]

1. Prepare zymosan suspension in a 15- or 50-mL polypropylene tube by mixing 20 mg of zymosan with 8 mL, 20 mL, and 40 mL of 4.25 % PDF for intraperitoneal administrations on days 0, 1, and 2–4, respectively (*see* **Note 5**). The zymosan suspension needs to be prepared just before each injection.
2. Scrape the rat peritoneum, as described above (*see* **steps 1–8** in Subheading 3.1).
3. When the incised peritoneum is closed by suturing, intraperitoneal injection of zymosan solution (5 mg/2 mL) (*see* **Note 6**) is performed as day 0 treatment just before closing the peritoneum. Tight suturing of the incision is necessary to prevent leakage of the zymosan suspension.
4. Following intraperitoneal injection on day 0, zymosan suspension is injected intraperitoneally with 5 mg/5 mL on day 1 and 10 mg/10 mL on days 2–4 [6] (*see* **Note 6**). The injected area is

not suitable for subsequent examination of peritoneal pathologies. We therefore recommend that each intraperitoneal injection be performed through the unscraped side of the abdomen.

3.3 Collection of Tissue for Analysis of Pathologies

1. Kill animal with an overdose of anesthetic.
2. Perform a laparotomy via a midline incision.
3. Before harvesting peritoneal samples, carefully check the scraped side of the parietal peritoneum macroscopically to confirm site.
4. The peritoneal membrane is then dissected and divided into four tissue blocks (5 mm × 10 mm each) or more for light microscopy and immunofluorescent microcopy analysis (*see* **Note 7**).

3.4 Tissue Processing

1. Fix harvested peritoneal tissue blocks in 10 % buffered formalin overnight for light microscopic analysis. Process to paraffin wax and embed. Cut 3-μm-thick sections from paraffin-embedded blocks for light microscopy.
2. Freeze tissue blocks in a cryogenic embedding compound using liquid nitrogen. Cut 4–5-μm frozen sections with a cryostat.

3.5 Histochemical Analysis

1. The degree of peritoneal fibrosis fibrin exudation can be evaluated using Masson's trichrome stain under LM [6].
2. Peritoneal thickness is used as a marker of damage to the parietal peritoneum. Using light microscopy, measure the peritoneal thickness at a magnification of ×100, at five independent points, in each of 20 fields/a rat.
3. To estimate total cell accumulation, the total number of infiltrated cells is counted in 20 fields at ×200 magnification under LM [6]. Results are expressed as an average per field.
4. Using IF staining, the LCA-positive (leukocyte) and ED1-positive (monocyte-macrophage) cells can be qualitatively and quantitatively evaluated. Again we routinely express cell infiltrate as the average number of LCA or ED1-positive cells in a field under ×200 magnification.

4 Notes

1. Suitable alternatives include povidone-iodine solutions such as the Isodine™ (10 % v/v) (Meiji Seika Pharm. Co., Tokyo, Japan).
2. Alternative approved methods of anesthesia, such as inhalation anesthesia, can be used.
3. Constant performance of the scraping technique may be difficult. It is recommended that the scraping technique be performed by

the same operator for all rats in a single series of experiments. It is recommended that a second person subsequently randomizes animals to individual experimental groups.

4. When incisions are sutured, the recommended interval for sutures is approximately 5 mm for the peritoneum and muscles to prevent leakage of the intraperitoneal dwell. In contrast, more spread out suturing of the skin is acceptable when closing the skin incision. In both cases, running sutures are convenient.
5. In our original description of this method, the zymosan suspension was prepared in an acidic peritoneal dialysate [6]. We have also used zymosan suspension prepared in neutral PDF to induce a slightly milder form of inflammation and peritonitis [12]. Using a neutral PDF results in less cell death than a more acidic solution, with a corresponding less peritoneal injury.
6. After each injection, it is necessary to rub and massage the abdomen to spread zymosan suspension throughout the abdominal cavity.
7. In order to avoid histological artifacts, when harvesting tissue samples, it is important to avoid sampling from the area around both the sutured incision and injection sites. When peritoneal tissues are harvested and dissected, it is important not to touch the peritoneal surface directly.

Acknowledgments

This work was supported in part by Grants-in-Aid 24591227 and 24591228 for Scientific Research from the Ministry of Education, Science, and Culture of Japan and by a Japanese association of dialysis physician grants-in-aid (JADP Grant 2012-07).

References

1. Kawanishi H, Moriishi M (2005) Epidemiology of encapsulating peritoneal sclerosis in Japan. Perit Dial Int 25:S14–S18
2. Nishimura M, Ito Y, Mizuno M et al (2008) Mineralocorticoid receptor blockade ameliorates peritoneal fibrosis in new rat peritonitis model. Am J Physiol Renal Physiol 294: F1084–F1093
3. Sawai A, Ito Y, Mizuno M et al (2011) Peritoneal macrophage infiltration is correlated with baseline peritoneal solute transport rate in peritoneal dialysis patients. Nephrol Dial Transplant 26:2322–2332
4. Mizutani M, Ito Y, Mizuno M et al (2009) Connective tissue growth factor (CTGF/CCN2) is increased in peritoneal dialysis patients with high peritoneal solute transport rate. Am J Physiol Renal Physiol 298:721–733
5. Aguilera A, Yáñez-Mo M, Selgas R et al (2005) Epithelial to mesenchymal transition as a triggering factor of peritoneal membrane fibrosis and angiogenesis in peritoneal dialysis patients. Curr Opin Investig Drugs 6:262–268
6. Mizuno M, Ito Y, Mizuno T et al (2009) Zymosan, but not LPS, developed severe and progressive peritoneal injuries accompanied with complement activation in peritoneal dialysate fluid in a rat peritonitis model with mechanical scraping. J Immunol 183:1403–1412

7. Mizuno M, Ito Y, Mizuno T et al (2012) Membrane complement regulators protect against fibrin exudation in a severe peritoneal inflammation model in rats. Am J Physiol Renal Physiol 2012(302):F1245–F1251
8. Ishii Y, Sawada T, Shimuzu A et al (2001) An experimental sclerosing encapsulating peritonitis model in mice. Nephrol Dial Transplant 16:1262–1266
9. Wieczorowska-Tobis K, Korybalska K et al (1997) Long-term effects of glycylglycine peritoneal dialysis solution with neutral pH on peritoneum in rats. Adv Perit Dial 13:42–46
10. Nakamoto H, Imai H, Ishida Y et al (2001) New animal models for encapsulating peritoneal sclerosis: role of acidic solution. Perit Dial Int 21:S349–S353
11. Sorenson WG, Shahan TA, Simpson J (1998) Cell wall preparations from environmental yeasts: effect on alveolar macrophage function in vitro. Ann Agric Environ Med 5:65–71
12. Kim H, Mizuno M, Furuhashi K et al (2014) Rat adipose tissue-derived stem cells attenuate peritoneal injuries in rat zymosan-induced peritonitis accompanied by complement activation. Cytotherapy 16:357–368
13. Margetts PJ, Kolb M, Yu L et al (2001) A chronic inflammatory infusion model of peritoneal dialysis in rats. Perit Dial Int 21:S368–S372

Chapter 5

Renal Sympathetic Denervation in Rats

Masahiro Eriguchi and Kazuhiko Tsuruya

Abstract

Experimental renal sympathetic denervation is a well-established technique. Classically, renal sympathetic denervation is achieved by dorsal rhizotomy. While more recently, direct renal sympathetic denervation is typically applied by stripping all visible renal nerve bundles followed by painting with a solution of 10 % phenol in ethanol to remove the remaining nerves. In clinical settings, a reliable marker of renal sympathetic denervation or renal sympathetic overactivity has not been established. However, in experimental models, successful renal sympathetic denervation is validated by a decrease in renal norepinephrine content levels. This facilitates the assessment of incomplete denervation by technical failure and reinnervation for long-term experimental models. In this chapter, we introduce comprehensive methods for direct renal sympathetic denervation and measurement of renal norepinephrine content levels.

Key words Sympathetic nervous activity, Renal sympathetic denervation, Norepinephrine, Phenol, Rat model, Reinnervation

1 Introduction

Activation of the sympathetic nervous system is reported in hypertension as well as in heart and renal failure. Although, catheter-based renal sympathetic denervation has recently been applied on a clinical trial basis, the mechanisms involved in its efficacy were not fully elucidated. Nevertheless, this technique is widely considered by clinicians to be an attractive new therapeutic strategy. Experimental studies may therefore provide important information on the mechanisms of renal sympathetic driving antihypertensive and tissue protective effects.

As summarized in Table 1, renal denervation has been applied in various experimental settings [1–18]. The majority of studies use rat models with study durations ranging from 10 days to 24 weeks. However, not all of these disease models of renal denervation exhibit a blood lowering effect. Most experiments confirm the validity of renal denervation by persistent reduction of renal noradrenaline levels to approximately 10 ng/g tissue at the end of the studies.

Tim D. Hewitson et al. (eds.), *Kidney Research: Experimental Protocols*, Methods in Molecular Biology, vol. 1397, DOI 10.1007/978-1-4939-3353-2_5, © Springer Science+Business Media New York 2016

Table 1
Blood lowering effect and study duration of renal denervation in different models

Reference	Published year	Animal	Model	Treatment	Study duration	Blood pressure
1	2015	Wistar rat	NO synthase inhibition	DNx	10 weeks	↓
2	2014	Female SHR	UNx	DNx	3 weeks	↓
3	2013	SHRSP	8 % NaCl diet	DNx	2–4 weeks	↓
4	2013	129S1/SvImJ mouse	UUO	DNx	10 days	N/A
5	2013	SHR/cp	Metabolic syndrome	DNx	19 weeks	↓
6	2012	Sprague-Dawley rat	UNx + aortic regurgitation	DNx	24 weeks	→
7	2011	SHR	DM induced by STZ	DNx	45 days	→
8	2010	Dahl salt-sensitive rat	UNx + 8 % NaCl diet	DNx	6 weeks	→
9	2008	Sprague-Dawley rat	Anti-Thy-1.1 nephritis	DNx	1 week	N/A
10	2008	SPRD-Cy/+ rat	ADPKD	DNx	4 weeks	↓
11	2006	Sprague-Dawley rat	Chronic AII infusion	DNx	5 weeks	↓
12	2005	Sprague-Dawley rat	UNx + DOCA-salt	DNx	47 days	↓
13	2004	Sprague-Dawley rat	DM induced by STZ	DNx	2 weeks	→
14	2002	Sprague-Dawley rat	UNx + cigarette smoke	DNx	12 weeks	→
15	2000	Sprague-Dawley rat	UNx + cyclosporine A	DNx	3 weeks	→
16	1995	Sprague-Dawley rat	5/6 nephrectomy	DR	6 weeks	↓
17	1995	Sprague-Dawley rat	5/6 nephrectomy	DR	6 weeks	↓
18	1994	Sprague-Dawley rat	NO synthase inhibition	DNx	4 weeks	↓

ADPKD autosomal dominant polycystic kidney disease, *AII* angiotensin II, *DM* diabetes mellitus, *DNx* renal denervation, *DOCA* deoxycorticosterone acetate, *DR* dorsal rhizotomy, *NO* nitric oxide, *N/A* not available, *SHR* spontaneously hypertensive rat, *SHRSP* stroke-prone spontaneously hypertensive rat, *SHR/cp* spontaneously hypertensive/NIH-corpulent rat [SHR/NDmc-cp (fat/fat)], *STZ* streptozotocin, *UNX* uninephrectomy, *UUO* unilateral ureteral obstruction

Some studies have also addressed the issue of reinnervation after renal denervation [19, 20], although this remains controversial. For example, complete renal sensory and sympathetic reinnervation was reported at 12 weeks after renal denervation [20]. By contrast, the effects of renal denervation were verified by persistent reduction of renal norepinephrine levels lasting at least 6–9 months [6, 21]. Thus, recovery of renal innervation varies in different models, and verification of renal denervation must be confirmed by measurement of renal norepinephrine at the end of the study.

2 Materials

2.1 Animals and Induction of Anesthesia

1. Rats: More than 200 g of rat weight (6 weeks old and over) is suitable to perform the procedure easily. We use 8-week-old male Wistar rats for renal denervation [1], but any other strain, size, age, or gender is acceptable [2, 3, 5–15, 18].
2. Anesthetic: inhalational anesthetic (e.g., sevoflurane), 10 % sodium pentobarbital in saline (v/v).
3. Warming plate (heated pad) to keep animal warm during surgery.

2.2 Approach to Renal Sympathetic Nerve

1. Stereoscopic microscope. We use a Wild M10 (Leica, Wetzlar, Germany), but any other stereoscopic microscopes are acceptable.
2. Surgical instrument and additional materials: fine blunt-tipped microsurgical tweezers, scissors, two microsurgery hooks (consisting of a clip and an elastic band), cotton swabs, and gauze dressings (Fig. 1).

2.3 Renal Sympathetic Denervation

1. Surgical instruments: fine blunt-tipped microsurgical tweezers and fine-tipped cotton swabs (*see* **Note 1** and Fig. 1).
2. Denervation solution: phenol in ethanol (10 % v/v). Phenol and absolute ethanol should be high purity grade chemicals. Melt phenol with hot water, pipette it, and add in ethanol quickly to avoid re-solidification in a pipette tip (*see* **Note 2**).

2.4 Verification of Renal Sympathetic Denervation (a Measurement of Kidney Norepinephrine Content)

1. Anesthetic: inhalational anesthetic and 10 % v/v of sodium pentobarbital in saline.
2. Surgical instrument and additional materials: tweezers, scissors, cotton swabs, and 21-gauge butterfly needles (Terumo, Tokyo, Japan).
3. 50 mL plastic syringe.
4. Ice-cold phosphate-buffered saline (PBS; pH 7.4).
5. Liquid nitrogen.

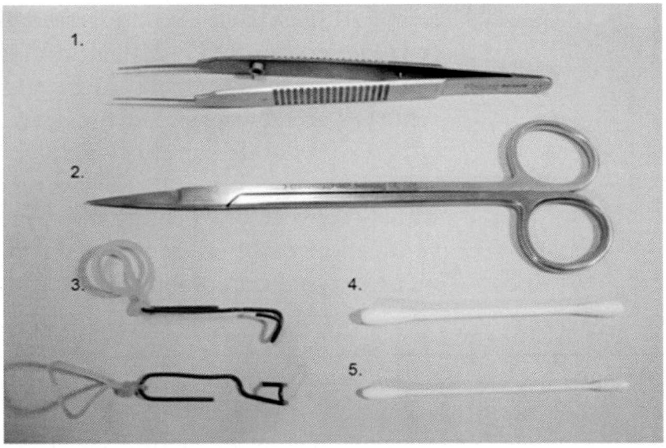

Fig. 1 Surgical instrument and additional materials for renal denervation. (*1*) Fine blunt-tipped microsurgical tweezers, (*2*) scissors, (*3*) two microsurgery hooks (consisting of a clip and elastic band), (*4*) cotton swabs, (*5*) and fine-tipped cotton swabs

6. Homogenizer. We use a TissueLyser™ (Qiagen, Venlo, the Netherlands), but other homogenizers are acceptable.
7. Solution A, a solution for renal tissue homogenization: ice-cold saline containing 1 mM ethylenediaminetetraacetic acid (EDTA), 4 mM sodium disulfite, and 0.01 N hydrochloric acid (*see* **Note 3**).
8. Enzyme-linked immunosorbent assay (ELISA) for measurement of norepinephrine levels: Noradrenaline ELISA kit (Labor Diagnostika Nord, Nordhorn, Germany) (*see* **Note 4**).

3 Methods

3.1 Induction of Anesthesia

1. After brief anesthesia with an inhalational anesthetic, 10 % pentobarbital in saline is injected at 5 mg/100 g body weight intraperitoneally for maintenance anesthesia (*see* **Note 5**). Deep anesthesia will be obtained after 10 min and last for approximately 1 h. During operation the rat should be kept warm on a heated surgical bed to avoid hypothermia.

3.2 Approach to Renal Sympathetic Nerve

1. Position the rat in gently a prone-lateral position. Perform a lateral flank incision and open the retroperitoneal space along the outer side of the iliopsoas muscle using cotton swabs (*see* **Note 6**).
2. After exposing the kidney, a wide operative field is obtained with the help of two microsurgery hooks. At this time, move

aside the kidney with the help of gauze dressings and one side of the microsurgery hook. This facilitates visualization of kidney vessels.

3.3 Renal Sympathetic Denervation

1. With assistance of stereoscopic microscope (magnification ×20–30), tear off as much visible renal nerve bundles as possible using fine blunt-tipped microsurgical tweezers (*see* **Note 7** and Fig. 2).
2. Gently strip all around the renal artery using fine-tipped cotton swabs and fine blunt-tipped microsurgical tweezers (Fig. 3), then paint a solution of 10 % phenol in ethanol for 2 min using new fine-tipped cotton swabs (*see* **Note 8**). Pay attention not to touch the solution to the kidney, because only

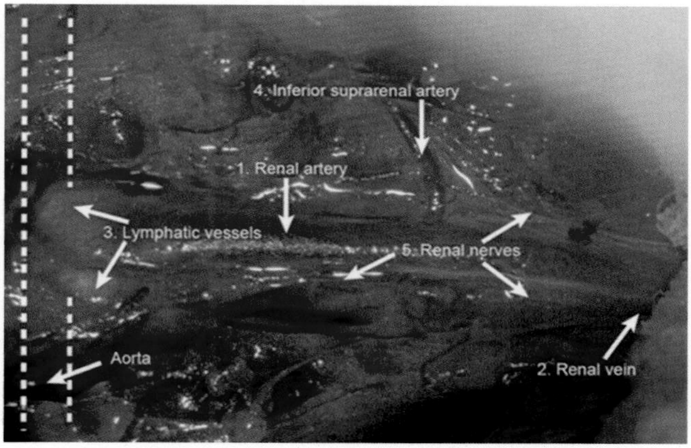

Fig. 2 Anatomy of the renal stalk. (*1*) Renal artery, (*2*) renal vein, (*3*) lymphatic vessel, (*4*) inferior suprarenal artery and, (*5*) renal nerves

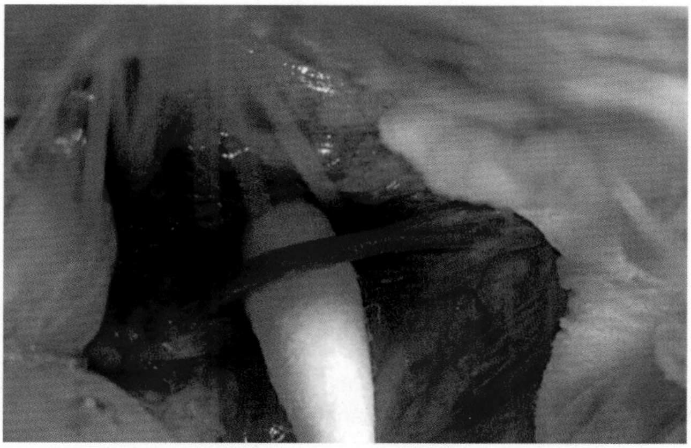

Fig. 3 A successfully stripped renal artery. Ensure to strip the renal artery gently to avoid vasoconstriction and injury to the artery and lymphatic vessel

a small amount of phenol induces kidney injury and renal sympathetic overactivity [22].

3. Wipe the excess solution of phenol in ethanol from around the kidney vessels with another set of new cotton swabs and close the surgical incision.

3.4 Verification of Renal Sympathetic Denervation (a Measurement of Kidney Norepinephrine Content)

1. At the end of the study, measure the kidney norepinephrine content for verification of denervation. Anesthetize the rat as previously described (*see* **step 1** in Subheading 3.1), perform an abdominal midline incision, and then expose the abdominal aorta using cotton swabs.
2. Immediately after blood removal from the cannulated abdominal aorta with 21-gauge butterfly needles, cut the inferior vena cava and slowly perfuse 70 mL of ice-cold PBS (pH 7.4) to harvest the kidneys. Snap-freeze collected renal cortical tissue in liquid nitrogen and store at −80 °C until analysis.
3. Homogenize the renal cortex in Solution A for 3 min at 30 Hz using a tissue homogenizer. Centrifuge the homogenate at 10,000 × *g* for 10 min at 4 °C and collect the supernatant.
4. Measure the norepinephrine level of the supernatant by ELISA kit according to the manufacturer's manual (*see* **Note 4**). Determine the kidney norepinephrine content by correcting the supernatant norepinephrine for the weight of renal cortex measured. As shown in Fig. 4 [1], confirm the validity of denervation by a decrease in norepinephrine content of the denervated kidney to approximately 5 % of that seen in the innervated kidney (*see* **Note 9**).

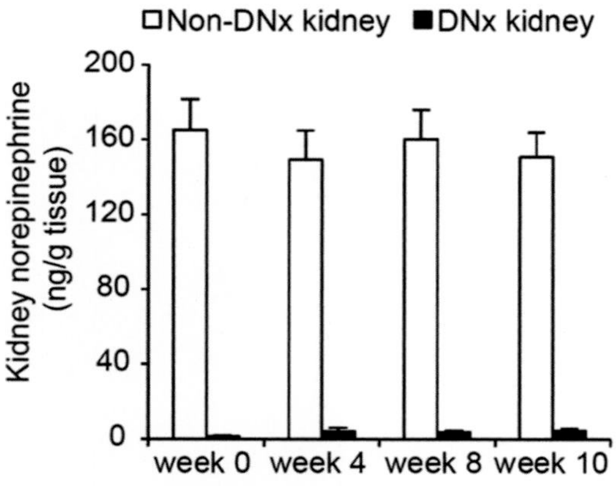

Fig. 4 Renal norepinephrine content levels in the unilateral renal denervation model. Renal cortical norepinephrine content in the denervated kidneys is markedly decreased compared with those from the innervated kidney, within the same rats, showing the validity of renal denervation throughout the study. Legend: *DNx* renal denervation. Reproduced from [1] with permission

4 Notes

1. Fine-tipped but blunt microsurgical tweezers will avoid injuring renal vessels. Fine-tipped cotton swabs for babies are suitable for treating renal vessels gently.
2. The melting point of phenol is 40.5 °C. Phenol in ethanol is stable for months at room temperature in a PTFE microtube.
3. EDTA is very insoluble at low pH and low temperature. Firstly, dissolve EDTA and sodium disulfite in hot water and add HCl just before use. The solution is stable in a plastic tube, but not in a glass bottle, and the solution should be freshly prepared for each use.
4. Alternatively, kidney norepinephrine content levels can be measured by commercially available high-performance liquid chromatography analyses.
5. Ensure to obtain high enough anesthesia to avoid pain-induced vasoconstriction. Therefore, no reaction by pricking the tail should be confirmed. If necessary, add additional pentobarbital intraperitoneally.
6. During opening of the retroperitoneal space, ensure not to perforate the peritoneal membrane. This leads to influx ascites and anesthetic into the retroperitoneal space, making it difficult to keep the operative field dry and to paint phenol around the kidney vessels.
7. Make sure not to injure the inferior suprarenal artery that often branches from the renal artery and the lymphatic vessel that runs around the renal artery. Note that treating the renal artery roughly induces vasoconstriction.
8. Instead of 10 % phenol in ethanol, 95 % ethanol can be applied for 10 days in a short-term mouse model, as reported by Kim et al. [4].
9. Norepinephrine content levels of innervated kidneys are approximately 160–170 ng/g tissue (can vary between 120 and 230 ng/g tissue in various reports), while those of denervated kidneys are approximately 10 ng/g tissue as previously described [1–3, 5, 11, 12, 17, 23].

References

1. Eriguchi M, Tsuruya K, Haruyama N et al (2015) Renal denervation has blood pressure-independent protective effects on kidney and heart in a rat model of chronic kidney disease. Kidney Int 87:116–127
2. Maranon RO, Lima R, Mathbout M et al (2014) Postmenopausal hypertension: role of the sympathetic nervous system in an animal model. Am J Physiol Regul Integr Comp Physiol 306:R248–R256
3. Nakagawa T, Hasegawa Y, Uekawa K et al (2013) Renal denervation prevents stroke and brain injury via attenuation of oxidative stress in hypertensive rats. J Am Heart Assoc 2(5):e000375

4. Kim J, Padanilam BJ (2013) Renal nerves drive interstitial fibrogenesis in obstructive nephropathy. J Am Soc Nephrol 24:229–242
5. Katayama T, Sueta D, Kataoka K et al (2013) Long-term renal denervation normalizes disrupted blood pressure circadian rhythm and ameliorates cardiovascular injury in a rat model of metabolic syndrome. J Am Heart Assoc 2(4):e000197
6. Rafiq K, Noma T, Fujisawa Y et al (2012) Renal sympathetic denervation suppresses de novo podocyte injury and albuminuria in rats with aortic regurgitation. Circulation 125:1402–1413
7. Dias LD, Casali KR, Leguisamo NM et al (2011) Renal denervation in an animal model of diabetes and hypertension: impact on the autonomic nervous system and nephropathy. Cardiovasc Diabetol 10:33
8. Nagasu H, Satoh M, Kuwabara A et al (2010) Renal denervation reduces glomerular injury by suppressing NAD(P)H oxidase activity in Dahl salt-sensitive rats. Nephrol Dial Transplant 25:2889–2898
9. Veelken R, Vogel EM, Hilgers K et al (2008) Autonomic renal denervation ameliorates experimental glomerulonephritis. J Am Soc Nephrol 19:1371–1378
10. Gattone VH II, Siqueira TM Jr, Powell CR et al (2008) Contribution of renal innervation to hypertension in rat autosomal dominant polycystic kidney disease. Exp Biol Med (Maywood) 233:952–957
11. Hendel MD, Collister JP (2006) Renal denervation attenuates long-term hypertensive effects of angiotensin II in the rat. Clin Exp Pharmacol Physiol 33:1225–1230
12. Jacob F, Clark LA, Guzman PA et al (2005) Role of renal nerves in development of hypertension in DOCA-salt model in rats: a telemetric approach. Am J Physiol Heart Circ Physiol 289:H1519–H1529
13. Luippold G, Beilharz M, Muhlbauer B (2004) Chronic renal denervation prevents glomerular hyperfiltration in diabetic rats. Nephrol Dial Transplant 19:342–347
14. Odoni G, Ogata H, Viedt C et al (2002) Cigarette smoke condensate aggravates renal injury in the renal ablation model. Kidney Int 61:2090–2098
15. Elzinga LW, Rosen S, Burdmann EA et al (2000) The role of renal sympathetic nerves in experimental chronic cyclosporine nephropathy. Transplantation 69:2149–2153
16. Campese VM, Kogosov E, Koss M (1995) Renal afferent denervation prevents the progression of renal disease in the renal ablation model of chronic renal failure in the rat. Am J Kidney Dis 26:861–865
17. Campese VM, Kogosov E (1995) Renal afferent denervation prevents hypertension in rats with chronic renal failure. Hypertension 25: 878–882
18. Matsuoka H, Nishida H, Nomura G et al (1994) Hypertension induced by nitric oxide synthesis inhibition is renal nerve dependent. Hypertension 23:971–975
19. Kline RL, Stuart PJ, Mercer PF (1980) Effect of renal denervation on arterial pressure and renal norepinephrine concentration in Wistar-Kyoto and spontaneously hypertensive rats. Can J Physiol Pharmacol 58:1384–1388
20. Mulder J, Hokfelt T, Knuepfer MM et al (2013) Renal sensory and sympathetic nerves reinnervate the kidney in a similar time dependent fashion following renal denervation in rats. Am J Physiol Regul Integr Comp Physiol 304:R675–R682
21. Grisk O, Grone HJ, Rose HJ et al (2001) Sympathetic reinnervation of rat kidney grafts. Transplantation 72:1153–1155
22. Ye S, Gamburd M, Mozayeni P et al (1998) A limited renal injury may cause a permanent form of neurogenic hypertension. Am J Hypertens 11:723–728
23. Veelken R, Hilgers KF, Porst M et al (2005) Effects of sympathetic nerves and angiotensin II on renal sodium and water handling in rats with common bile duct ligature. Am J Physiol Renal Physiol 288:F1267–F1275

Chapter 6

Decellularization of Rat Kidneys to Produce Extracellular Matrix Scaffolds

Mei Jin, Yu Yaling, Wang Zhibin, and Zhang Jianse

Abstract

The extracellular matrix (ECM) retains three-dimensional structures for the stimulation of cell growth, with components of the ECM relatively conserved between species. Interest in the use of decellularized scaffold-based strategies for organ regeneration is increasing rapidly. Decellularized scaffolds derived from animal organs are a promising material for organ engineering, with a number of prominent advances having been reported in the past few years.

In this article we describe a simple and robust methodology for generating decellularized rat kidneys. To obtain these scaffolds, we perfuse rat kidneys with detergents through the abdominal aorta. After decellularization, kidney scaffolds are harvested for evaluation of vascular structure and histology. Qualitative evaluation involves vascular corrosion casting, transmission electron microscopy, and several different histological and immunofluorescent methods. SDS residue levels are assessed by ultraviolet-visible spectrophotometer (UV-VIS).

Key words Kidney, Decellularized scaffolds, Tissue/organ engineering, Regeneration, Extracellular matrix

1 Introduction

Chronic kidney disease is an increasing public health issue, affecting between 8 and 16 % of the global adult population [1]. Although end stage renal disease (ESRD) can be managed with dialysis, transplantation remains the only available curative treatment. However, there is a long-term requirement for immune suppression. Furthermore, the supply of donor kidneys is inadequate to meet demand, with less than 20 % of US patients [2, 3] and only 1 % of patients in China [1] with ESRD expected to receive a kidney transplant. Tissue-engineered kidney substitutes generated in vitro or in vivo could therefore offer new strategies for ESRD. Possibilities include implantation of a bioactive kidney scaffold with a three-dimensional structure for the stimulation of cell growth, with and without implanting cells. The former needs a

Tim D. Hewitson et al. (eds.), *Kidney Research: Experimental Protocols*, Methods in Molecular Biology, vol. 1397, DOI 10.1007/978-1-4939-3353-2_6, © Springer Science+Business Media New York 2016

bioreactor for the cultivation of cells under monitored and controlled environmental and operational conditions [4], while the latter will use organisms as an autologous bioreactor. Both concepts require a kidney scaffold with a three-dimensional (3D) structure to mimic the real kidney and stimulate cell growth.

Meezan et al. [5] first described a method for the isolation of extracellular matrix from tissues in 1975. This has allowed the exploration of decellularization. With the development of tissue engineering, decellularized (DC) scaffolds have become a sensational basic material, as it can potentially retain the 3D architecture, including the microvasculature, of the original tissue. Notable advances in the development of tissue engineering have been made in multiple organs including the kidney [6], heart [7], trachea [8, 9], liver [10], and lung [11]. Since the first report by Nakayama et al. [12] of a decellularized rhesus monkey kidney, decellularized kidneys have been obtained from a number of species including the pig [13–15], rodent [16, 17], and human [6]. In addition, based on our previous experience on transplanting kidney DC scaffolds [16], DC kidney scaffolds are able to act as inductive template for functional organ recovery, allowing the damaged area to recellularize with autologous stem cells or differentiated cells. Given the importance of maintaining renal function, the decellularization of kidney tissue has received much attention.

We have recently developed a reliable protocol to generate DC rat kidney scaffolds, which is described here. These scaffolds retain intact vascular trees and overall architecture, such as a continuous Bowman's capsule, the basement membrane of the glomerular capillaries, and mesangial matrix, but lose all cellular components. Importantly, the detergent residue is controlled within nontoxic levels.

2 Materials

2.1 Preparation of Decellularized (DC) Kidney Scaffolds

1. Deionized water (dH_2O).
2. 0.01 M phosphate-buffered saline, pH 7.4 (PBS).
3. 5 % chloral hydrate.
4. 24-gauge cannula.
5. Peristaltic pump.
6. Antibiotics: 100 U/mL penicillin and 100 ng/mL streptomycin.
7. 50 U/mL heparin in 0.01 M PBS.
8. 0.8 % sodium dodecyl sulfate (SDS) in dH_2O (w/v) (*see* **Notes 1, 2**).
9. 0.1 % Triton X-100 in dH_2O (v/v).
10. Sprague-Dawley (SD) rats, approximately 2 months of age and 200–250 g in weight.

2.2 Assessing SDS Residue Levels in DC Kidney Scaffold

1. SDS standards: Prepare eight different concentrations of SDS ranging between 0 and 200 mg/mL in dH_2O.
2. 100 mg/mL proteinase K (Biomiga, San Diego, CA, USA) in protein lysis buffer (Biomiga).
3. Ultraviolet-visible spectrophotometer.
4. High-speed centrifuge.

2.3 Vascular Corrosion Casting of DC Kidney Scaffolds

1. Acetone.
2. ABS Sudan solvent mixture: 0.5 % w/v Sudan red dye in 10 % solution of acrylonitrile butadiene styrene (ABS) in acetone (w/v).
3. ABS blue pigment mixture: 0.1 % w/v blue pigment in 10 % solution of acrylonitrile butadiene styrene (ABS) in acetone (w/v).
4. 50 % hydrochloric acid in dH_2O (v/v).
5. Stereomicroscope (with camera).

2.4 Histology and Immunofluore scence Analysis

1. Dimethylbenzene.
2. Ethanol: 100, 90, 80, 70, and 50 % v/v solutions of ethanol in dH_2O.
3. Harris hematoxylin.
4. 1 % hydrochloric acid in ethanol (v/v).
5. 1 % ammonia in dH_2O (v/v).
6. 1 % eosin in dH_2O (w/v).
7. Neutral balsam.
8. Periodic acid.
9. Schiff reagent.
10. 0.5 % sodium metabisulfite: Combine 10 mL of a 10 % aqueous sodium metabisulfite stock, 10 mL 1 N hydrochloric acid, and 180 mL dH_2O.
11. Sulfuric acid.
12. Ponceau acid magenta.
13. 1 % phosphomolybdic acid in dH_2O (v/v).
14. Aniline blue.
15. 3 % solution of hydrogen peroxide in dH_2O (v/v).
16. Sodium citrate buffer: 10 mM sodium citrate, 0.05 % Tween 20 (v/v), pH 6.0.
17. Pressure cooker.
18. 2 % acetic acid in dH_2O (v/v).
19. Block: 5 % bovine serum albumin (BSA) in PBS.

20. Primary antibodies diluted in 1 % BSA in PBS: rabbit anti-collagen IV antibody (1:200, Sigma-Aldrich, St. Louis, MO, USA), rabbit anti-laminin antibody (1:200, Sigma-Aldrich), and rabbit anti-fibronectin antibody (1:200, Sigma-Aldrich).
21. Secondary antibody diluted in 1 % BSA in PBS: goat anti-rabbit IgG-FITC antibody (1:250, Sigma-Aldrich).
22. 4′,6-Diamidino-2-phenylindole (DAPI).
23. Antifade mounting medium.
24. Bright-field microscope.
25. Fluorescent microscope with filter sets for DAPI and FITC.

2.5 Transmission Electron Microscope Observation

1. 2.5 % glutaraldehyde in 0.1 M phosphate buffer: Combine 10 mL of 25 % stock glutaraldehyde and 50 mL 0.2 M phosphate buffer, pH 7.4. Make up to 100 mL with dH_2O.
2. 1 % osmium tetroxide in dH_2O (w/v).
3. 3 % uranyl acetate in dH_2O (w/v).
4. Acetone.
5. Embedding media: Epon 812.
6. Ultramicrotome.
7. EM grids.
8. Lead citrate.
9. Electron microscope.

3 Methods

3.1 Preparation of DC Kidney Scaffolds

1. Anesthetize rats with 5 % chloral hydrate (0.6 ml/100 g) via intraperitoneal injection.
2. Open abdominal cavity by a ventral midline incision, extending from the pubis to the xiphoid process.
3. Ligate suprarenal abdominal aorta, suprarenal inferior vena cava, lumbar arteries, testicular arteries of male rats, and ovarian arteries of female rats (*see* **Note 3**).
4. Insert a 24 G cannula into the infrarenal abdominal aorta.
5. Remove bilateral kidneys with blood vessels connected.
6. Connect the cannula with the peristaltic pump to allow continuous rinsing with various detergents (*see* **Note 2**). Perfuse at approximately 8 mL/min (*see* **Note 4**) in the following order (*see* **Note 5**):
   (a) 50 U/mL heparin in 0.01 M PBS for 30 min (*see* **Note 6**)
   (b) 0.1 % Triton X-100 for 3 h (*see* **Note 7**)

(c) dH_2O for 30 min

(d) 0.8 % SDS in dH_2O (v/v) for 3 h (*see* **Notes 1** and 7)

(e) dH_2O containing 100 U/mL penicillin and 100 mg/mL streptomycin for 24 h (*see* **Note 5**)

7. Once perfusion is complete, DC kidney scaffolds can be stored in 50 mL of dH_2O containing penicillin and streptomycin at 4 °C until use (*see* **Note 9**).

3.2 Assessing SDS Residue Levels in DC Kidney Scaffold

Residual SDS can be assessed using an ultraviolet-visible spectrophotometer (UV-VIS).

A standard curve for SDS is generated from the absorbance of eight different known concentrations of SDS. This curve is then used to calculate SDS residue in samples.

1. Prepare standard solutions of SDS in dH_2O (0–200 mg/mL).
2. Read absorbance by UV-VIS at 499 nm.
3. Plot to create a standard curve and use statistical software to calculate a regression line and equation.
4. Sponge up water inside the DC kidney scaffolds, and digest scaffolds with 100 mg/mL proteinase K (*see* **Note 10**) in protein lysis buffer at 50 °C overnight.
5. Centrifuge at 14,000 × g for 5 min to remove the precipitate.
6. Measure light absorption of the supernatant at 499 nm.
7. Calculate content of SDS in the supernatant by reference to the calibrated standard curve prepared above.

3.3 Vascular Corrosion Casting of DC Kidney Scaffolds

To confirm the integrity of microvasculature in the DC kidney scaffolds, we perform vascular corrosion casting at the end of perfusion in a representative normal and DC kidney.

1. At the end of perfusion, catheterize the inferior vena cava and abdominal aorta.
2. Inject 1–2 ml of acetone into DC kidney scaffolds through the inferior vena cava.
3. Pass 5 ml of 10 % acrylonitrile butadiene styrene (ABS) Sudan solvent mixture (*see* **Note 11**) through the abdominal aorta.
4. Simultaneously perfuse 10 mL of 10 % ABS blue pigment mixture via the inferior vena.
5. Harvest bilateral kidneys.
6. Cool samples in running water and corrode in 50 % hydrochloric acid in dH_2O for 1–3 days.
7. The morphology and distribution of vasculature can be observed under stereomicroscope and recorded by imaging.

3.4 Histology and Immunofluorescence Analysis

Samples are prepared for histological and immunofluorescence analyses by following standard protocols for paraffin embedding and sectioning (*see* **Note 12**).

3.4.1 Hematoxylin-Eosin Staining

1. Dry kidney sections at 60 °C, 8 h.
2. Dewax with dimethylbenzene for 30 min × 2.
3. Rehydrate with 100, 100, 90, 80, 70, and 50 % ethanol consecutively over 5 min.
4. Wash in dH_2O, 5 min.
5. Immerse in hematoxylin, 2 min.
6. Rinse under running water, 5–10 s.
7. Color separate with 1 % hydrochloric acid in ethanol for 3 s.
8. Rinse under running water, 1–2 s.
9. Promote blue with 1 % ammonia, 5–10 s.
10. Rinse under running water, 1–2 s.
11. Immerse in 1 % Eosin in dH_2O, 10 s.
12. Rinse under running water, 1–2 s.
13. Dehydrate with 80 %, 90 %, and 100 % alcohol for 2 min each.
14. Immerse in dimethylbenzene for 5 min × 2.
15. Mount with neutral balsam.
16. Examine and image slides with a bright-field microscopy.

3.4.2 Periodic Acid-Schiff Staining

1. Dry kidney sections at 60 °C, 8 h.
2. Dewax with dimethylbenzene for 30 min × 3.
3. Rehydrate with 100, 100, 90, 80, 70, and 50 % v/v ethanol in distilled water for a total of 5 min.
4. Wash in distilled water, 5 min.
5. Immerse in 0.5 % periodic acid solution, 5 min.
6. Rinse under 70 % alcohol, 5–10 s.
7. Immerse in Schiff reagent, 20 min.
8. Immerse in 0.5 % sodium metabisulfite solution, 2 min.
9. Rinse under running water, 10 min.
10. Immerse in hematoxylin, 2 min.
11. Promote blue with 1 % ammonia, 5–10 s.
12. Rinse under running water, 30 s.
13. Sequentially dehydrate with 80 %, 90 %, and 100 % v/v ethanol in distilled water for 5 min.
14. Immerse in dimethylbenzene for 10 min × 2.
15. Mount with neutral balsam.
16. Observe slides and capture images using a bright field.

3.4.3 Masson's Staining

1. Dry kidney sections at 60 °C, 8 h.
2. Dewax with dimethylbenzene for 30 min twice.
3. Rehydrate with 100, 100, 90, 80, 70, and 50 % v/v ethanol in dH_2O sequentially for a total of 5 min.
4. Wash in dH_2O, 5 min.
5. Immerse in hematoxylin, 5 min.
6. Wash in dH_2O, 10 min.
7. Immerse in Ponceau acid magenta, 5 min.
8. Immerse in 2 % acetic acid, 30 s.
9. Immerse in 1 % phosphomolybdic acid, 3 min.
10. Dye with aniline blue, 5 min.
11. Immerse in 0.2 % acetic acid, 30 s.
12. Dehydrate with 90 % and 100 % alcohol for 3 min.
13. Immerse in dimethylbenzene for 5 min × 2.
14. Mount with neutral balsam.
15. Observe slides with a bright-field microscope and capture images.

3.4.4 Immunofluorescent Staining

1. Dry kidney sections at 60 °C, 8 h.
2. Dewax with dimethylbenzene for 30 min × 2.
3. Rehydrate with 100, 100, 90, 80, 70, and 50 % ethanol sequentially for 5 min each.
4. Wash in PBS, 3 × 5 min.
5. High-pressure antigen retrieval: Heat slides to 90 kpa in citrate buffer antigen retrieval solution for 2 min, and then naturally cool to room temperature.
6. Quench with 0.3 % hydrogen peroxide for 10 min to eliminate endogenous peroxidase activity.
7. Wash in PBS, 3 × 5 min.
8. Incubate in block (5 % BSA in PBS) for 1 h.
9. Drain and wipe off excess serum and incubate in primary antibody overnight at 4 °C.
10. Wash in PBS, 3 × 5 min.
11. Incubate in secondary antibody at room temperature for 2 h.
12. Wash in PBS, 3 × 5 min.
13. Counterstain nuclei in DAPI for 10 min.
14. Wash in PBS, 3 × 5 min.
15. Mount with antifade mounting medium.
16. Observe and image slides with a fluorescent microscope using filter sets appropriate for DAPI and FITC.

3.5 Transmission Electron Microscope Observation

Transmission electron microscopy is used to characterize the microstructure of the DC kidney scaffolds, with very high spatial resolution.

1. Fix samples with 2.5 % glutaraldehyde in 0.1 M phosphate buffer overnight at 4 °C.
2. Wash in PBS, 3 × 15 min.
3. Postfix with 1 % osmium tetroxide for 1 h at 37 °C.
4. Wash in PBS, 2 × 15 min.
5. Stain with 2 % uranyl acetate, 1 h.
6. Dehydrate with 70, 80, 90, 100, and 100 % ethanol in 0.1 M PBS for 15 min each.
7. Infiltrate with a 1:1 mixture of acetone-embedding fluid, 1 h at 37 °C.
8. Infiltrate with acetone-embedding (1:4) fluid overnight at 37 °C.
9. Infiltrate with embedding fluid, 1 h at 45 °C.
10. Solidify 3 h at 45 °C and 48 h at 65 °C.
11. Use an ultramicrotome to cut ultrathin sections (80 nm) and mount on grids.
12. Double stain with a 3 % aqueous solution of uranyl acetate and lead citrate for 2 min each, with a wash in dH_2O in-between.
13. Examine sections using an electron microscope at 70 kV, and image by high-resolution CCD digital camera.

3.6 Analysis and Interpretation

After decellularization with continuous detergent perfusion, DC kidney scaffolds have a somewhat transparent appearance (Fig. 1a). Vascular corrosion casting shows that the vascular tree in the DC kidney scaffolds is well maintained compared with intact kidney (Fig. 1b, c). Standard H&E staining reveals that blue-stained nuclei are not observed, but pink-stained components are present in the DC kidney scaffolds. Control tissue is shown in Fig. 1d, e. As pink-stained components include both cytoplasm and extracellular matrices, this may reflect the morphological difference in H&E staining between intact kidney and DC kidney scaffolds (Fig. 1d, e). In addition, PAS and Masson's staining can be used to examine the ECM (e.g., scaffolds lose renal cells but keep normal vascular tree and continuous extracellular matrix). Immunofluorescence analysis shows that protein components (notably collagen IV, laminin, and fibronectin) of DC kidney scaffolds remain intact, while DAPI staining was negative in our scaffolds (Fig. 2). Our previous electron microscopy studies have shown that DC kidney scaffolds maintain an intact Bowman's capsule, glomerular capillary basement membrane, and mesangial matrix (Fig. 1h).

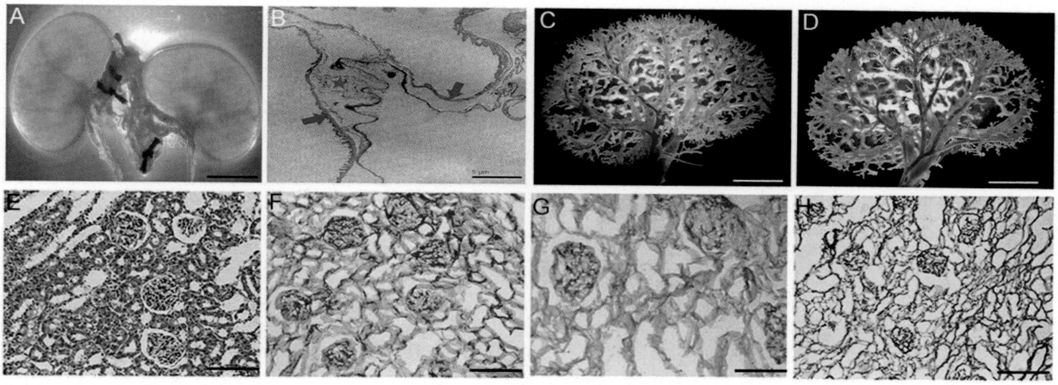

Fig. 1 Characterization of the DC kidney scaffolds. (**a**) Gross appearance of harvested DC kidney scaffolds. (**b**) Electron microscopy observation shows intact extracellular matrix in DC kidney scaffold. *Blue arrows* indicate the membrane of Bowman's capsule, a *green arrow* indicates the basement membrane of the glomerular capillaries, and a *red star* points to the mesangial matrix. (**c** and **d**) Vascular corrosion casting shows a normal vascular tree of DC kidney scaffold (**d**) compared with intact kidney (**c**). (**e** and **f**) H&E staining shows the existence of *blue*-stained nuclei in intact kidney (**e**) but not DC kidney scaffold (**f**). (**g**) Masson's staining shows that green-stained collagenous fibers in DC kidney scaffold. (**h**) PAS staining shows the presence of the ECM (e.g., basement membranes) in DC kidney scaffold. Note that the capillary loops of the glomeruli are clearly displayed. Scale bars = 5 mm (**a**, **c**, and **d**), 100 μm (**e**–**h**) and 5 μm (**b**) (Reproduced from YL Yu et al. Biomaterials 2014; 35: 6822–6828 with permission of Elsevier)

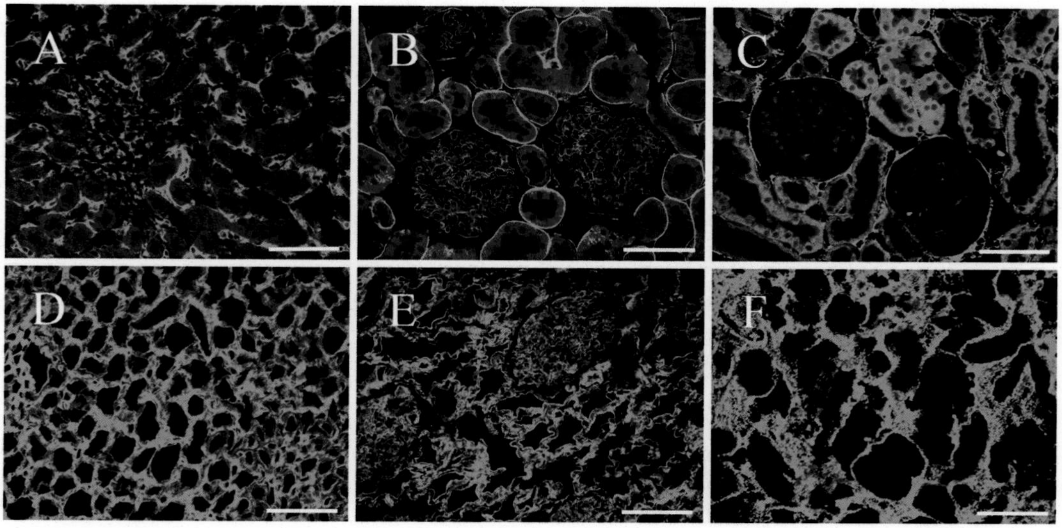

Fig. 2 Immunofluorescence of collagen IV, laminin (LN), and fibronectin (FN). (**a**–**c**) Native kidney. (**d**–**f**) Decellularized kidney scaffolds. (**a** and **d**) Scale bars = 100 μm. (**b**, **c**, **e**, and **f**) Scale bars = 50 μm. In all panels, cell nuclei stain was *blue* with DAPI, while fluorescent immunohistochemical staining for specific markers appears *green*. These indicate the intact kidney architecture and matrix proteins are retained and undisturbed, while cells and nuclear material are removed compared with the native

Given that SDS is toxic to cells, it is important to confirm the absence of SDS residues in the scaffolds. In our studies, the concentration of SDS residues in DC kidney scaffolds is typically 50.0 ± 1.7 μg/g, which is well below the toxic level of 133.3 μg/g.

4 Notes

1. In the decellularization process, it is very important that the detergent used is sodium dodecyl sulfate, not sodium dodecyl sulfonate.
2. For decellularization, we don't use trypsin, as it will damage the structural proteins severely.
3. Before cannulation, make sure that ligated vessels include the suprarenal vena cava, suprarenal abdominal aorta, lumbar arteries, testicular arteries of male, and ovarian arteries of female animals.
4. A higher or lower flow velocity is not suitable.
5. All of the fluids for decellularization should be preheated to 25–30 °C.
6. After abdominal aorta cannulation, perfuse with heparin in PBS for 30 min to prevent blood clotting.
7. The concentration of SDS and Triton X-100 must be checked, as elevated SDS and Triton X-100 will damage the extracellular matrix microstructure. Conversely lower detergent concentrations do not work well.
8. Given that SDS is toxic to cells, it is necessary to ensure that there is sufficient perfusion of deionized water to remove all traces of SDS.
9. The prepared DC scaffolds can be preserved for 1 week at 4 °C and 2 months at −80 °C.
10. To detect SDS residue, the samples must be digested completely.
11. Avoid over injection of pigments into the scaffolds when preparing vascular corrosion casts.
12. Sections for histology and immunofluorescent analysis should be approximately 5 μm in thickness.

References

1. Jha V, Garcia-Garcia G, Iseki K et al (2013) Chronic kidney disease: global dimension and perspectives. Lancet 382(9888):260–272
2. Centers for Disease Control and Prevention. National chronic kidney disease fact sheet: general information and national estimates on chronic kidney disease in the United States, 2010. http://www.cdc.gov/DIABETES//pubs/factsheets/kidney.htm. Accessed 2 Mar 2015
3. US Department of Health and Human Services (2013) National Transplantation Data

Report. OPTN: Data Organ Procurement and Transplantation Network. http://optn.transplant.hrsa.gov/latestData/step2.asp? Accessed 2 Mar 2015

4. Portner R, Nagel-Heyer S, Goepfert C, Adamietz P, Meenen NM (2005) Bioreactor design for tissue engineering. J Biosci Bioeng 100:235–245
5. Meezan E, Hjelle JT, Brendel K, Carlson EC (1975) A simple, versatile, nondisruptive method for the isolation of morphologically and chemically pure basement membranes from several tissues. Life Sci 17:1721–1732
6. Song JJ, Guyette JP, Gilpin SE, Gonzalez G, Vacanti JP, Ott HC (2013) Regeneration and experimental orthotopic transplantation of a bioengineered kidney. Nat Med 19:646–651
7. Ott HC, Matthiesen TS, Goh SK et al (2008) Perfusion-decellularized matrix: using nature's platform to engineer a bioartificial heart. Nat Med 14:213–221
8. Gonfiotti A, Jaus MO, Barale D et al (2014) The first tissue-engineered airway transplantation: 5-year follow-up results. Lancet 383:238–244
9. Macchiarini P, Jungebluth P, Go T et al (2008) Clinical transplantation of a tissue-engineered airway. Lancet 372:2023–2030
10. Uygun BE, Soto-Gutierrez A, Yagi H et al (2010) Organ reengineering through development of a transplantable recellularized liver graft using decellularized liver matrix. Nat Med 16:814–820
11. Petersen TH, Calle EA, Zhao L et al (2010) Tissue-engineered lungs for in vivo implantation. Science 329:538–541
12. Nakayama KH, Batchelder CA, Lee CI, Tarantal AF (2010) Decellularized rhesus monkey kidney as a three-dimensional scaffold for renal tissue engineering. Tissue Eng Part A 16:2207–2216
13. Orlando G, Farney AC, Iskandar SS et al (2012) Production and implantation of renal extracellular matrix scaffolds from porcine kidneys as a platform for renal bioengineering investigations. Ann Surg 256:363–370
14. Park KM, Woo HM (2012) Porcine bioengineered scaffolds as new frontiers in regenerative medicine. Transplant Proc 44:1146–1150
15. Sullivan DC, Mirmalek-Sani SH, Deegan DB et al (2012) Decellularization methods of porcine kidneys for whole organ engineering using a high-throughput system. Biomaterials 33:7756–7764
16. Yu YL, Shao YK, Ding YQ et al (2014) Decellularized kidney scaffold-mediated renal regeneration. Biomaterials 35:6822–6828
17. Ross EA, Abrahamson DR, St John P et al (2012) Mouse stem cells seeded into decellularized rat kidney scaffolds endothelialize and remodel basement membranes. Organogenesis 8:49–55

Part II

Imaging Techniques

Chapter 7

Use of Cationized Ferritin Nanoparticles to Measure Renal Glomerular Microstructure with MRI

Kevin M. Bennett, Scott C. Beeman, Edwin J. Baldelomar, Min Zhang, Teresa Wu, Bradley D. Hann, John F. Bertram, and Jennifer R. Charlton

Abstract

Magnetic resonance imaging (MRI) is becoming important for whole-kidney assessment of glomerular morphology, both in vivo and ex vivo. MRI-based renal morphological measurements can be made in intact organs and allow direct measurements of every perfused glomerulus. Cationic ferritin (CF) is used as a superparamagnetic contrast agent for MRI. CF binds to the glomerular basement membrane after intravenous injection, allowing direct, whole-kidney measurements of glomerular number, volume, and volume distribution. Here we describe the production, testing, and use of CF as an MRI contrast agent for quantitative glomerular morphology in intact mouse, rat, and human kidneys.

Key words Kidney, Glomerulus, Morphology, Chronic kidney disease, Magnetic resonance imaging, Cationic ferritin, Contrast agent

1 Introduction

The total number of nephrons in the kidneys predicts long-term susceptibility of a patient to chronic kidney and cardiovascular disease. Nephron number can also predict allograft success and may be more sensitive than standard clinical blood or urinary markers in predicting renal response to therapy. Cationized ferritin (CF), a derivative of the 24-subunit iron storage protein ferritin, can be used in native and modified form to detect, count, and measure functioning glomeruli with magnetic resonance imaging (MRI) [1–7]. This technique of CF-enhanced MRI (CFE-MRI) is versatile and can be used in conjunction with other imaging techniques to map kidney microstructure in healthy and diseased rodent kidneys and in humans. These protocols describe the formation of CF and the use of CFE-MRI for preclinical MRI-based histology in rats, mice, and humans (*see* **Note 1**).

Tim D. Hewitson et al. (eds.), *Kidney Research: Experimental Protocols*, Methods in Molecular Biology, vol. 1397, DOI 10.1007/978-1-4939-3353-2_7, © Springer Science+Business Media New York 2016

2 Materials

2.1 Cationized Ferritin Synthesis, Cationization, and Validation

1. Doubly pure water (diH_2O).
2. Better Bradford Assay™ kit (Thermo Scientific, Rockford, Il, USA).
3. Horse spleen apoferritin (Sigma Aldrich, St. Louis, MO. USA).
4. Horse spleen ferritin (Sigma Aldrich).
5. 2-(*N*-morpholino)ethanesulfonic acid (MES).
6. $Fe(II)Cl_2$.
7. 0.15 M NaCl solution in diH_2O.
8. 8000 Da molecular weight cut-off (MWCO) dialysis bags.
9. *N*,*N*-dimethyl-1,3,propanediamine (DMPA).
10. 1-ethyl-3(3-dimethyl-aminopropyl) carbodiimide hydrochloride (EDC).
11. 2 N HCl solution in diH_2O.
12. 2 N NaOH solution in diH_2O.
13. Phosphate buffered saline (PBS), pH 7.4.
14. Peristaltic pump.
15. Bruker minispec™ (Bruker Biospin Corporation, Billerica, MA, USA).
16. Paraffin film.
17. Syringe filter.
18. Chemistry glassware including manifold, three neck flask, coil condenser, and water bath.
19. Nitrogen gas cylinder and regulator (50 psi).

2.2 Magnetic Resonance Imaging and Computational Requirements for Running Image Analysis

1. MRI system and hardware (recommended field strength of 7T or greater).
2. Intel CPUs or Intel ×86/×64 compatible CPUs like AMD.
3. Memory: 4 GB or above (12 GB recommended for human kidney segmentation).
4. Matlab (The Mathworks, Inc. Natick, MA. USA): Matlab version R2011b or newer (*see* **Note 2**).

2.3 Kidney Perfusion

1. 23–26 gauge needle for penile or tail vein infusion (rat).
2. A 30 gauge needle attached to a glass syringe for retro-orbital infusion (mice).
3. 18–21 gauge and 25 gauge needles for cardiac perfusion in rat and mouse respectively.
4. Approved animal anesthetic.

5. Saline.
6. Fixative: 10 % neutral buffered formalin (NBF) and/or 2 % solution of 1 M glutaraldehyde solution in 0.1 M sodium cacodylate (*see* **Note 3**).
7. Phosphate buffered saline (PBS), pH 7.4.
8. Peristaltic pump for perfusion.

3 Methods

3.1 Ferritin Metal Core Formation (Adapted from Refs. 8–10)

The following protocol describes the synthesis of an iron-oxide core in the apoferritin protein (*see* **Note 4**). Apoferritin concentrations should be calculated using a Better Bradford™ assay. The nanoparticle molar concentration is directly calculated from the protein concentration obtained by the Bradford assay by dividing the protein concentration by the molecular weight of apoferritin 24-mer (457,000 Da). These procedures describe small-scale synthesis of ~10 ml ferritin in ~2 mM concentrations. Synthesis can be scaled up as desired, though we have not tested this protocol on large-scale syntheses.

All synthesis should be performed inside a laminar flow hood or chemical safety hood, preferably under sterile conditions. A recommended chemistry setup for filling apoferritin is shown in Fig. 1. The components are (a) double manifold importing nitrogen gas and (b) a gas bubbler, (c) a three neck flask fitted with a coil condenser to prevent solvent loss and an infusion needle to inject the iron, sealed with sleeve septa, (d) $FeCl_2$ solution degassing separately, (e) water bath at 60–65 °C warming the three neck flask and apoferritin solution, (f) syringe pump to inject the iron solution at 16 μL/min.

1. Prepare a round-bottom flask containing 20 nmol (9.15 mg) apoferritin in 10 mL 0.05 M MES, adjusted to pH 8.5.
2. Prepare a second round-bottom flask containing excess of 50 mM $Fe(II)Cl_2$.
3. Place both flasks into a water bath or heat blanket to maintain a temperature of 60–65 °C.
4. Degas both flasks by attaching a sleeve septum to the flask opening and sealing around the edges with paraffin film. Insert a needle attached to a tube leading to a regulated nitrogen tank into one septum and another through a nitrogen bubbler. Add nitrogen to the container at 50 psi. Slowly reduce the pressure of the gas until no bubbles are formed in the liquid of the flask.
5. Carefully draw excess de-aerated $FeCl_2$ solution into a 5 mL syringe fitted with an infusion needle, and then quickly transfer

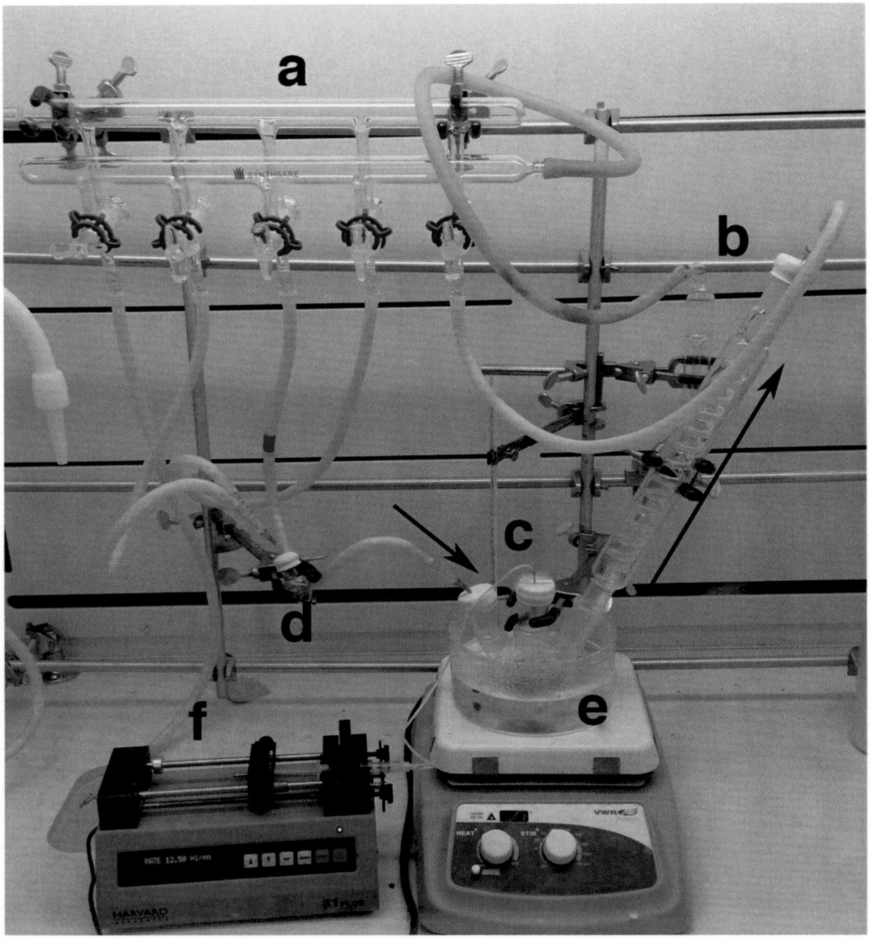

Fig. 1 Recommended chemistry setup for filling apoferritin. (**a**) Double manifold importing N_2 gas and expelling nitrogen through a gas bubbler (**b**). (**c**) Three neck flask fitted with a coil condenser to prevent solvent loss and an infusion needle to inject the iron, sealed with sleeve septa. (**d**) $FeCl_2$ solution degassing separately. (**e**) Water bath at 60–65 °C warming the three neck flask and apoferritin solution. (**f**) Syringe pump to inject the iron solution at 16 μL/min. *Arrows* show the direction of N_2 flow

the needle to the septum on the center neck of the flask. Using a syringe pump, inject the $FeCl_2$ into the apoferritin at a rate of 16 μL/min for 75 min (3000 Fe/ferritin). You may adjust this time between 25 and 125 min (1000–5000 Fe/ferritin); increasing the time typically increases transverse MR relaxivity (r_2), but decreases yield.

6. Dialyze the ferritin-containing solution overnight (12 h) against 0.15 M NaCl using dialysis tubing (8000 Da MWCO).
7. Measure final protein concentration using the Better Bradford Assay™ by UV/visible spectrophotometer absorbance at 595 nm.

8. Measure MR relaxivity using a Bruker minispec™ or equivalent, at the desired field strength and MR acquisition parameters. Relaxivity can be measured in solution or in low melting point agar, with the temperature kept below 50 °C to avoid denaturing the ferritin protein.
9. Store at 4 °C.

3.2 Cationized Ferritin Synthesis

We synthesize CF in our lab using a protocol adapted from Danon et al. [11]. All procedures should be performed in a chemical safety cabinet, following all safety protocols for the chemicals. This synthesis is for 27.5 mg, approximately 4 mg/mL. Volumes can be scaled as needed (*see* **Note 5**).

1. Add 1.5 mL of deionized water in a 20 mL beaker.
2. Slowly add 0.5 mL of stock 8 M (~99 %, ~200 DMPA/COO-) *N*,*N*-dimethyl-1,3,propanediamine (DMPA) to the beaker under stirring.
3. Add 4 mL 2N HCl, then adjust to pH 6.0 with 0.2 N HCl and 0.2 N NaOH.
4. Add 0.5 mL 55 mg/mL native ferritin to the solution of DMPA.
5. Add 200 mg 1-ethyl-3(3-dimethyl-aminopropyl) carbodiimide hydrochloride (EDC, ~50 DMPA/COO-).
6. Carefully maintain pH at 6 for 1 h. The addition of EDC will make the solution slightly basic, approaching 6.5, so it can be brought back down by addition of 0.2 N HCl. Take care to avoid adding too much HCl, as a pH below 4.0 can denature ferritin.
7. Once pH is stabilized, cover with paraffin film and stir at room temperature overnight (12 h).
8. Dialyze at least two times in 8000 Da MWCO dialysis bags against 3.5 L of PBS at 4 °C for 12–24 h each time.
9. Store CF at 4 °C. The solution can be filtered through a syringe filter to remove aggregates before use.
10. Measure concentration using a Better Bradford™ assay or equivalent.

3.3 Validating Cationization

To test the effectiveness of the cationization reaction, determine the number of carboxylate residues before and after the reaction by titration [11]. In our experience the titrated sample can be brought back to pH 7.2 with 0.1 N HCl, dialyzed into PBS, and used as normal.

1. Add 1–10 mg of ferritin or cationized ferritin to 5 ml diH_2O (using more ferritin will give a more accurate result). For the buffer standard, add the same volume of buffer (typically saline or PBS) to 5 ml diH_2O (*see* **Note 6**).

2. Make a stock solution of aqueous 10 mM HCl.
3. Slowly decrease the pH of the ferritin or buffer solution to 5.2 with HCl.
4. Continue adding HCl solution in 1 μL increments, measuring the total volume (V_{ferritin} or V_{buffer}) of HCl solution required to decrease the pH from 5.2 to 3.0.
5. To calculate the number of titratable carboxylate groups per mg of ferritin molecule:

$$\left[\text{COO}^-\right] = \frac{\left(V_{\text{fn}} - V_{\text{b}}\right) \times \left[\text{HCl}\right]}{M_{\text{fn}}}. \tag{1}$$

Variables: $[\text{COO}^-]$ = titratable carboxylates (μmol/mg ferritin), M_{fn} is the mass of ferritin being titrated (mg), V_{fn} and V_{b} = volume of HCl solution required to decrease the pH from 5.2 to 3.0 (mL), [HCl] = concentration of the HCl solution (10 μmol/mL).

The average surface charge per ferritin, $\tilde{Z}$, can be estimated from the known amino acid sequence of horse spleen ferritin [NCBI, PDB: 1IER_A] (MW = 476.4 kDa, Asp 288, Glu 360, Lys 216, Arg 264).

$$\text{SF} = \left|\frac{\left[\text{COO}^-\right]_{\text{M,NF}} \times \text{MW}_{\text{NF}}}{A_{\text{C,NF}}}\right|, \tag{2}$$

where

$$\tilde{Z}_{\text{M,NF}} = Z_{\text{C,NF}} \times \text{SF} \tag{3}$$

and

$$\tilde{Z}_{\text{CF}} = \tilde{Z}_{\text{NF}} + 2 \times \left(\left[\text{COO}^-\right]_{\text{M,NF}} - \left[\text{COO}^-\right]_{\text{M,CF}}\right) \times \text{MW}_{\text{NF}} \tag{4}$$

Variables: SF = surface fraction (unitless), $[\text{COO}^-]$ = carboxylate/mg (μmol/mg), MW = molecular weight (476.4 mg/μmol), A = calculated carboxylates per ferritin (−648 q), $\tilde{Z}$ = average surface charge per ferritin (q), Z = calculated total charge per ferritin (−168 q) (*see* **Note 7**). Subscripts: M = measured, C = Concentration.

3.4 Cationized Ferritin Labeling and MRI in Rat Kidneys

For all experiments, obtain regulatory approval from the relevant institutional animal care and use committee.

1. Weigh and prepare rats for intravenous injections.
2. Inject CF intravenously (e.g., by tail vein or penile vein). Typical bolus injections are performed with a total of 5.75 mg of CF per 100 g per body weight, with injected concentrations of 11 mg/mL. It is common to deliver the boluses in three

equal injections of 1 ml spaced 1.5 h apart (2 × the blood half-life of CF). Toxicity manifests as systemic hypoxia, characterized by a blue skin tone in the feet and tail.

3. 1.5 h after the last injection, anesthetize the rat with either isoflurane or an equivalent (e.g., ketamine/xylazine cocktail), matched to the weight of the animal.
4. Incise the chest and abdomen of the rat to expose the heart, liver, and kidneys. Excise the right lung or make an incision in the right ventricle of the heart. Insert a large-gauge needle (18–21 gauge) into the left ventricle of the heart and systemically perfuse the rat using a peristaltic pump to drive the perfusate through the needle and into the vasculature. Perfuse first with saline at a rate of at least 100 mL/min (*see* **Note 8**) for a total volume of 400–500 mL. Confirm no visible blood in the kidneys, then perfuse with 10 % neutral buffered formalin at a rate of ~100 mL/min for a total of 100 mL of formalin (*see* **Note 9**). Remove the kidneys and place into fixative (*see* **Note 3**).
5. Store kidneys in fixative at 4 °C for at least 6 h.
6. Kidneys can be imaged with MRI immediately if fixed in glutaraldehyde fixative. If they were initially fixed in formalin, they must be transferred to a large volume of PBS (at least 250 mL) and stored overnight before imaging. Tissue fixation typically dehydrates tissue. MRI signal is directly proportional to the number of water molecules in a sample. Therefore, it is preferable to rehydrate tissue in PBS prior to imaging.
7. MRI systems and associated hardware vary widely between institutions. Two major factors influencing image quality are the MRI field strength and quality and suitability of the radiofrequency (RF) coils. We have obtained excellent results in kidney glomerular morphology at field strengths of 7T and greater.
8. Place the excised kidney(s) inside a tube filled with PBS into the RF coil, and insert the RF coil into the bore of the magnet. The glomeruli can be readily detected using a gradient-recalled echo (GRE) pulse sequence with an appropriate TE/TR. For example, at 7T our typical resolution for 3D MRI is 70 μm in each dimension, with TE/TR = 20/80 and a spoiler gradient, using GRE-MRI. Averaged acquisitions can be used to improve signal-to-noise.

CFE-MRI of rat kidneys is shown in Fig. 2.

3.5 Cationized Ferritin Labeling and MRI in Mouse Kidneys

1. Weigh mice.
2. Inject CF intravenously (tail vein) or alternatively using a retroorbital injection as outlined by Yardeni et al. [12]. Bolus injections are performed using a total of 5.75 mg/100 g of body weight CF, with a concentration ~11 mg/mL. This total dose

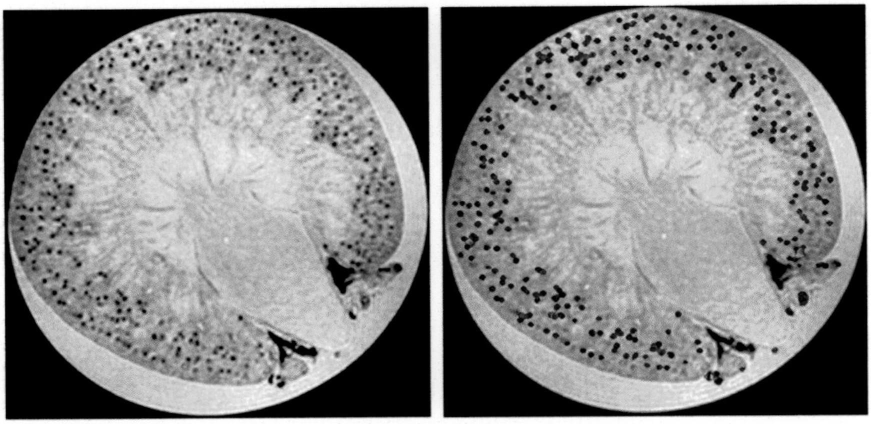

Fig. 2 Identification of individual glomeruli in whole rat kidney with MRI enhanced by intravenously injected cationized ferritin (CFE-MRI). A representative axial kidney image from a 3-dimensional (3D) magnetic resonance imaging (MRI) dataset from a cationic ferritin (CF)-injected rat is shown (*left*). The data were analyzed with a 3D counting algorithm to identify glomeruli. Regions defined as glomeruli by the computational 3D counting algorithm were assigned an arbitrary color exclusively for visualization (*right*). Reproduced with permission from ref. 2

is administered in two injections separated by 90 min (2 × the blood half-life of CF) to minimize toxicity (which typically manifests as systemic hypoxia and pulmonary hemorrhage).

3. Retro-orbital injections are effective and technically easier than tail vein injections to ensure the adequate delivery of CF in mice. In brief, the animal is anesthetized on a warming device using an inhalant anesthetic delivered initially in a Plexiglas chamber followed by a funnel-shaped nose cone attached to a scavenger device, in addition to topical anesthetic prior to the injection. The mouse is positioned in a lateral recumbency. Use the first finger and thumb of the non-injecting hand to protrude the mouse's eyeball from the socket, taking care not to occlude the trachea or cervical vessels. A 30 gauge needle attached to a glass syringe is slowly introduced at an angle of 30 °C into the medial canthus with the bevel of the needle facing downward to reduce potential damage. The administrator should slowly and smoothly inject the CF avoiding an initial aspiration. This technique has been published to work for volumes up to 150 μL [12].
4. 1.5 h after the last injection, anesthetize the mouse with either isoflurane or euthanasia solution appropriate for the weight of the animal.
5. A Y-shaped incision is made caudal to the pelvis and the skin is reflected to expose the abdomen and chest. The sternum is reflected to expose the heart and the abdominal contents are reflected to the animal's left side to observe the right kidney.

A 25 gauge needle attached to either a 60 mL syringe or an automated pump is inserted into the left ventricle after a large incision in the right atrium has been made. Perfuse with normal saline or PBS until the kidneys are clear of visible blood and uniformly pale in color.

6. Perfuse with 10 % neutral buffered formalin at a rate of ~10 mL/min for a total of 10 mL of formalin. Remove the kidneys and place into fixative (*see* **Note 3**).
7. Store kidneys in fixative at 4 °C for at least 6 h.
8. Kidneys can be imaged with MRI immediately if fixed in glutaraldehyde. If they were initially fixed in formalin, they must be transferred to a large volume of PBS (at least 250 mL) and stored overnight before imaging.
9. Place the excised kidney(s) inside a tube filled with PBS into the RF coil, and insert the RF coil into the bore of the magnet. The glomeruli can be readily detected using a gradient-recalled echo (GRE) pulse sequence with an appropriate TE/TR. For example, at 7T our typical imaging parameters for 3D GRE-MRI with a spoiler gradient are: Resolution—50 μm in each direction; TE/TR = 20/80. The number of averaged acquisitions is dependent upon the desired signal-to-noise.

3.6 Cationized Ferritin Labeling and MRI in Human Kidneys

Human donor organs can be obtained through an organ procurement program, such as the International Institute for the Advancement of Medicine (IIAM, Edison, NJ, USA). The organs are obtained at death and can be preserved as if for transplant. Organs used for research must not be viable for transplant and must be donated specifically to research by the donor or family.

1. Receive the kidneys within 24 h of cross-clamp in University of Wisconsin solution.
2. Catheterize the renal artery and perfuse with 120 mL of PBS over ~1 min.
3. Follow the PBS perfusion with a perfusion of 60 mL of CF (11 mg/mL solution).
4. Wash out the excess CF by perfusing with 60 mL of PBS over ~30 s.
5. Fix tissue by perfusing with 60 mL of NBF over ~30 s.
6. Store the fixed kidney in formalin at 4 °C.
7. 48 h prior to imaging, wash and hydrate tissue in a large volume of PBS (500–1000 mL). Replace PBS every ~12 h. Tissue fixation typically dehydrates tissue. MRI signal is directly proportional to the number of water molecules in a sample. Therefore it is preferable to rehydrate tissue in PBS prior to imaging.

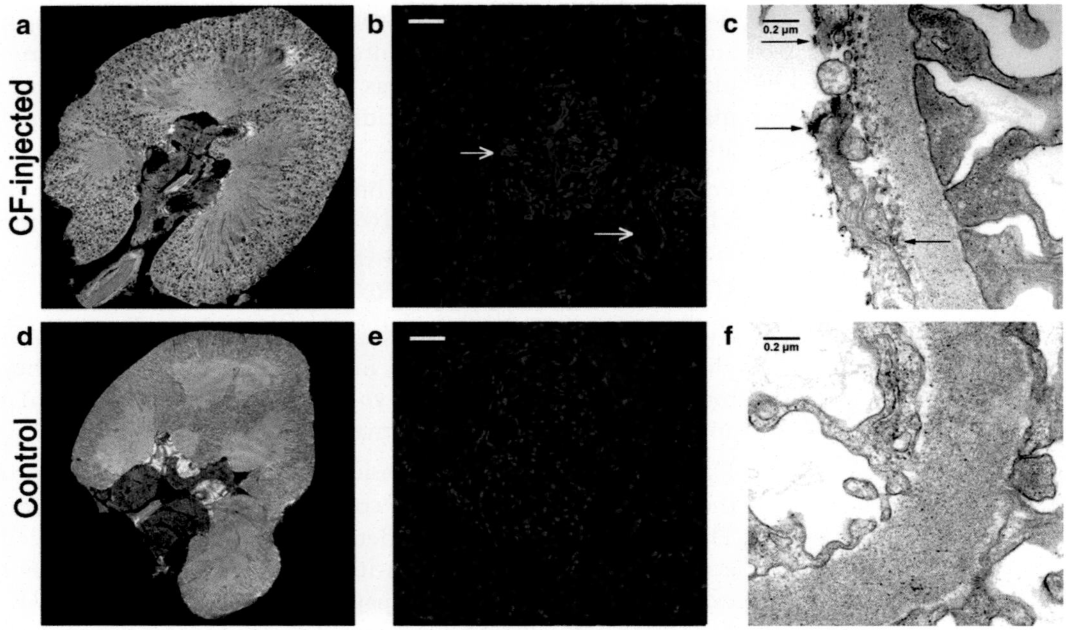

Fig. 3 Intravenously injected cationized ferritin specifically labels glomeruli in perfused human donor kidneys, making them visible with 7T MRI. MR images of CF-labeled human kidneys reveal punctate dark spots throughout the cortex (**a**). An unlabeled control kidney shows no such spots (**d**). Immunofluorescence confirms the accumulation of CF (*red*) in glomeruli (**b**, *upper arrow*) and leakage of CF into tubules of CF-perfused kidneys (**b**, *lower arrow*). Unlabeled control glomeruli remain clear of CF-related immunofluorescence (**e**). Transmission electron microscopy (TEM) confirms the accumulation of CF in the glomerular basement membrane (*arrows*) (**c**). The glomerular capillary walls of the naive control kidney were clear of CF (**f**). White scale bars = 50 μm. Black scale bars = 0.2 μm. Reproduced with permission from ref. 3

8. The donor kidney should be sealed in a plastic bag with a small amount of PBS. Air should be evacuated from the bag. The PBS and evacuation are done to ensure that the kidney does not dry out during MRI scanning.
9. The kidney should be placed in an appropriately sized RF coil (we have found that a 72-mm quadrature coil is a good coil for imaging donor human kidneys). Glomeruli may be detected using a GRE pulse sequence. At 7T, we have found that a resolution of 117 μm in each dimension, a TE/TR of 20/39 ms, and 5 averages are sufficient to detect glomeruli.

CFE-MR images of human kidneys, with immunofluorescence and transmission electron microscopy images, are shown in Fig. 3.

3.7 Image Processing to Measure Glomerular Morphology

A computer algorithm can be used to analyze the MR images and measure number and volume of glomeruli. This protocol describes the technique our group has developed [13] and used for glomerular morphology [2, 3].

A three-phased pipeline, termed Hessian based multi-Features Clustering (HmFC), can be used to perform fast, reliable, and accurate measurements and counts of CF-labeled glomeruli in MR images.

Based on this pipeline, a semi-automated Quantitative Glomeruli Assessment Toolkit (qGAT) is available through our group. At the time of this publication, the Toolkit, manual, and full description of the software are available online at

http://swag.engineering.asu.edu/HmFC.htm.

The workflow of the Toolkit is as follows:

1. Load the MR images (current version only supports Analyze 7.5 format).
2. For the raw images, the kidney boundary is first defined. This can be done either by using the software or loading the boundary mask, which can be predefined offline.
3. Identify Hessian matrix of the image, using the positive semi-definite property of the Hessian matrix; the non-glomerular regions (non-convex regions) are filtered out resulting in a number of glomerulus candidates (*see* **Note 10**).
4. For each glomerulus candidate, six features are extracted: average intensity, divergence, distance to the kidney boundary (provided by domain expert), region volumes (provided by domain expert), shape index (SI), and Laplacian of Gaussian (LoG). A dataset with six features for each glomerulus candidate is generated.
5. The Variational Bayesian Gaussian Mixture model is applied to the dataset to group the glomerulus candidates. Here, the initial number of clusters needs to be assigned. Based on the empirical experiments, 10 clusters are recommended for a rodent kidney and 30 clusters are recommended for a human kidney for optimal detection, assuming a human imaging matrix size of $512 \times 512 \times 896$ compared to a rat kidney image matrix size of $256 \times 256 \times 256$. In addition, the number of glomeruli from human kidney (500,000 ~ 2,000,000) is at least ten times that of a rodent model.
6. A glomerulus cluster is picked by visually checking against other clusters. For rodent, the majority of glomeruli are covered by one cluster. For human kidneys, the glomeruli may fall into several clusters. The user may pick multiple clusters to get combined segmentation results.

4 Notes

1. CFE-MRI can be performed with commercial-grade CF, followed by MRI. Subheadings 3.1–3.3 are a cost-effective, flexible option but are not required for CFE-MRI.

2. Since qGAT only depends on the Matlab platform, it has no OS requirement as long as the OS supports Matlab.
3. Fixative can be either 2 % glutaraldehyde in 0.1 M sodium cacodylate or 10 % formalin. Retrieve biopsy cores before fixation if a different fixative is required for histology or electron microscopy. For example, electron microscopy typically requires glutaraldehyde fixation of small samples (<1 mm^3), while immunohistochemistry typically requires formalin fixation or paraffin embedment.
4. Apoferritin can be recombinantly expressed or purchased for use in this synthesis scheme.
5. CF derived from horse spleen can be obtained commercially from Sigma Aldrich (St. Louis, MO, USA).
6. Because PBS has a high concentration of phosphate (p*K*a of 7.2 and 2.1), having an accurate buffer standard is crucial. The dialysate from cationization can be used as a standard.
7. The molecular weight is based on 100 % H ferritin. The resultant calculations should be considered estimates, with uncationized control protein measured for comparison.
8. A higher rate—as much as 200 ml/min—may be necessary to remove all blood from kidneys. It is best to find the lowest flow rate necessary to remove all blood from the kidneys, as excess pressure may damage tissue.
9. Formalin causes spontaneous twitching movement in the animal. Therefore, effective delivery of fixative should be obvious.
10. This reduces the size of the dataset by a factor of ~10.

Acknowledgements

This work was funded by a grant from the NIH Diabetic Complications Consortium, The American Heart Association, The US National Institutes of Health grant NIH DK-091722, and a grant from the Hartwell Foundation.

References

1. Bennett KM, Zhou H, Sumner JP, Dodd SJ, Bouraoud N, Doi K, Star RA, Koretsky AP (2008) MRI of the basement membrane using charged nanoparticles as contrast agents. Magn Reson Med 60:564–574
2. Beeman SC, Zhang M, Gubhaju L, Wu T, Bertram JF, Frakes DH, Cherry BR, Bennett KM (2011) Measuring glomerular number and size in perfused kidneys using MRI. Am J Physiol Renal Physiol 300:F1454–F1457
3. Beeman SC, Cullen-McEwen LA, Puelles VG, Zhang M, Wu T, Baldelomar EJ, Dowling J, Charlton JR, Forbes MS, Ng A, Wu Q-Z, Armitage A, Egan GF, Bertram JF, Bennett

KM (2014) MRI-based glomerular morphology and pathology in whole human kidneys. Am J Physiol Renal Physiol 306: F1381–F1390

4. Bennett KM, Bertram JF, Beeman SC, Gretz N (2013) The emerging role of MRI in quantitative renal glomerular morphology. Am J Physiol Renal Physiol 304:F1252–F1257
5. Charlton JR, Beeman JF, Bennett KM (2013) MRI-detectable nanoparticles: the potential role in the diagnosis of and therapy for chronic kidney disease. Adv Chronic Kidney Dis 20:479–487
6. Heilmann M, Neudecker S, Wolf I, Gubhaju L, Sticht C, Schock-Kusch D, Kruz W, Bertram JF, Schad LR, Gretz N (2012) Quantification of glomerular number and size distribution in normal rat kidneys using magnetic resonance imaging. Nephrol Dial Transplant 27: 100–107
7. E. J. Baldelomar, J.R. Charlton, S.C. Beeman, L. Cullen-Mcewen, V.M. Pearl, J.F. Bertram, T. Wu, M. Zhang, K.M. Bennett. Phenotyping by magnetic resonance imaging nondestructively measures glomerular number and volume distribution in mice with and without nephron reduction. Kidney Int (In Press)
8. Clavijo-Jordan MV, Caplan MR, Bennett KM (2010) Simplified synthesis and relaxometry of magnetoferritin for magnetic resonance imaging. Magn Reson Med 64:1260–1266
9. Uchida M, Terashima M, Cunningham CH, Suzuki Y, Willits DA, Yang PC, Tsao PS, McConnell MV, Young MJ, Douglas T (2008) A human ferritin iron oxide nano-composite magnetic resonance contrast agent. Magn Reson Med 60:1073–1081
10. Clavijo Jordan MV, Beeman SC, Baldelomar EJ, Bennett KM (2014) Disruptive chemical doping in a ferritin-based iron oxide nanoparticle to decrease r2 and enhance detection with T1-weighted MRI. Contrast Media Mol Imaging 9:323–332
11. Danon D, Goldstein L, Marikovsky Y, Skutelsky E (1972) Use of cationized ferritin as a label of negative charges on cell surfaces. J Ultrastruct Res 38:500–510
12. Yardeni T, Eckhaus M, Morris HD, Huizing M, Hoogstraten-Miller S (2011) Retro-orbital injections in mice. Lab Anim (NY) 40:155–160
13. M. Zhang, T. Wu*, and K.M. Bennett. Small blob identification in medical images using regional features from optimum scale. IEEE Trans Biomed Eng. 2015. 62(4): 1051-62

Chapter 8

Biopsychronology: A Method Using Live Tissue Staining to Image Cell Function in the Kidney

Muhammad Imtiaz Ashraf, Dietmar Fries, Werner Streif, Felix Aigner, Paul Hengster, Jakob Troppmair, and Martin Hermann

Abstract

Methods to monitor the status of a graft prior to transplantation are highly desirable to avoid unnecessary surgical interventions and follow-up treatments and to optimize the clinical outcome as delayed graft function may lead to costly and lengthy follow-up treatments or even organ loss. As a promising step in this direction we present a method which combines the use of fine needle biopsies, the staining of living cells with dyes suitable to monitor mitochondrial status/cellular integrity, and live confocal real-time analysis.

This approach provides information about the functional and structural intactness of an organ within a few minutes. To confirm the feasibility of this approach, we recently published a pilot study using rodent kidneys. The results demonstrated that this method is suitable to monitor organ damage caused by ischemia or short periods of reperfusion. This procedure required minimal time for sample preparation and data acquisition and is suitable for recording damage resulting from unphysiological stress to the organ.

Key words Biopsy, Kidney, Real-time live confocal microscopy

1 Introduction

In solid organ transplantation, one eminent challenge is to assess rapidly and in a minimally invasive procedure donor organ quality, which in addition to extended ischemia times is also affected by donor age, medical history, and other variables. Due to organ shortage, donor criteria have been extended and borderline quality organs are increasingly included, stressing the need for the assessment of organ function prior to transplantation. In kidney transplantation, histological analyses of the pretransplant donor renal biopsies (PTDB) are often performed in case of marginal donors [1, 2]. However, they are time-consuming and delay transplantation, resulting in further deterioration of the graft. Here we describe a novel minimally invasive approach, which monitors the

Tim D. Hewitson et al. (eds.), *Kidney Research: Experimental Protocols*, Methods in Molecular Biology, vol. 1397, DOI 10.1007/978-1-4939-3353-2_8, © Springer Science+Business Media New York 2016

preservation of key cellular functions such as mitochondrial membrane potential as well as viability and structural integrity. Our approach is based on the acquisition of biopsies, incubation with appropriate reporter dyes, which monitor the functional status of viable cells, followed by real-time live confocal analysis.

In our pilot work we used the following four different stains: (1) tetramethyl rhodamine methyl ester (TMRM), a dye that is readily sequestered by functioning mitochondria with intact mitochondrial membrane potential [3], (2) propidium iodide (PI), which is only taken up by nuclei of dead cells, (3) Syto16, which stains nuclei in dead and living cells [4], and (4) FITC-coupled wheat germ agglutinin conjugate (WGA), which binds to *N*-acetyl-d-glucosamine and *N*-acetyl-d-neuraminic acid residues, thereby staining cell membranes as well as matrix components [3, 5, 6] and thus can be used to monitor the intactness of the tissue. By using appropriate fluorophores, these dyes may be combined for simultaneous detection of various parameters. Moreover, a wealth of additional fluorescent markers is available, which can be adopted for this purpose to further increase clinically relevant information. A flow scheme of our approach is depicted in Fig. 1. Following kidney biopsy, samples are incubated with the live stains Syto 16, PI, and WGA and immediately processed for imaging. Serial images are taken along the whole length of the biopsy allowing for the coverage of a cross section through a large area of the organ (Fig. 1). An inset depicting tubuli in the cortical area of the kidney is shown in the upper left corner of Fig. 1 and one showing a glomerulus is inserted in the lower right corner.

In order to confirm that our approach is suitable to monitor changes in the cellular environment, we analyzed rat kidney biopsies after a 24-h incubation step at 37 °C in the tissue culture incubator (24 h 37 °C) or following permeabilization of the cell membrane by digitonin treatment, which results in the collapse of cellular homeostasis and cell death. The pictures shown in Fig. 2 focus on the corticular area of the kidney and comprise several tubuli. Double staining with either Syto 16 (green) or PI (red; Fig. 2a–c) or TMRM (red) and WGA (green; Fig. 2b–d) documents the cell viability. Despite immediate processing of the samples (i.a.b) a heterogeneity in the staining pattern can be observed which documents the presence of viable and nonviable regions in the tubular area. After the 24-h incubation step at 37 °C, the number of PI-positive (dead) cells increases substantially, while only a fraction of Syto 16 positive/PI negative cells is still present at the end of the incubation period. After the digitonin addition all nuclei are

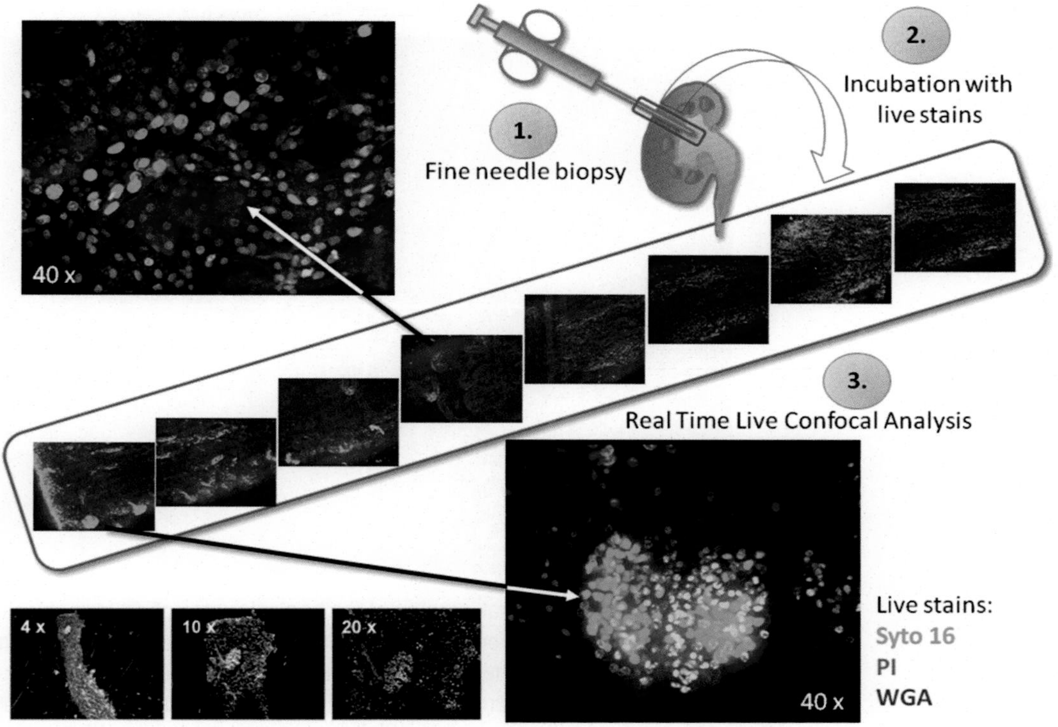

Fig. 1 The three steps to "biopsychronology." In a first step a fine needle biopsy is taken (shown is an example of a rat kidney). The second step consists of an incubation of the kidney biopsy with live stains. Stains can vary, depending on the cell parameters, being analyzed. In this case the biopsy was stained with Syto16 (in *green*, staining all the nuclei), propidium iodide (PI, in *red*, staining the nuclei of the nonvital cells), and wheat germ agglutinin (WGA, in *violet*, staining the tissue morphology of the kidney sample). In the third step real-time live confocal analysis with a spinning disc system is performed. Due to the spinning disc technology, the confocal analysis can be performed with non-fixed biopsy samples. If a whole kidney biopsy is taken, serial pictures at low magnification can be assembled for overview purposes. From low magnifications (see *bottom left* corner showing images acquired with a 4×, 10×, and 20× objective) to a more detailed analysis of regions of interest using higher magnifications as exemplified with a 40× water immersion objective (see two examples at the *top left* and *bottom right* corner of the figure) can easily be performed. (Reproduced from Transplant International 2014 27:868–876 with permission)

PI/Syto 16 positive resulting in a yellow overlay color. A similar picture emerges when looking at the mitochondrial membrane potential. At the earliest time point after the biopsy areas with decreased mitochondrial function are already detectable (Fig. 2b). Finally, the mitochondrial transmembrane potential has completely collapsed after the permeabilization of the cell membrane by digitonin (Fig. 2f).

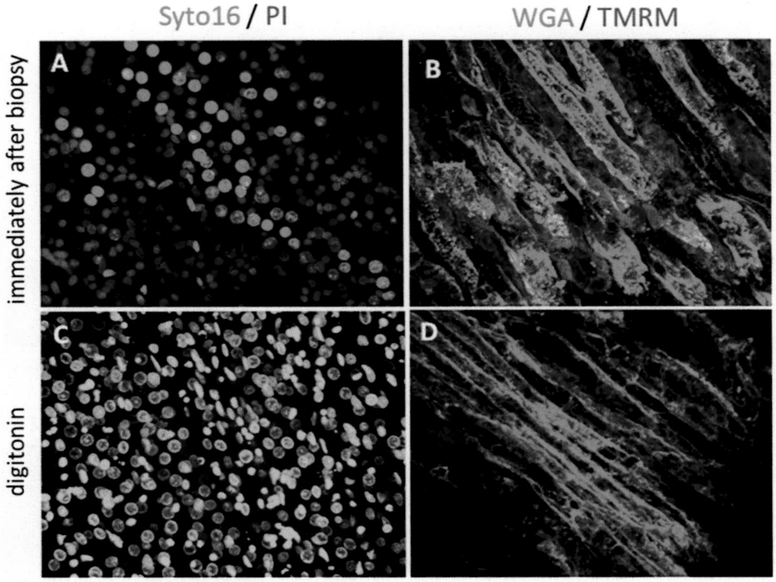

Fig. 2 Real-time live confocal imaging of the corticular area of a kidney biopsy showing several tubuli. Following left renal ischemia/reperfusion (IR) (35 min) fine needle biopsies of murine kidneys were taken and incubated with live stains. (**a** and **c**) Syto 16 (*green*), propidium iodide (PI; *red*) visualizing nuclei of living (Syto 16) and dead cells (PI), and (**b** and **d**) wheat germ agglutinin (WGA, *green*) and TMRM (*red*), staining cell membranes and functional mitochondria. Images were acquired with a 40× water immersion objective. One representative example is shown consisting of a z-stack of 25 planes with a spacing of 1 μm

The staining pattern obtained with WGA showed the same heterogeneity as did the staining with TMRM. After cultivation for 24 h at 37 °C, the structural integrity of the biopsy was lost and increased cell death was apparent as documented by Syto16/PI staining. Permeabilization with digitonin finally resulted in massive cell death (Fig. 2c and d) but has no influence on the staining pattern of WGA (Fig. 2d).

In order to demonstrate the suitability of our method to monitor the status of a graft prior to transplantation, murine kidneys were exposed to 24 h of cold ischemia. Following this, longitudinal biopsies were taken and incubated with Syto 16 (green), propidium iodide (PI; red), and WGA. Figure 3 shows the result of such a staining documenting the viability in a tubular area before (left image) and after 24 h of cold ischemia (right image).

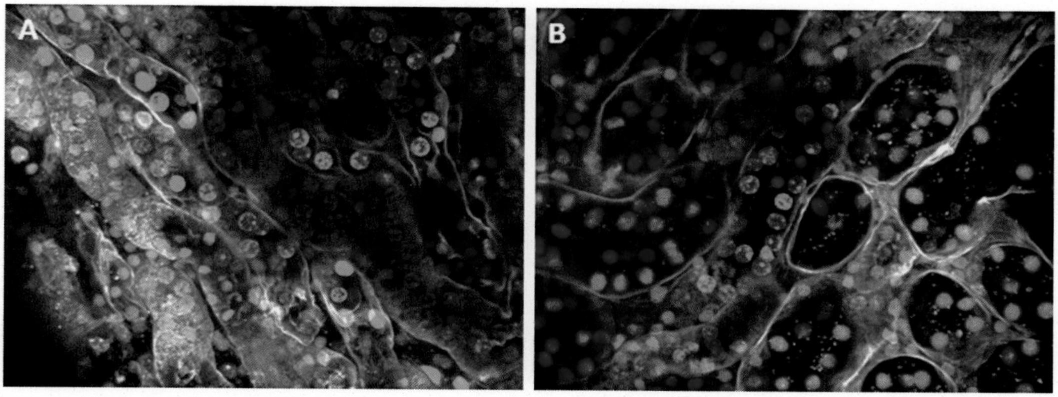

Fig. 3 Real-time live confocal imaging of the corticular area of a kidney biopsy. Following left renal IR (35 min) fine needle biopsies of murine kidneys were taken and incubated with the live stains Syto 16 (*green*), propidium iodide (PI; *red*), and wheat germ agglutinin (WGA, *white*), to stain all nuclei, nuclei of the dead cells, and cell membranes, respectively. *Left image* shows a tubular area, the *right image* a tubular area after 24 h of cold ischemia. Images were acquired with a 40× water immersion objective. Shown is a z-stack of 25 planes with a spacing of 1 μm

2 Materials

2.1 Animal Experimentation

1. Animals (mouse, rat) (*see* **Note 1**).
2. Surgical microscope.
3. Anesthetic gas evaporator.
4. Isoflurane.
5. Sterilized surgical instruments (forceps, scissors, surgical clamp holder).
6. Bipolar coagulation forceps.
7. Bipolar coagulation generator.
8. Nontraumatic vascular clamps.
9. Skinsept.
10. Underpads.
11. Gauze.
12. Transpore tape.
13. Autoclave unit.
14. Syringes.
15. Needles.
16. Preservative solution, e.g., HTK (Histidine-tryptophan-ketoglutarate).
17. Falcon™ tubes.
18. Surgical gloves.
19. Saline.
20. Hair trimmer.

2.2 Biopsy Acquisition and Culture

1. Super-core semi-automatic biopsy needle (18 g×9 cm) (Angiotech, Stenlose, Denmark).
2. Scalpels/blades.
3. Forceps.
4. Fine scissors.
5. Petri dishes.
6. Pipette set.
7. Pipet aid.
8. Eight-well chambered cover glass (Nalge Nunc International, Rochester, NY, USA).
9. Medium: Dulbecco's Modified Eagle's Medium/Nutrient Mixture F-12 containing 15 mM HEPES, 15 mM pyridoxine and $NaHCO_3$, supplemented with 5 % fetal calf serum (FCS), penicillin (100 U/ml) (PAA), streptomycin (100 μg/ml) (PAA), and 2 mM l-glutamine (*see* **Note 2**).
10. Water bath.
11. Laminar flow.
12. Tissue culture incubator (37 °C, 5 % CO_2).

2.3 Live Cell Staining

1. Fluorescent probes (propidium iodide, Syto-16, WGA, TMRM, *see* **Note 3**).
2. Aluminum kitchen foil.
3. Pipette and Tips.

2.4 Confocal Microscopy

1. Microlens enhanced Nipkow disk-based UltraVIEW™ RS confocal scanner (Perkin Elmer, Wellesley, MA, USA) mounted on an Olympus IX-70 inverse microscope (Olympus, Nagano, Japan; *see* **Note 4**).
2. UltraVIEW™ LCI software version 5.4 (Perkin Elmer).
3. Eight-well chambers (Nalge Nunc International).

3 Methods

3.1 Organ Preparation

To start biopsychronology, perfuse the organ with a preservative solution (e.g., HTK) to make sure that the organ is completely free of blood cells. Here, we present a perfusion procedure for rat kidney. The method, however, can be easily adapted for other organs (*see* **Note 5**).

1. Anesthetize the animal with isoflurane (oxygen flow 350–400 ml/min, isoflurane 2.5–3 %). Other anesthetics may also be used, ketamine (100 mg/kg BW) and xylazine (10 mg/kg BW).

2. Remove hair from the abdominal region of the animal with the help of a hair trimmer.
3. Place the animal ventrally on a sterile pad under the surgical microscope.
4. Open the abdominal cavity of the animal by a long midline incision and two additional cuts ranging diagonally in a 45° angle from the pubic symphysis.
5. Move all gastrointestinal organs to the right side, exposing the left kidney and the abdominal vessels.
6. Place a saline wet gauze on the retracted intestines.
7. Liberate left renal pedicels from the surrounding tissue and vessels by blunt dissection and cauterization.
8. Expose aorta approximately 2 mm proximal and 5 mm distal of the left renal pedicles by blunt dissection.
9. Place two nontraumatic vascular clamps on the proximal and distal ends of the exposed region of aorta.
10. Puncture the inferior vein cava with scissors to allow the outflow of the blood.
11. For perfusion, insert a cannula into the aorta distal of the left renal pedicles and flush the kidney slowly with 6–8 mL of ice-cold HTK preservative solution.
12. Procure the kidney by cutting the renal pedicles and ureter with scissors.
13. Kidney can be used immediately for biopsy acquisition or subjected to a desired treatment, e.g., ischemia.

3.2 Fine Needle Biopsy

To obtain renal biopsies perform below mentioned steps under the flow hood with sterilized instruments.

1. Fill 50 mL of culture medium in a petri dish and place the harvested kidney in the medium.
2. Carefully remove the capsular layer of the kidney by using a pair of forceps and scissors.
3. Set the needle biopsy to the first throw, thus creating a specimen slot of 10 mm penetration depth.
4. Insert cannula of the biopsy needle across the kidney from the outer cortex to the hilum in a way that the specimen slot of the biopsy needle is in the desired region (Fig. 1).
5. Push the trigger to fire cutting cannula.
6. Withdraw biopsy needle.
7. Pull the cutting cannula to the second throw, exposing the biopsied tissue in the specimen slot.
8. Using forceps, pick up the biopsy at the hilum end and immerse into the medium in the petri dish.

9. To obtain further biopsies repeat **steps 3–8**.
10. To prepare for staining, transfer biopsies into the eight-chambered cover slips (2 biopsies/well), containing 200 μL Hepes-buffered culture medium.

3.3 Acquisition of Longitudinal Biopsies

To mimic longitudinal biopsies taken in the clinic in routine practice, sharp blades/scalpel can also be used instead of needle biopsies (*see* **Note 6**).

1. Following the first two steps (see above), carefully shave out the thin surface slices (4–6) of the kidney with the sharp blade, similar to skin shave biopsy.
2. To prepare for staining, transfer biopsies into the eight-chambered cover slips (2 biopsies/well), containing 200 μL Hepes-buffered culture medium (*see* **Note 7**).

3.4 Biopsy Staining and Live Confocal Imaging

1. Place 200 μL medium in each well of the eight-chambered cover slips.
2. Add fluorescent dyes (final concentrations as recommended by the manufacturer) (*see* **Note 8**) and mix to allow homogenous distribution.
3. Transfer the biopsies (needle/longitudinal) into the medium. Make sure that the biopsies are submerged in the medium.
4. Wrap the eight-chambered cover slip with aluminum foil to prevent photobleaching of the fluorophores.
5. Incubate the biopsies for 15 min in tissue culture incubator at 37 °C, 5 % CO_2.
6. Wash the biopsies by replacing the staining medium with fresh medium. Use a pipette and do not touch the biopsy while washing.
7. Add 200 μL medium to each well of the eight-chambered cover slips containing the biopsy (*see* **Note 9**).
8. Perform microscopy with an inverted confocal microscope. Start with a low magnification (e.g., 10× objective) to obtain a general overview and use higher magnifications (e.g., 40× water immersion objective) to acquire detailed images of the biopsy (*see* **Note 10**).
9. Image acquisition can be performed by using only one optical plane or by adding several optical planes to a z-stack (*see* **Note 11**).

4 Notes

1. Though the procedure has been originally established for the rat/mice kidneys, it can be adapted for every organ or tissue of all organisms, including human.

2. Depending on the tissue and the live stains the usage of other buffer solutions may be required. For biopsychronology the cells and tissues should be placed in buffer solutions which do not alter cell viability or functional parameters. When using cell culture media, keep in mind the pH rises as soon as the incubation is not performed in a CO_2 incubator. Therefore the use of Hepes-buffered media is strongly recommended in order to avoid unnecessary pH-induced stress to cells.
3. Several other live stains can be applied which are suitable to visualize radical oxygen species (e.g., MitoTracker™ Red CM-H2XRos), calcium levels (e.g., Rhod-2), or other parameters.
4. Although we used a confocal microscope allowing live cell imaging, other microscopes (even nonconfocal setups) may also allow a quality assessment of the biopsy. Depending on the systems used, damage due to photobleaching has to be considered and specially addressed by using appropriate controls.
5. Care should be taken to minimize handling time of the biopsy before imaging.
6. Longitudinal biopsies mentioned here are more appropriate for real-time imaging of the cells at the surface of the kidney.
7. The biopsies should be analyzed within 20 min after harvesting the organ.
8. Depending on the size and type of biopsy, concentration as well as incubation times can be adapted.
9. Place the cut slice in the eight-chambered cover slips with the surface side of the kidney facing the bottom of the chamber, containing 200 μL medium in each well. The biopsy should be small enough to freely sink to the bottom of the well. Otherwise the distance between the objective and the sample might be bigger than the working distance of the objective. For samples with a high content of fat cells, floating of the biopsy may present a problem. In such a case aspirate the staining medium until the biopsy makes contact with the bottom of the cover slip.
10. Upright microscopes may also be used. For such a purpose, the biopsy has to be placed in a setting allowing for upright microscopy.

    Imaging should be performed as fast as possible, especially when several biopsies are analyzed. Stability of the dye in the specific setting should be tested in pilot experiments. Depending on the dyes, the staining might change with time. After imaging the last sample, it is recommended to go back to the first sample and check whether the staining is still as it was when starting the imaging.
11. Keep in mind that the mechanical stress while taking/handling the biopsy can also lead to cell damage.

References

1. Esposito C, Migotto C, Torreggiani M et al (2012) Pretransplant and protocol biopsies may help in defining short and mid-term kidney transplant outcome. Transplant Proc 44:1889–1891
2. Mazzucco G, Magnani C, Fortunato M, Todesco A, Monga G (2010) The reliability of pre-transplant donor renal biopsies (PTDB) in predicting the kidney state. A comparative single-centre study on 154 untransplanted kidneys. Nephrol Dial Transplant 25:3401–3408
3. Kuznetsov AV, Troppmair J, Sucher R, Hermann M, Saks V, Margreiter R (2006) Mitochondrial subpopulations and heterogeneity revealed by confocal imaging: possible physiological role? Biochim Biophys Acta 1757:686–691
4. Wlodkowic D, Skommer J, Darzynkiewicz Z (2011) Rapid quantification of cell viability and apoptosis in B-cell lymphoma cultures using cyanine SYTO probes. Methods Mol Biol 740:81–89
5. Hermann M, Pirkebner D, Draxl A, Margreiter R, Hengster P (2005) "Real-time" assessment of human islet preparations with confocal live cell imaging. Transplant Proc 37:3409–3411
6. Wright CS (1984) Structural comparison of the two distinct sugar binding sites in wheat germ agglutinin isolectin II. J Mol Biol 178:91–104

Part III

In Vivo Analysis of Metabolism in Ischemia-Reperfusion

Chapter 9

Prolonged and Continuous Measurement of Kidney Oxygenation in Conscious Rats

Maarten P. Koeners, Connie P.C. Ow, David M. Russell, Roger G. Evans, and Simon C. Malpas

Abstract

A relative deficiency in kidney oxygenation, i.e., renal hypoxia, may contribute to the initiation and progression of acute and chronic kidney disease. A critical barrier to investigate this is the lack of methods allowing measurement of the partial pressure of oxygen in kidney tissue for long periods in vivo. We have developed, validated, and tested a novel telemetric method that can do this. Here we provide details on the calibration, implantation, implementation for data recording, and reuse of this telemetry-based technology for measurement of medullary tissue oxygen tension in conscious, unrestrained rats. This technique provides an important additional tool for investigating the impact of renal hypoxia in biology and pathophysiology.

Key words Kidney, Tissue oxygen concentration, Medulla, Telemetry

1 Introduction

Clinical and experimental data indicate that multiple forms of acute and chronic kidney disease are associated with a relative deficiency in oxygen tension (PO_2) within kidney tissue (i.e., renal hypoxia). These observations have formed part of the evidence underlying the hypothesis that tissue hypoxia is a critical driver of the development and progression of both acute kidney injury and chronic kidney disease [1–11]. Yet the precise roles of renal hypoxia in the underlying mechanism(s) remain unresolved. A major technical limitation has been the absence of methods allowing long-term measurement of kidney tissue PO_2 in conscious, unrestrained animals.

The advent of implantable radio-telemetry has been pivotal within cardiovascular research for recording of physiological signals, like arterial blood pressure, over extended periods of time in unstressed animals (i.e., unhindered by anesthesia or restraint). Recent advancements in telemetric technology, particularly the

Tim D. Hewitson et al. (eds.), *Kidney Research: Experimental Protocols*, Methods in Molecular Biology, vol. 1397,
DOI 10.1007/978-1-4939-3353-2_9, © Springer Science+Business Media New York 2016

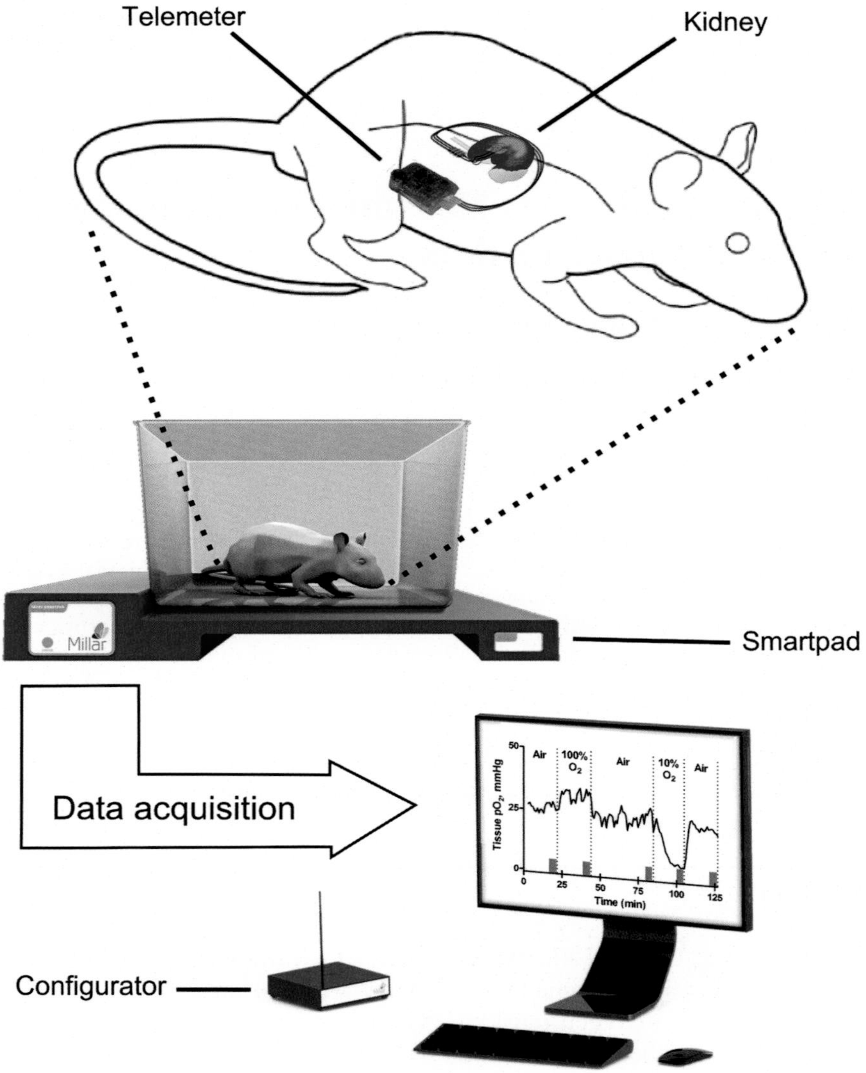

Fig. 1 Schematic diagram of the fully implantable telemetry system for measurement of kidney tissue PO_2. After implantation surgery each rat is placed on a SmartPad that enables transcutaneous recharging and continuous recording with a data acquisition system. Control of the telemeter settings is facilitated wirelessly by the Configurator

ability to recharge the batteries of telemetric transmitters in vivo, in a completely noninvasive manner [12, 13], have enabled us to combine radio-telemetry with an electrochemical oxygen sensor.

Herein we report the methods required for use of this exciting new technology for measurement of renal tissue PO_2, with high temporal resolution, in conscious, unrestrained rats (Fig. 1) [14]. Previously we have validated this technology (Fig. 2) and observed stable renal medullary tissue PO_2 over a 19-day period (Figs. 3 and 4)

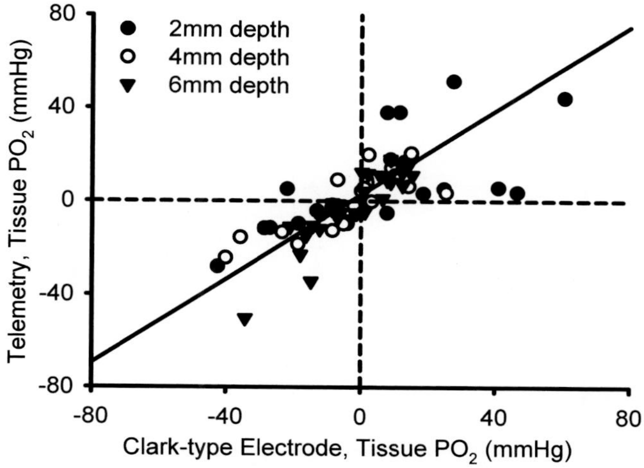

Fig. 2 Validation of the telemetry system. Changes in kidney PO_2 in anesthetized rats, measured by Clark electrode and the carbon paste oxygen electrode-telemetry system. Kidney PO_2 was changed by altering inspired oxygen. Measurements were made sequentially at different depths below the cortical surface. *Symbols* represent individual data points. The *solid line* represents the line of best fit, calculated by ordinary least products regression analysis ($r = 0.74$, $P < 0.001$). Reproduced from ref. 14 with permission from the American Physiological Society

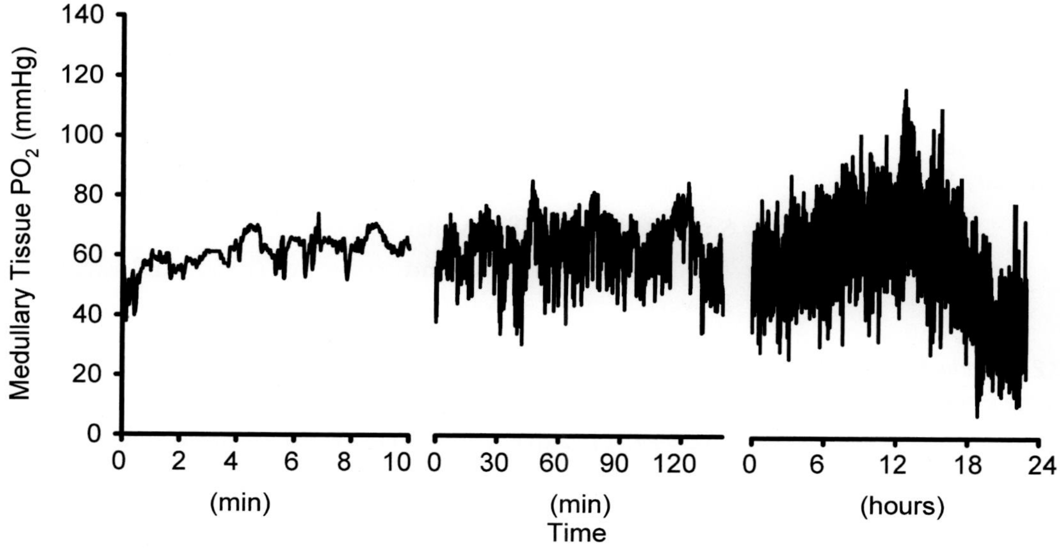

Fig. 3 A typical recording of renal medullary tissue PO_2 measured over minutes, hours, and one day using a telemeter that had been implanted for a week. Values are derived using the post-implantation calibration relationship and subtraction of zero offset values

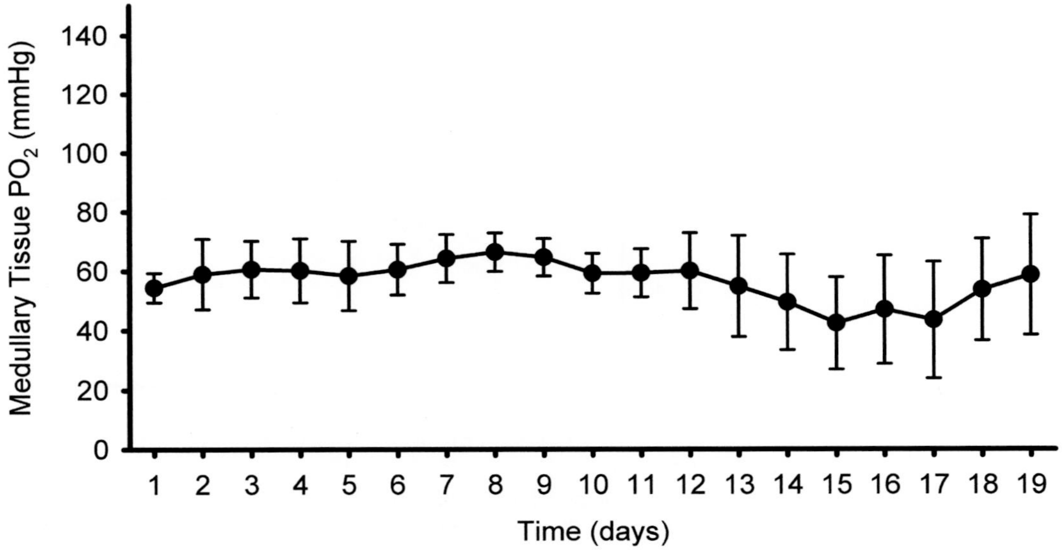

Fig. 4 Mean daily medullary tissue PO_2 across a 19-day period. Values are derived using the post-implantation calibration relationship and subtraction of zero offset values and are shown as mean ± SEM of six rats (*see* **Note 11**). Reproduced from ref. 14 with permission from the American Physiological Society

and reproducible responses to hypoxia and hyperoxia (Fig. 5). The telemetric method we have developed represents the only existing approach for measurement of renal oxygenation in conscious, unrestrained rats.

2 Materials

2.1 Oxygen Telemetry System

1. Telemeter (TR57Y, Millar Inc., Houston, TX, USA).
2. Configurator and ConfigSoft™ software (TR190, Millar Inc.).
3. SmartPad™ (TR181, Millar Inc.).
4. Acquisition hardware and software (e.g., ADInstruments PowerLab and Lab Chart Data).

2.2 Connection of the Electrodes

1. Oxygen electrode: carbon paste electrode (Millar Inc.).
2. Teflon (Polytetrafluoroethylene, PTFE) coated silver wire (AG549511, Advent Research Materials, Suffolk, UK) or reference and auxiliary electrodes (Millar Inc.).
3. FluoroEtch™ (Acton Technologies, Pittston, PA, USA).
4. Acetic acid.
5. 70 % ethanol.
6. Scalpel.

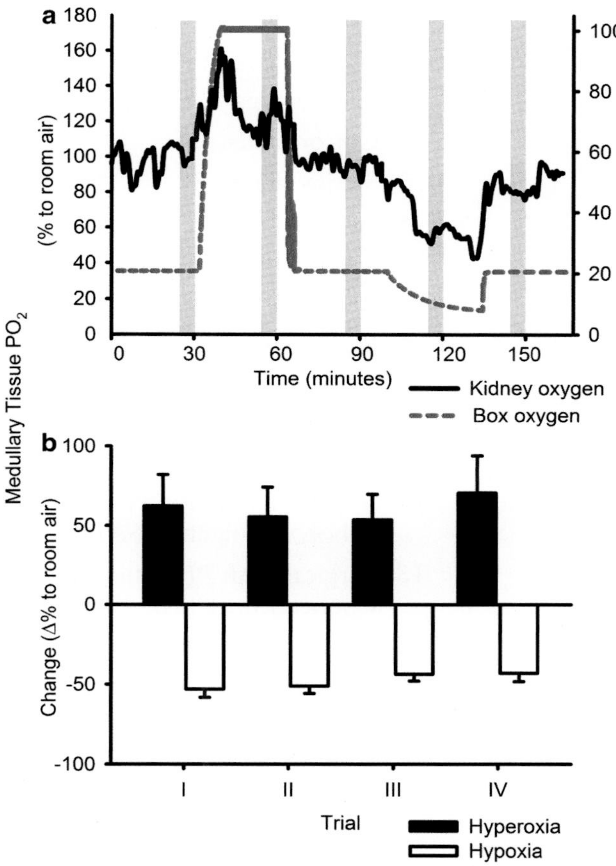

Fig. 5 Responses of medullary tissue PO_2 to altered inspired oxygen content. (**a**) Typical response in one rat (*solid line*, 0.01 Hz low-pass filtered). The *dashed line* indicates the oxygen levels measured in the chamber. The *shaded area* shows the 5-min periods over which average data were obtained during the four trials over the 3-week period of telemeter implantation. (**b**) Mean changes in medullary tissue oxygen in response to hyperoxia (*closed bars*) and hypoxia (*open bars*) of 14 rats. *I–IV* represent the successive trials performed at 3–4 day intervals. Reproduced from ref. 14 with permission from the American Physiological Society

7. Fine (long) nose pliers.
8. Needles for applying epoxy adhesives.
9. Silver conductive epoxy adhesive (8331-14G, MG-Chemicals, Ontario, Canada).
10. Clear nail varnish.
11. Polyurethane tubing (internal diameter of 1.5 mm).
12. Generic (slow setting) 2-part epoxy adhesive, Selleys Super Strength Araldite™ (Selleys Australia & New Zealand, Padstow, NSW, Australia).

2.3 Implantation of the Telemeter

1. Isoflurane.
2. Scalpel.
3. Fine forceps.
4. Cotton buds.
5. Sterile saline.
6. Gauzes.
7. Scissors.
8. Retractor for laparotomy.
9. Hemostats.
10. Suture to suture the leads to the adventitia of the aorta (non-absorbable, 6-0).
11. Needle to pre-puncture the kidney (30G).
12. *n*-Butyl-2 cyanoacrylate tissue adhesive (TissueSeal, Ann Arbor, Michigan, USA).
13. SurgicalMesh™ (Textile Development Associates Inc, Brookfield, Connecticut, USA).
14. Sutures and staples (or sutures) to close abdominal muscle and skin.
15. Buprenorphine.
16. Sodium pentobarbitone (or alternative method of euthanasia).
17. Terg-a-zyme enzyme detergent (Sigma-Aldrich, St. Louis, MO, USA).

3 Methods

3.1 System Overview

The Millar Tissue Oxygen Telemeter (TR57Y) is an implantable telemeter for measurement of tissue PO_2 in small animals (>200 g) such as the rat. The telemeter incorporates a potentiostat and is encased in silicone (Fig. 1). The unit has three stainless steel coiled lead wires that must be attached, by the user, to the three electrodes required for measurement of tissue PO_2 (working, reference, and auxiliary). The oxygen electrode (working) is a carbon paste electrode [15–19], while the reference and auxiliary electrodes are silver electrodes. The telemeter units come in a fully implantable form, i.e., fitted with all three electrodes. Although the electrodes used for measurement of PO_2 are not suitable for reuse (stabilization of the oxygen electrode in vivo requires deposition of proteins at the tip of the electrode), the great advantage of the system is that the user can attach new electrodes before the next experiment.

3.2 Connection of the Carbon Paste Oxygen Electrode to a Telemeter

These are the recommended procedures for attaching new electrodes to the leads of the telemeter (Fig. 6). The process of attachment of the electrodes to the telemeter leads requires at least 3 days. An additional 2 days is required for the connection to be completely cured, prior to calibration and implantation (*see* **Note 1**).

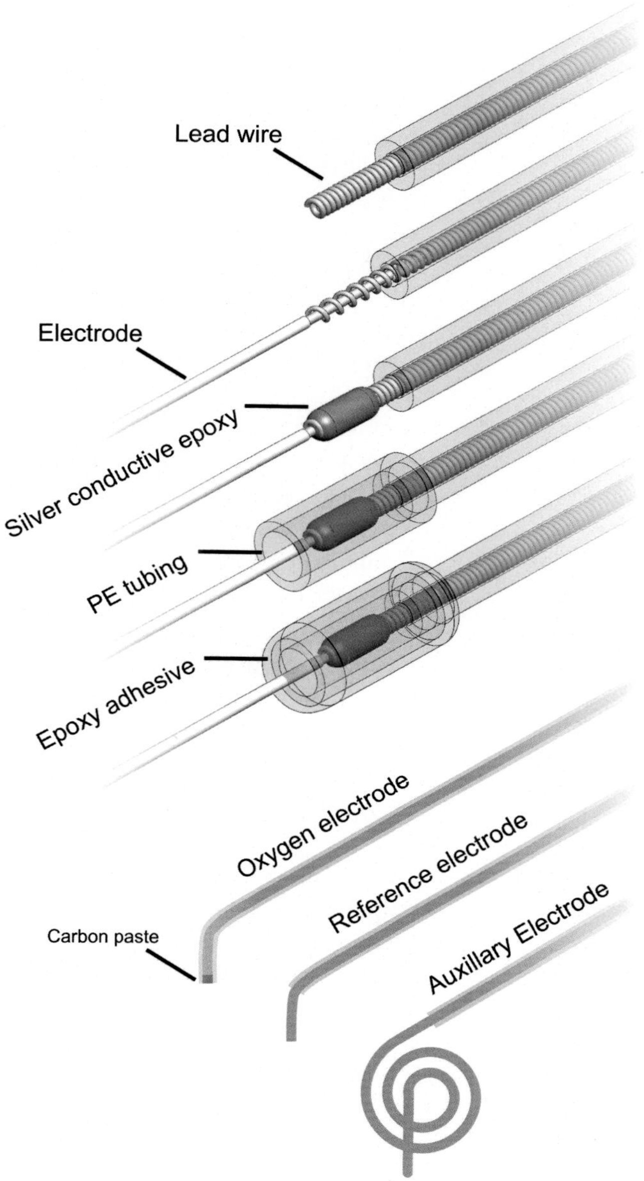

Fig. 6 Electrode lead preparation. In order to reuse the telemeter, new electrodes are attached to the telemeter leads. Silver conductive epoxy will facilitate conduction between electrodes and lead wire. Polyethylene (PE) tubing in combination with epoxy adhesive will provide a waterproof seal. Before use ~5 mm and 8 cm of the Teflon coat is removed from the reference and auxiliary electrode, respectively. Both the carbon paste oxygen and reference electrode are bent at a 90° angle ~5 mm from the electrode tip while the auxiliary electrode is coiled

3.3 Electrode Lead Preparation

The oxygen electrode supplied by Millar Inc. is pre-etched by the supplier. The etching process enhances adhesion of the polyurethane tubing (used to cover the connection) to the electrodes. Approximately 5 mm of the Teflon-coating of the electrodes should be removed at one end of the electrode to facilitate insertion of the electrodes into the stretched stainless steel coils of the telemeter leads. A properly etched electrode will have a light brown color on the Teflon coating. For best results we strongly recommend purchasing the electrodes from Millar Inc. for use with the TR57Y telemeters. Although Millar Inc. can supply the reference and auxiliary electrodes, these can be easily made and etched by the researcher from Teflon-coated silver wire.

3.3.1 Etching of the Reference and Auxiliary Electrodes

1. Cut ~5 cm and ~8 cm of the Teflon-coated silver wire for the reference and auxiliary electrode, respectively.
2. Heat FluoroEtch, acetic acid, and distilled water in a water bath at 80 °C. FluoroEtch deactivates rapidly once exposed to air. Therefore the sealed bottle should be placed inside the water bath.
3. Immerse ~3 cm of the end of the electrodes to be etched into the etchant for 1 min before immersing them into 70 % ethanol and distilled water for 30 s each in succession.
4. Finally, immerse the electrodes into acetic acid for 1 min.
5. The etched end of the electrode will appear brown in color once it is successfully etched.

3.3.2 Preparation of the Telemeter for Reattachment of Electrodes

The telemeters can be reused after they are explanted. Using a scalpel carefully remove the used electrodes from the explanted telemeter, as close to the original joint as possible to preserve as much telemeter lead length as possible.

3.3.3 Lead Wire Preparation

1. Place the telemeter leads on a flat surface so that the tubing is not stretched. Cut off any used and exposed stainless steel coiled wire.
2. Cut the leads so that all three of the telemeter leads are of the same length.
3. For each lead, hold the polyurethane tubing upright. Using a pair of long nose pliers, extend and stretch the unexposed stainless steel coiled wire ~1 cm past the end of the polyurethane tube.
4. Carefully remove any colored silicone (blue, black, or red coverings around the coiled wires) from within the center of the extended coiled wire using forceps or by rolling the silicone coverings on the lead between two fingers.

5. Cut the stretched lead wire so that approximately 3 mm is extending out of the end of the tubing for each lead.
6. The leads are color coded as follows:
   (a) Blue lead = carbon paste electrode (oxygen electrode)
   (b) Black lead = reference electrode
   (c) Red lead = auxiliary electrode

3.3.4 Electrode Attachment (Fig. 6)

1. Insert the exposed wire of the electrode into the coiled telemeter lead wire (at least 2 mm).
2. Mix a small amount of the two component silver conductive epoxy (1:1 ratio). The working time of this product is about 10 min. After that time it becomes too thick to apply. Using a needle apply the silver epoxy between and around the areas where the electrode comes into contact with the telemeter leads. Ensure that the silicone tubing of the electrode is free of silver epoxy.
3. Using a fresh (or cleaned) needle, remove excess epoxy. The joint should be able to fit inside the polyurethane tubing, so the connection should be no wider than the coiled wire itself.
4. Leave the silver epoxy to dry overnight.

3.3.5 Sealing the Connection (Fig. 6)

1. Coat the silver conductive epoxy with two layers (30 min in between each) of clear lacquer (nail varnish).
2. After the second coating is dry (>30 min), mix a small amount of the two component adhesive epoxy in a 1:1 ratio.
3. Apply a layer of adhesive epoxy over the silver epoxy layer. Cut a piece of polyurethane tubing (1.5 mm inner diameter) and carefully slide it along the electrode, ensuring that the tubing covers the connection and fits around the telemeter lead. Take care not to damage the tip of the carbon paste electrode.
4. Leave this to dry overnight. Keep the leads and electrodes vertical to prevent uneven setting of the glue.
5. Mix a small amount of the two component adhesive epoxy in a 1:1 ratio. Allow it to cure for 15 min before applying to the connection.
6. Apply the epoxy adhesive to each of the connections so that all sides are covered and that some of the etched (brown) part of the electrode coating is covered. Do not apply excess epoxy or it will run down the electrode. Ensure that both ends of the polyurethane tubing covering the silver conductive epoxy layer are completely sealed.

Leave to dry/cure for a minimum of 2 days. Once the epoxy adhesive is completely dry, the electrodes can be prepared for calibration and subsequent implantation.

3.4 Preparation before Telemeter Implantation

All telemeters are shipped switched off ("safe mode") to maintain battery life. They must be activated via the following process prior to the first use.

1. *Note the Serial Number of Your Telemeter*
   Each telemeter has a unique serial number printed on the casing. The default channel setting of all shipped telemeters is Channel "setup."
2. *Set up the Wireless TR181 SmartPad™*
   The TR181 SmartPad™ is a universal receiver and recharging pad for all small animal telemeters. The SmartPad™ automatically detects the type of telemeter it is communicating with and all analogue outputs are low-pass filtered at 1 kHz. Each SmartPad™ has a serial number printed on the side above the status indicator light. Note that the SmartPad™ should not be operated on a metal bench or within 10 cm of a metal object such as a filing cabinet as this may cause the detection of a high current and thus automatically disable the charging field.
3. *Set up the TR190 Configurator*
   The Configurator System includes the TR190 Configurator hardware and ConfigSoft™ software (*see* **Note 2**), which are used together to wirelessly control any 5* series telemeters and TR181 SmartPads™. Install the ConfigSoft™ software on your computer. Connect the antenna to the rear of the TR190 Configurator and then connect the T190 Configurator to your computer by the USB connection provided.
4. *Pair the Telemeter with a SmartPad™*
   Place the telemeter to be used on the SmartPad™ you wish to pair it with. In the telemeter configuration and diagnostics menu of the TR190 ConfigSoft™ software, you can set the channel number of a telemeter by entering the serial number in the box and then selecting a channel number to change it to. In the SmartPad™ configuration and diagnostics menu, enter the serial number of the SmartPad™ and set it to the same channel as the telemeter to pair the two devices and enable both inductive charging and data transmission. The SmartPad™ indicator will turn green, indicating successful pairing, charging, and data transmission. Battery level and charging status can also be monitored with the ConfigSoft™ software.
5. *Connect the SmartPad™ to a Data Acquisition System*
   Connect the analogue BNC connectors located on the side of the SmartPad™ to a data acquisition system (Inputs 1–3) according to the following output allocation:

Telemeter model	Output 1	Output 2	Output 3
TR57Y	Oxygen	NA	Temperature

The temperature of the telemeter body is monitored to regulate the charging field in order to prevent overheating of the device. It is not a necessity to connect the temperature output to the data acquisition system.

3.5 Pre-implantation Calibration

The current flowing between the auxiliary electrode and the oxygen electrode is measured by the telemetry circuit. This is converted into a voltage output by the SmartPad™ to provide a voltage which varies between 0 and 4 V. Specifically 1 V is equal to 0 nA and 2 V is equal to −200 nA. The system has a maximum range of up to −600 nA. The working electrode current in nanoamperes is calibrated at the factory so as to be directly proportional to the concentration of oxygen. A calibration factor is provided with each electrode for use in the conversion of the measured current to PO_2 (and/or oxygen concentration). We suggest that the telemeters are calibrated both before implantation and after explanation, using the same procedure (Fig. 7).

1. Add 20 mL of saline into a glass cell into which you place the three electrodes. Add a mini stirring bead and place the cell on a magnetic stirrer/heater. Set the temperature of the saline to 37 °C. Partially seal the cell and saturate the 20 ml saline with N_2 for 30 min (Fig. 7a).
2. Meanwhile, in a second cell saturate 20 mL of saline with O_2 for at least 30 min to create a 100 % O_2 saturated solution. Keep this solution O_2-saturated during the entire calibration procedure.
3. Stop the infusion of N_2 by pulling the tube out the saline solution so that it is just above the fluid surface.
4. Add incrementing aliquots of this O_2-saturated solution (+416, +425, +434, +443, and +452 μL) into the cell in which the electrodes have settled to generate typical responses which have a linear relationship (Fig. 7b). Addition of these solutions to the calibration cell will result in PO_2 of 15.8, 31.5, 47.3, 63.1, and 78.9 mmHg, respectively (equating to concentrations of O_2 of 25, 50, 75, 100, and 125 μM) (*see* **Note 3**).

In the field of biomedical science, tissue oxygenation is typically measured as partial pressure [mmHg]. The carbon paste oxygen electrodes used in this system directly measures the number of moles of oxygen in the vicinity of the electrode tip. It can be thought of as "counting" dissolved oxygen ions in solution, where the measured current in the wire is proportional to the number counted. We can use molar concentration of oxygen [mol/m^3] in our physiological experiments but it becomes difficult when we need to compare our results with those of other researchers, who mostly present data in mmHg.

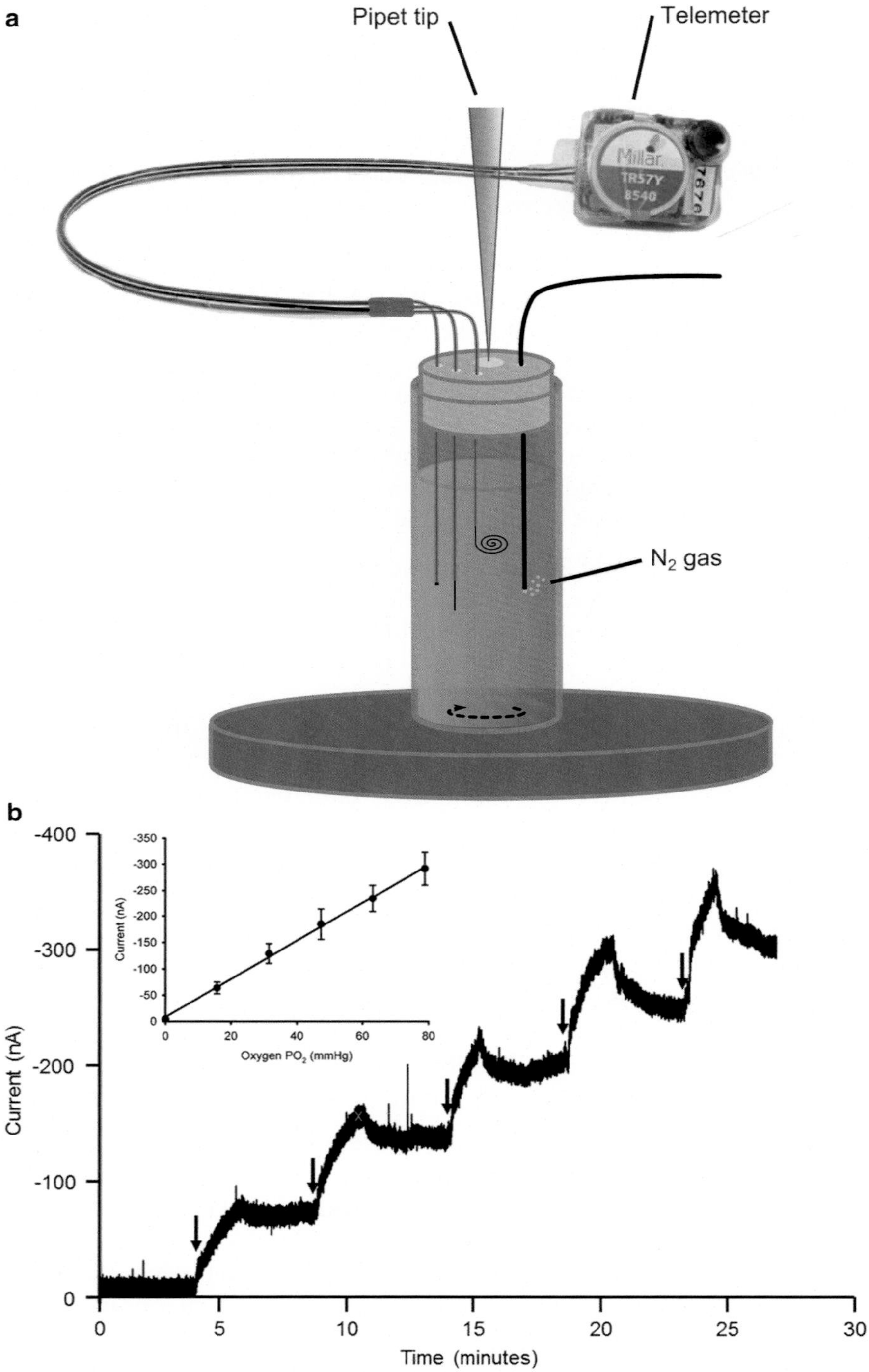

Fig. 7 Calibration of the telemetry system. (**a**) Schematic diagram of a cell with the electrodes placed in saline. Before calibration the saline within the cell is saturated with N_2. (**b**) Typical response upon cumulative addition of aliquots of saline saturated with 100 % O_2 to the cell. *Arrows* show when aliquots were added. The *inset* displays the calibration relationship between current measured by the telemeter (nA) and PO_2

It is possible to calibrate the sensors for use in [mmHg] but there exists no analytical solution. The conversion is simply derived from empirical evidence. Bolger et al. use the following conversion factors [16]:

O_2%	*P* (mmHg)	*c* (μM)
0	0	0
Air [21]	152	240
100	760	1200

Which results in the following formula:
$P = c.k$
where:
c is the molar concentration [mol/m^3]
k is a constant, 0.631 [mmHg.m^3/mol]
P is the partial pressure of oxygen in solution [mmHg].

3.6 Preparation of Electrode for Implantation in the Kidney

Before the electrodes can be implanted, a length of the insulating Teflon coating of the auxiliary and reference electrode must be removed. For the reference and auxiliary electrode, remove ~5 mm and 8 cm (increases the surface area) of the Teflon coat respectively, by rolling a scalpel gently along the Teflon-coated silver wire, ensuring you do not damage the exposed silver wire in the process. In addition, the 10 cm long auxiliary electrode must be bent into a coil to generate a compact implantable configuration (Fig. 6).

As renal tissue PO_2 is spatially heterogeneous, the depth of electrode implantation will affect the measured PO_2. For inner medullary implantation, it is recommended that you bend both the carbon paste oxygen and reference electrode at a 90° angle ~5 mm from the electrode tip (*see* **Note 4**). The position of the bend can be altered to allow the tip of the probe to be positioned at the required depth below the surface of the renal cortex (Fig. 6).

3.7 Sterilization

1. The effects of ethylene oxide gas sterilization on the carbon paste electrodes have not been determined. We currently recommend that the radio transmitter of the telemeter and the lead wires, but not the attached electrodes, are cold sterilized using 2 % glutaraldehyde for at least 1 h before implantation surgery.
2. At the time of surgery, the surface of the electrodes, with the exception of the carbon paste oxygen electrode, can be sterilized using 70 % alcohol. Do not wipe the tip of the carbon paste oxygen electrode as this will ruin the carbon paste which is absolutely necessary for the functionality of the oxygen telemeter.

3.8 Telemeter Implantation

Telemeters should be fully charged before implanting. Use the ConfigSoft™ software to monitor battery level. Once fully charged, a telemeter can be placed in Safe Mode to deactivate it and maintain battery charge. To reactivate the telemeter simply place it on the SmartPad™, or in case the telemeter is reactivated after implantation, place the animal in its home cage after the surgery and place the home cage back on the SmartPad™ it was originally paired with.

As the implantation surgery can take 1.5–2 h, we recommend the use of isoflurane anesthesia as this avoids the need for multiple administration of an injectable anesthetic.

1. Prepare the rat for a laparotomy by shaving the skin overlying the abdomen.
2. Disinfect the skin with iodine and inject the analgesic buprenorphine s.c. (3 μg/100 g rat).
3. Under isoflurane anesthesia place the rat in dorsal recumbence, make a ~2 cm incision ventral midline abdominal incision, and separate the skin from the underlying muscle by blunt dissection.
4. Extend the abdominal incision through the linea alba into the abdominal cavity taking care to avoid the underlying organs. Use a retractor to gain appropriate access and visibility.
5. Deflect the intestines to the rat's right with saline soaked gauzes to expose the left kidney (*see* **Note 5**). Make sure that during the procedure the intestine in general and the exposed kidney in particular remain moist by applying drips of sterile saline.
6. After identifying the abdominal aorta and the kidney, dissect the aorta free from the surrounding fatty tissue (Fig. 8).
7. Suture the telemeter joints (i.e., the connections between the electrodes and the telemeter leads) to the adventitia of the aorta with 6-0 suture (non-absorbable) in the following order: reference, oxygen carbon paste, and auxiliary electrode (*see* **Note 6**).
8. Make a small puncture with a 30 gauge needle in the kidney where the reference electrode will be inserted.
9. Insert the reference electrode and secure it to the kidney with n-Butyl-2 cyanoacrylate tissue adhesive.
10. Insert the carbon paste oxygen electrode in the same way.
11. Cover both electrodes with SurgicalMesh™ and secure the mesh to the renal capsule with some additional tissue adhesive.
12. Secure the auxiliary electrode to the covered electrodes and surrounding kidney tissue with mesh and tissue adhesive on top (Fig. 8).

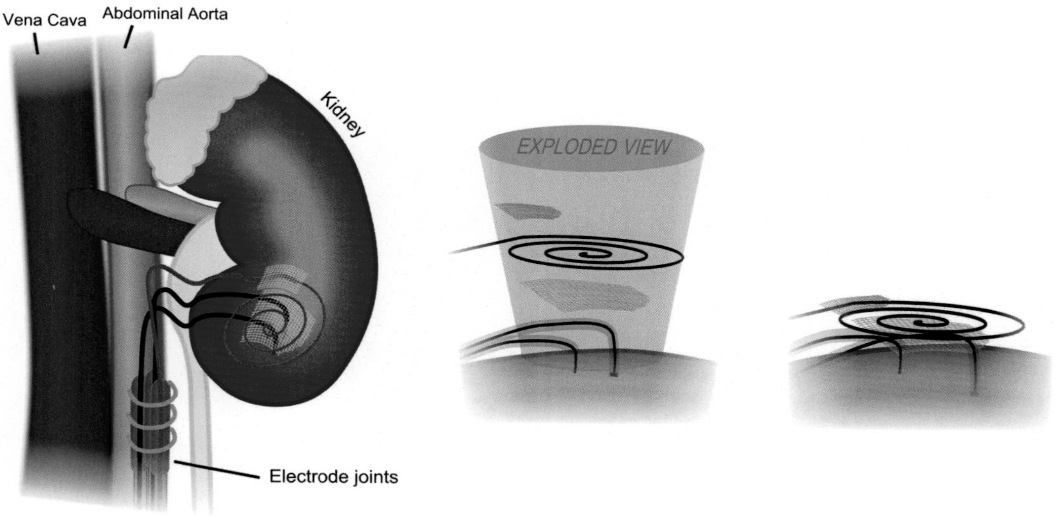

Fig. 8 Implantation of the electrodes in the kidney. To anchor the telemeter leads, the joints of all electrodes are sutured on the abdominal aorta. After pre-puncture of the capsula and kidney tissue, the oxygen and reference electrodes are inserted until a depth of ~5 mm (for medulla) and anchored with tissue glue and mesh. On top the auxiliary electrode is anchored using glue and a smaller piece of mesh

13. Remove the gauze from the abdomen and replace the abdominal contents.
14. Insert the telemeter under the abdominal muscles, on the right side of the abdomen. Ensure the telemetry leads remain close to each other and parallel. Maneuver the telemeter into a stable position.
15. Insert the suture needle individually through each of the two suture tabs on the telemeter to fix it to the underside of the abdominal muscle (Fig. 9).
16. Pour ~1 mL of saline in the abdominal cavity. Close the ventral midline incision in two layers. If appropriate sterile procedures are followed, there should be no need for antibiotics. However, you may choose to administer an agent such as enrofloxacin s.c. (0.5 mg/100 g rat).
17. Finally, place the cage on a heated platform (30–37 °C) for 3–24 h to facilitate the rat's recovery.
18. For postoperative analgesic care inject buprenorphine s.c. (3 μg/100 g rat) every 8–14 h for up to 3 days or as required. Supply soft food (e.g., "recovery gel") for 1–3 days. Monitor bodyweight daily for at least 1 week. If bodyweight 5–7 days after surgery is >15 % less than its preoperative weight, appropriate action should be taken (typically the rat must be humanely killed).

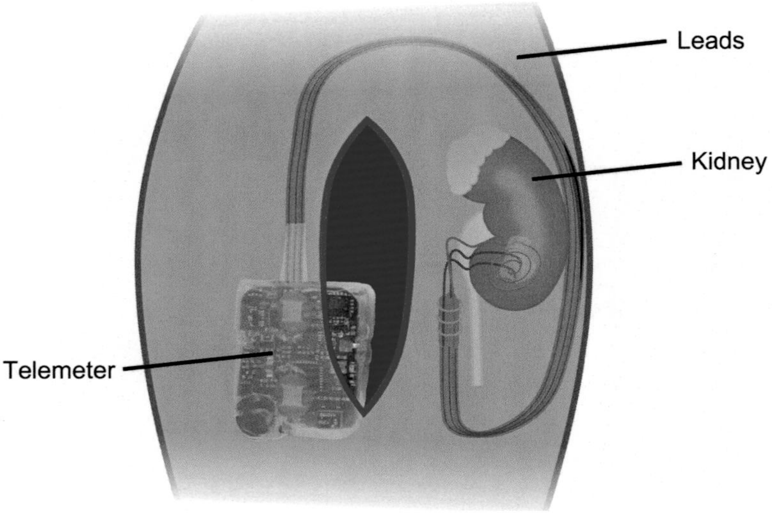

Fig. 9 Placement of the telemeter in the abdominal cavity. The telemeter leads are placed counterclockwise in the abdominal cavity and the telemeter body in the lower right quadrant. The telemeter is secured to the abdominal muscles by non-absorbable sutures through the suture tabs

3.9 Data Recording

Data can be collected 24 h a day, continuously as the telemeter batteries never need to be replaced. The animal's home cage is constantly on the SmartPad™ which charges the telemeter and receives and transmits data to your data acquisition device (*see* **Note 7**). To verify that the telemetry system can measure differences in kidney oxygenations, the rat can be exposed to different concentrations of oxygen in the inspired air and/or vasoactive drugs can be given. For example, we measured the changes in medullary tissue PO_2 when rats were subjected to 30-min periods in which they inspired hypoxic (10 % O_2 to 90 % N_2) and hyperoxic (100 % O_2) gas (Fig. 5).

3.10 Post-explantation Calibration and Cleaning and Storage

1. In order to assess the zero offset of the telemeter system in vivo, measure the PO_2 after cardiac arrest induced by humanely killing the animal with an overdose of anesthetic (e.g., >200 mg/ml sodium pentobarbitone). Subtract this offset value from data collected in vivo before subsequent data analysis.
2. One you have established the offset value, open the abdomen to explant the telemeter (*see* **Note 8**). Depending on the duration of implantation, the telemeter may be considerably encapsulated in surrounding tissue.
3. Carefully separate away all tissue, taking care not to damage the telemeter body, the connected leads or the electrode tips.
4. Cut the sutures and slowly pull the telemeter out.

5. Cut out the kidney and aortic tissue to which the leads are connected. Carefully free the leads and the electrodes from any tissue and gently dissect and pull the electrodes out of the kidney (*see* **Note 9**).
6. The whole system can now be removed for post-explantation calibration (*see* **Note 10**) as described under "Pre-implantation calibration."
7. Thereafter cut the electrodes just below the joints (i.e., the connection between sensor and telemeter lead). This will enable the connection of new electrodes to the end of the telemeter leads for use in further experiments.
8. Place the whole telemeter in an enzymatic detergent such as Terg-a-zyme enzyme detergent (Sigma-Aldrich) for cleaning.

4 Conclusion

In conclusion the telemetric method described herein represents an exciting approach for measurement of kidney tissue PO_2 in conscious, unrestrained rats. The high accuracy and temporal resolution of the data generated using this method allows long-term investigation of kidney PO_2, i.e., 24 h/day for many days. This opens up the opportunity for the examination of changes in PO_2 during the development and progression of both acute kidney injury and chronic kidney disease. Such studies may help elucidation of the roles of renal hypoxia in the development of kidney disease.

Telemetric measurement of tissue oxygen tension can also be applied to other parts of the kidney (other than the medulla) or other organs (e.g., the liver). Furthermore, combining measurement of PO_2 with telemetric measurement of blood pressure and blood flow would allow investigation of the causes and consequences of renal hypoxia. Finally the method we describe could (and should) be combined with techniques that can demonstrate spatial resolution, like functional magnetic resonance imaging [17, 20], laser speckle [21], fluorescence probes [22], or immunohistochemistry [23]. Taken together, we believe this technique provides an important additional tool for investigating the impact of tissue oxygenation in biology and pathophysiology.

5 Notes

1. The connection between telemeter joints and electrodes must be waterproof. The correct procedure should be followed, including allowing several days for the epoxy adhesive to cure so it can reach its maximum bonding strength. To test whether the joints are waterproof, they can be individually submerged

in saline while recording PO_2. Leakage will be observed as a maximum, saturated signal of −600 nA.

2. Configurator hardware and ConfigSoft™ software do not work on a Mac computer.
3. Calibrations can be modified to accommodate the physiological range of the tissue under investigation.
4. For implantation in renal cortical tissue, it is recommended that you bend the tip of both the carbon paste and reference electrode at a 90° angle, ~2 mm from the electrode tip.
5. Electrodes can be implanted in the right kidney (or other tissues).
6. The electrode leads can be sutured to the back muscle instead of the aorta.
7. Some data acquisition software allow for video recording which can be used to test for the possibility of locomotor activity-induced variations in the PO_2. Previously we have not observed any overt correlation between locomotor activity and the signal from the telemeter [14].
8. Be careful not to cut the telemeter leads while explanting.
9. For histology purposes (e.g., fibrosis around the electrodes), the kidney can be fixed (e.g., with formalin) while the electrodes are still within the kidney. However this precludes post-explantation calibration.
10. There appears to be some drift in calibration of the probes over the first few days of implantation. We recommend calibration of the probes both before and after implantation [14].
11. Values of medullary tissue PO_2 generated by these telemetric probes appear to exceed those generated previously in anesthetized animals [14]. This may represent some level of inaccuracy of the system, the confounding effects of anesthesia in previous studies, or both.

References

1. Evans RG, Gardiner BS, Smith DW, O'Connor PM (2008) Intrarenal oxygenation: unique challenges and the biophysical basis of homeostasis. Am J Physiol Renal Physiol 295(5): F1259–F1270
2. Heyman SN, Khamaisi M, Rosen S, Rosenberger C (2008) Renal parenchymal hypoxia, hypoxia response and the progression of chronic kidney disease. Am J Nephrol 28(6):998–1006
3. Legrand M, Mik EG, Johannes T, Payen D, Ince C (2008) Renal hypoxia and dysoxia after reperfusion of the ischemic kidney. Mol Med 14(7-8):502–516
4. Nangaku M (2006) Chronic hypoxia and tubulointerstitial injury: a final common pathway to end-stage renal failure. J Am Soc Nephrol 17(1):17–25
5. Tanaka T, Kato H, Kojima I et al (2006) Hypoxia and expression of hypoxia-inducible factor in the aging kidney. J Gerontol A Biol Sci Med Sci 61(8):795–805
6. Evans RG, Ince C, Joles JA, Smith DW, May CN, O'Connor PM, Gardiner BS (2013) Haemodynamic influences on kidney oxygenation: clinical implications of integrative physiology. Clin Exp Pharmacol Physiol 40(2): 106–122

7. Hansell P, Welch WJ, Blantz RC, Palm F (2013) Determinants of kidney oxygen consumption and their relationship to tissue oxygen tension in diabetes and hypertension. Clin Exp Pharmacol Physiol 40(2):123–137
8. Singh P, Ricksten SE, Bragadottir G, Redfors B, Nordquist L (2013) Renal oxygenation and haemodynamics in acute kidney injury and chronic kidney disease. Clin Exp Pharmacol Physiol 40(2):138–147
9. Liss P, Cox EF, Eckerbom P, Francis ST (2013) Imaging of intrarenal haemodynamics and oxygen metabolism. Clin Exp Pharmacol Physiol 40(2):158–167
10. Nangaku M, Rosenberger C, Heyman SN, Eckardt KU (2013) Regulation of hypoxia-inducible factor in kidney disease. Clin Exp Pharmacol Physiol 40(2):148–157
11. Manotham K, Tanaka T, Matsumoto M et al (2004) Evidence of tubular hypoxia in the early phase in the remnant kidney model. J Am Soc Nephrol 15(5):1277–1288
12. McCormick D, Hu AP, Nielsen P, Malpas S, Budgett D (2007) Powering implantable telemetry devices from localized magnetic fields. Conf Proc IEEE Eng Med Biol Soc 2007:2331–2335
13. Budgett DM, Hu AP, Si P et al (2007) Novel technology for the provision of power to implantable physiological devices. J Appl Physiol 102(4):1658–1663
14. Koeners MP, Ow CP, Russell DM et al (2013) Telemetry-based oxygen sensor for continuous monitoring of kidney oxygenation in conscious rats. Am J Physiol Renal Physiol 304(12): F1471–F1480
15. Bolger FB, Bennett R, Lowry JP (2011) An in vitro characterisation comparing carbon paste and Pt microelectrodes for real-time detection of brain tissue oxygen. Analyst 136(19):4028–4035
16. Bolger FB, McHugh SB, Bennett R et al (2011) Characterisation of carbon paste electrodes for real-time amperometric monitoring of brain tissue oxygen. J Neurosci Methods 195(2):135–142
17. Lowry JP, Griffin K, McHugh SB, Lowe AS, Tricklebank M, Sibson NR (2010) Real-time electrochemical monitoring of brain tissue oxygen: a surrogate for functional magnetic resonance imaging in rodents. NeuroImage 52(2):549–555
18. Kealy J, Bennett R, Lowry JP (2013) Simultaneous recording of hippocampal oxygen and glucose in real time using constant potential amperometry in the freely-moving rat. J Neurosci Methods 215(1):110–120
19. Lowry JP, Fillenz M (1997) Evidence for uncoupling of oxygen and glucose utilization during neuronal activation in rat striatum. J Physiol 498(Pt 2):497–501
20. Niendorf T, Pohlmann A, Arakelyan K et al (2015) How bold is blood oxygenation level-dependent (BOLD) magnetic resonance imaging of the kidney? Opportunities, challenges and future directions. Acta Physiol (Oxf) 213(1):19–38
21. Scully CG, Mitrou N, Braam B, Cupples WA, Chon KH (2013) Detecting physiological systems with laser speckle perfusion imaging of the renal cortex. Am J Physiol Regul Integr Comp Physiol 304(11):R929–R939
22. Zhang S, Hosaka M, Yoshihara T, Negishi K, Iida Y, Tobita S, Takeuchi T (2010) Phosphorescent light-emitting iridium complexes serve as a hypoxia-sensing probe for tumor imaging in living animals. Cancer Res 70(11): 4490–4498
23. Rosenberger C, Rosen S, Paliege A, Heyman SN (2009) Pimonidazole adduct immunohistochemistry in the rat kidney: detection of tissue hypoxia. Methods Mol Biol 466:161–174

Chapter 10

Magnetic Resonance Imaging (MRI) Analysis of Ischemia/Reperfusion in Experimental Acute Renal Injury

Andreas Pohlmann, Karen Arakelyan, Erdmann Seeliger, and Thoralf Niendorf

Abstract

Imbalance between renal oxygen delivery and demand in the first hours after reperfusion is suggested to be decisive in the pathophysiological chain of events leading to ischemia-induced acute kidney injury. Here we describe blood oxygenation level-dependent (BOLD) magnetic resonance imaging (MRI) for continuous monitoring of the deoxyhemoglobin-sensitive MR parameter T_2* in the renal cortex, outer medulla, and inner medulla of rats throughout renal ischemia/reperfusion (I/R). Changes during I/R are benchmarked against the effects of variations in the fraction of inspired oxygen (hypoxia, hyperoxia). This method may be useful for investigating renal blood oxygenation of rats in vivo under various experimental (patho)physiological conditions.

Key words Magnetic resonance imaging (MRI), Blood oxygenation level dependent (BOLD), Oxygenation, Rats, Ischemia, Hypoxia, Hyperoxia

1 Introduction

Ischemic acute kidney injury (AKI) can have a variety of causes including kidney hypoperfusion or ischemia followed by restoration of renal perfusion. Reperfusion of the kidney is generally associated with further renal tissue injury and a profound inflammatory response [1–3] called "reperfusion injury." No specific therapeutic approaches for AKI are currently available, because the underlying pathophysiological mechanisms remain in completely understood. Imbalance between renal oxygen delivery and demand in the first hours after reperfusion is suggested to be decisive in the pathophysiological chain of events leading to AKI [4, 5]. However, it is quite early to make ultimate statements on the pathophysiology of renal ischemia/reperfusion (I/R), because the established "gold-standard" modalities for in vivo characterization of renal hemodynamics and oxygenation have major methodological limitations: they do not provide sufficient spatiotemporal coverage of the

Tim D. Hewitson et al. (eds.), *Kidney Research: Experimental Protocols*, Methods in Molecular Biology, vol. 1397, DOI 10.1007/978-1-4939-3353-2_10, © Springer Science+Business Media New York 2016

kidney and, moreover, being invasive, they are applicable only in animal models. Therefore, noninvasive in vivo imaging of the spatiotemporal changes of renal oxygenation is conceptually attractive for preclinical studies and essential for clinical studies. Magnetic resonance imaging (MRI) offers a noninvasive approach to obtain insight into renal oxygenation under physiological conditions and renal disorders. Blood oxygenation level-dependent (BOLD) MRI with quantitative parametric mapping of the magnetic resonance relaxation time T_2* is thought to provide a surrogate of renal blood oxygenation [6]. The BOLD effect relies on the paramagnetic property of deoxygenated hemoglobin. The effective transversal relaxation time T_2* is sensitive to the concentration of deoxygenated hemoglobin per tissue volume element (voxel), which depends on O_2-saturation of hemoglobin, but notably also on other factors, such as the blood (vessel) volume per tissue volume element and the hematocrit, i.e., the concentration of hemoglobin per blood volume unit [7–9].

By acquiring several MR images with different T_2* weighting a parameter map of renal T_2* can be calculated, which permits quantitative comparison of renal T_2* throughout the time course of an experiment, as well as between experiments and between different animals. Here we describe continuous monitoring of T_2* in the renal cortex, outer medulla, and inner medulla of rats with a temporal resolution of approximately 80 s. This technique is employed to assess the temporal change in renal T_2* throughout 45 min of renal ischemia followed by 100 min of reperfusion and is compared with the effects of variations in the fraction of inspired oxygen (FiO_2) (hypoxia, hyperoxia). The fast and pronounced decline of renal blood oxygenation during ischemia can be depicted, as well as the lasting reduction of outer medullary T_2* (blood oxygenation) after reperfusion, in contrast to cortical T_2*, which recovers back to baseline. This method may be useful for investigating renal blood oxygenation of rats in vivo under various (patho)physiological conditions.

2 Materials

2.1 Animals

Male rats (Wistar, Sprague–Dawley or Lewis) with a body mass of 320–340 g, as the spatial constraints of the MR environment require the use of relatively small rats (≤350 g). The rats are allowed ad libitum food (standard diet) and must be housed at standard conditions.

2.2 Surgical Preparation

1. Anesthesia: Urethane solution (Sigma–Aldrich, Steinheim, Germany; 20 % in distilled water).
2. Temperature-controlled operation table.

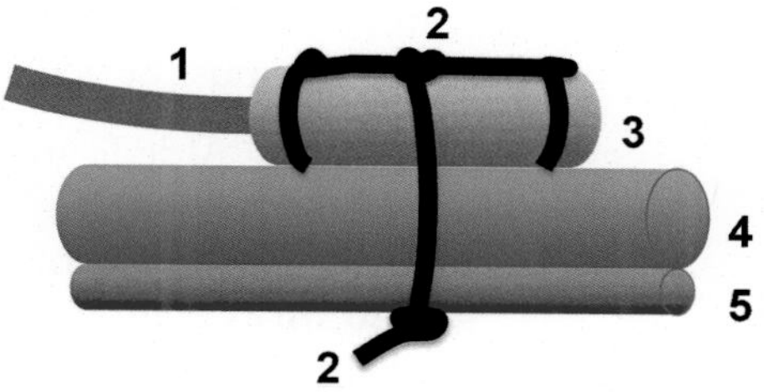

Fig. 1 Schematic view of the hydraulic occluder used for induction of renal ischemia during MRI. *1* = indistensible extension tube, *2* = sutures, *3* = distensible silicone tube, *4* = renal vein, *5* = renal artery. A water-filled syringe, connected to the indistensible extension tube, is used to create a hydraulic pressure, which leads to an inflation of the distensible tube. This causes a compression of the renal artery and vein and restricts the blood flow. Adapted from ref. 10 with permission from the Public Library of Science

3. Operation microscope (Leica MZ6; Leica Microsystems, Wetzlar, Germany) with magnification range between 6.3× and 40×.
4. Set of microsurgical instruments (including dissecting scissors, forceps, needle holders) and threads (including Vicryl polyglactin 910, 4/0; Prolene 6/0; Ethicon, Norderstedt, Germany).
5. Three occluders (*see* Fig. 1): The remotely controlled hydraulic occluder must be custom-made, as it was designed and prepared in our laboratory. The occluder consists of a distensible tube ("head" of the occluder) connected by an indispensable catheter to a syringe. The "head" of the occluder is made of a high grade silicone elastomer inner tube (Silikonkautschuk, Detakta Isolier- und Messtechnik GmbH & Co KG, Germany). For the connection between the "head" of the occluder and a regulative syringe, an indistensible extension catheter (Portex Polythene Tubing; Ref 800/110/200; inner diameter 0.58 mm) was used.
6. Interferometric temperature measurement system (ACS-P4-N-62SC, Opsens, Quebec City, Canada), including a fiber-optical temperature probe (OTP-M, AccuSens, Opsens). If you use a GaAs crystal-based system instead be aware of the offset caused by the magnetic field (approx. 4.7 °C at 9.4 T).

2.3 Magnetic Resonance Imaging

Magnetic resonance imaging (MRI) requires access to an ultra-high field MRI system including suitable accessories for the MR acquisition (radio frequency antennas), positioning, anesthesia, warming, and monitoring of physiological parameters, and trained personnel for operating the MRI system.

Due to the small size of rats in comparison with humans, a much higher spatial resolution is required to depict the kidney with adequate detail. This in turn demands a high signal-to-noise ratio (SNR), which must be achieved by use of tailored MR equipment.

1. MR system: A dedicated small animal MR system with a magnetic field strength of 7 T and higher is recommended. Here we describe the use of a 9.4 T 20 cm bore system (Biospec™ 94/20, Bruker Biospin, Ettlingen, Germany) equipped with a gradient system with integrated shim set (B-GA12S2, Bruker Biospin; gradient amplitude 440 mT/m, max. slew rate 3440 T/m/s).
2. Radio frequency (RF) coils: Use RF coils (antennas for RF transmission and reception) suitable for abdominal imaging, such as a transmit/receive rat body volume coil (72 mm inner diameter, quadrature; Bruker Biospin, Ettlingen, Germany) or preferably a transmit only *rat body volume coil* (72 mm inner diameter, linear; model T10325V3, Bruker Biospin) in combination with a receive only *rat heart coil array* (curved, 2×2 elements; model T12814V3, Bruker Biospin). Use of the latter coil setup is assumed here, as it allows for much higher spatial resolution due to its superior SNR when compared with the transmit/receive *volume coil.*
3. Animal holder: An animal holder (here model T11739, Bruker Biospin) designed for the size of the animals and the geometry of the RF coils is provided by the MR system/RF coil manufacturer (*see* **Note 1**).
4. Gases: O_2, N_2, and compressed air, as well as a gas-mixing system (FMI Föhr Medical Instruments GmbH, Seeheim-Ober Beerbach, Germany) to achieve required changes in the oxygen fraction of inspired gas mixture (FiO_2). The following gas mixtures are required during the experiment: (i) for hypoxia—10 % O_2/90 % N_2; (ii) for hyperoxia—100 % O_2; (iii) for normoxia—21 % O_2 (air).
5. Device for FiO_2 monitoring in gas mixtures: for example Capnomac AGM-103 (Datex GE, Chalfont St Giles, UK).
6. Device for warming of animal: Use a circulating warm-water-based heating system, consisting of a flexible rubber blanket with integrated tubing (part no. T10964, Bruker Biospin) connected to a conventional warm-water bath (SC100-A10, ThermoFisher, Dreieich, Germany). For alternative coil setups, water pipes may be built into the animal holder.
7. Monitoring of physiological parameters: For monitoring of respiration and core body temperature throughout the entire MR experiment, use a small animal monitoring system (Model 1025, Small Animal Instruments, Inc., Stony Brook, NY, USA), including a rectal temperature probe and pneumatic pillow.

8. Data analysis: Quantitative analysis of the data requires a personal computer and MATLAB software (R13 or higher; The Mathworks, Natick, MA, USA), ImageJ (Rasband, W.S., ImageJ, U.S. National Institutes of Health, Bethesda, Maryland, USA, http://imagej.nih.gov/ij/, 1997–2014), or a similar software development environment. Analysis steps described in the Subheading 3.7-3.13 can be performed manually by using the functions provided by the software development environment. Most of these steps benefit from (semi-) automation by creating software programs/macros—these steps are indicated by the computer symbol (🖳).

3 Methods

3.1 Preparation of MRI

1. Start the *ParaVision*™ *5* software and—only before the very first experiment—create and store the following MR protocols.
   (a) *Protocol_TriPilot* (pilot scan): conventional FLASH pilot with seven slices in each direction (axial, coronal, sagittal).
   (b) *Protocol_T2axl* (axial pilot scan): RARE sequence, repetition time (TR) = 560 ms, effective echo time (TE) = 24 ms, RARE factor 4, averages = 4. Define as geometry an axial field of view (FOV) = 70 × 52 mm², matrix size (MTX) = 172 × 128, eight slices with a thickness of 1.0 mm and distance of 2.2 mm, and an acquisition time of approximately a minute. Respiration trigger on (per phase step), flip-back on, fat saturation on.
   (c) *Protocol_T2corsag* (coronal/sagittal pilot scan): like *Protocol_T2axl*, but only one slice in coronal orientation.
   (d) *Protocol_PRESSvoxel* (shim voxel): conventional PRESS protocol, with a voxel size of = 9 × 12 × 22 mm³.
   (e) *Protocol_MGE* (T_2* mapping): multi-gradient echo (MGE) sequence, TR = 50 ms, echo times = 10, first echo = 1.43 ms, echo spacing 2.14 ms, averages = 4. Define as geometry a coronal oblique image slice with a FOV = (38.2 × 50.3) mm², MTX = 169 × 113 zero-filled to 169 × 215, and a slice thickness of 1.4 mm. Respiration trigger on (per slice), fat saturation on.
   (f) *Protocol_TOF* (angiography): FLASH sequence, TR = 11 ms, TE = 3 ms, flip angle = 80°, spatial in-plane resolution of 200 × 268 μm², with 15 slices of 1.0 mm thickness.
2. Switch on the gradient amplifiers of the MR system, which will also power on the automatic animal positioning system *AutoPac*™.
3. Install the transmit/receive *rat body volume coil* in the magnet bore.

4. Connect the animal holder to the animal positioning system (*AutoPac*™).
5. Install the *rat heart coil array* RF coil including its preamplifier on the animal bed.
6. Attach the face mask unit (commercial or custom-made, as described in **Note 1**) to the animal holder and connect it to the inspiratory gas providing system (luer tubing).
7. Place the flexible rubber mat of the warm-water-based heating system on top of the *rat heart coil array* and connect it to the warm-water circulation.
8. Switch on the water bath. Adjust the temperature to approximately 45 °C (*see* **Note 2**).
9. Attach the rectal temperature probe and pneumatic pillow to the small animal monitoring system and place the probes on the animal bed, at the lower abdominal position of the rat.
10. Attach all tubes and cables along the length of the animal bed using tape (*see* **Note 3**).

3.2 Surgical Preparation

Surgery must be performed parallel to MRI preparation outside the MR scanner room (in a neighboring preparation room) for safety reasons.

1. Anesthetize the animal by intraperitoneal injection of urethane (20 % solution, 6 mL/kg body mass) (*see* **Note 4**).
2. After reaching the required depth of anesthesia (determined by specific physiological signs such as muscle relaxation degree, absence of the paw withdrawal reflex, absence of the swallowing reflex, whisker movements, etc.), carefully shave the coat in the abdominal area of the rat (hair clipper Aesculap Elektra II GH2, Aesculap AG, Tuttlingen, Germany).
3. Place the rat in supine position on a warmed-up (39 °C) temperature-controlled operating table and fix the paws of the animal to the table by means of sticky tapes.
4. Open the abdominal cavity by a midventral incision (4–5 cm). Carefully dissect both renal arteries from the surrounding tissues.
5. Place the hydraulic occluder around the renal artery and vein. In an ideal case a simultaneous arteriovenous occlusion can be achieved (*see* **Notes 5** and **6**).
6. Place a fiber-optical temperature probe in close proximity to the kidney, in order to monitor the temperature of the kidney throughout the investigation.
7. Mark the localization of the investigated kidney's upper and lower pole on the skin of the abdomen. This is essential for

optimal positioning of the rat in the MRI scanner (i.e., optimal position of the rat's kidney relative to the MR coil).

8. Fill the abdominal cavity with warm saline (37 °C). For replenishment of abdominal saline, a catheter must be placed in the abdominal cavity.
9. Leave the tube/cable extensions of the occluder, of the catheter used for abdominal flushing, as well as the temperature probe, through the caudal cutting edge of the median abdominal incision.
10. Close the abdominal cavity by continuous suture.

3.3 Transport, Mask, Positioning

1. Transfer the animal into the MR scanner room (*see* **Note 7**).
2. Position the rat supine on the MRI animal bed, while aligning the kidney (pen markings on skin) with the center of the RF surface coil array.
3. Place a respiratory mask loosely around the muzzle of the spontaneously breathing rat. Open air supply to a rate of 1000 mL/min.
4. Switch on the small animal monitoring system. Insert the rectal temperature probe after cleaning it with alcohol and dipping it into Vaseline™.
5. Place the pneumatic pillow on the abdomen and cover the animal with the warming blanket. Watch the respiration trace on the monitor of the small animal monitoring system and adjust pillow position until the respiratory motion is captured well (*see* Fig. 2) (*see* **Note 8**).
6. Set the trigger options of the small animal monitoring system such that the trigger gate (indicated by white/red color horizontal bars parallel to the respiratory trace) opens for a maximum 280 ms (delay 30 ms) around the expiratory peak (*see* Fig. 2).

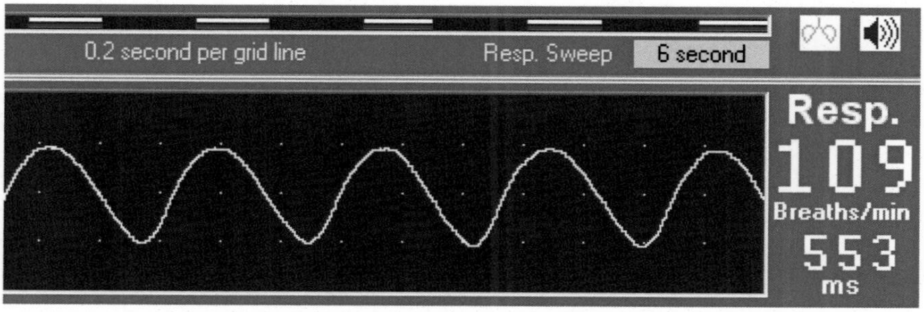

Fig. 2 Small animal monitoring system (SA Instruments, *see* Subheading 2). Setup of respiratory triggering for the MRI acquisition, with *begin delay* of 30 ms and *maximum width* of 280 ms for a typical respiratory rate of ~110 per minute. *Horizontal white bars* at top indicate trigger window

7. Press the *Out* button of the *AutoPac* system to make sure the animal bed is in the reference position. Switch on the *Laser* position marker and drive animal bed until the center of the RF surface array coil is aligned with the Laser position. Switch off the Laser (this point is now stored as the point of interest).
8. Double check that along the entire animal bed nothing protrudes such that it would not fit into the volume RF coil that is installed inside the magnet.
9. Press the *Work Position* button of the *AutoPac*™ system to drive the point of interest (point on the animal bed marked by the Laser) to the iso-center of the magnet.

3.4 MRI Prescans

1. Start the *ParaVision*™ *5* software and register a new subject and study (*see* **Note 9**). For the first scan select the *Protocol_TriPilot.*
2. Tune and match the *rat body volume coil* using the *Wobble* function.
3. Start the first pilot scan and verify on the acquired images that the kidneys are in the center of the magnet (field of view) and well positioned within the signal intensity profile of the surface coil. If necessary correct the animal and/or Laser marker position and then repeat the tuning/matching and this pilot scan (*see* **Note 10**).
4. Load the *Protocol_T2axl* (axial pilot scan), edit the geometry such that the eight axial slices cover the kidney (*see* **Note 11**), and run the scan.
5. Load the *Protocol_T2corsag* (coronal/sagittal pilot scan), edit the geometry of the slice such that it is perpendicular (oblique coronal) to the axial reference image and crosses the kidney centrally along its longest dimension, and run the scan.
6. Clone the *Protocol_T2corsag* and like *Protocol_T2axl*, edit the geometry of the slice such that it is perpendicular (oblique sagittal) to the oblique coronal reference image and crosses the kidney centrally along its longest dimension, and run the scan.
7. Load the *Protocol_PRESSvoxel* (shim voxel) and edit the geometry (including its size) of the voxel such that it just encloses the kidney.
8. Shim on the PRESS voxel: from the *Spectrometer Control Tool > Acq > Current Adjustments* run the *Method specific adjustment for the Local Field Homogeneity* followed by *Method specific adjustment for the Basic Frequency.*
9. Load the *Protocol_MGE* (T_2* mapping) and edit the geometry (using the axial, oblique coronal, and oblique sagittal T_2 scans as references) such that the slice is strictly coronal with regard to the kidney. Run the scan and check image geometry and quality.

10. Load the *Protocol_TOF* (angiography) and edit the geometry such that it crosses the kidney, sagittal with regard to the kidney. Run the scan and after acquisition locate the bright signal of the large vessels within the kidney.

3.5 Baseline, Interventions

1. Clone the *Protocol_MGE* 5 times and run these baseline scans with the *GOP* button of the *Spectrometer Control Tool* (not the *Traffic Lights* button of the *Scan Control Tool*).
2. Clone the *Protocol_MGE* 45 times.
3. Start of Hypoxia. Change the gas flowing through the respiratory mask to 10 % O_2/90 % N_2.
4. Start a *Protocol_MGE* (with GOP) immediately after the onset and another exactly 5 min after the start of hypoxia.
5. End of Hypoxia. Change the gas flowing through the respiratory mask back to air (21 % O_2).
6. Start a series of *Protocol_MGE* (with GOP; using the *Queued acquisition* macro) to run continuously for 20 min (recovery period).
7. Start of Hyperoxia. Change the gas flowing through the respiratory mask to 100 % O_2.
8. Start a *Protocol_MGE* (with GOP) immediately after the onset and another exactly 5 min after the start of hyperoxia.
9. End of Hyperoxia. Change the gas flowing through the respiratory mask back to air (21 % O_2).
10. Start a series of *Protocol_MGE* (with GOP; using the *Queued acquisition* macro) to run continuously for 20 min (recovery period).
11. Remove the remaining *Protocol_MGE* scans (in *ready* status) from the scan list. Clone *Protocol_TOF* once, and *Protocol_MGE* 45 times.
12. Start of Occlusion. Inflate the remotely controlled occluder.
13. Simultaneously start the *Protocol_TOF* and after acquisition verify the absence of the blood flow signal in the renal vessels (*see* **Note 12**).
14. Start a series of *Protocol_MGE* (with GOP; using the *Queued acquisition* macro) to run continuously for ~45 min (ischemia).
15. After ~43 min remove the excess *Protocol_MGE* scans from the scan list. Clone *Protocol_TOF* once, and *Protocol_MGE* 90 times.
16. After exactly 45 min exit the *Queued acquisition*.
17. End of Occlusion. Rapidly deflate the occluder.
18. Simultaneously start the *Protocol_TOF* and after acquisition verify the re-occurrence of the blood flow signal in the renal vessels.

19. Start a series of *Protocol_MGE* (with *GOP*; using the *Queued acquisition* macro) to run continuously for 100 min (recovery period).
20. After 100 min exit the *Queued acquisition* and remove excess *Protocol_MGE* scans from the scan list.

3.6 End of Experiment

1. Carefully remove the respiratory mask from the animal's muzzle.
2. Move the animal into preparation room and place in supine position on a warm operating table.
3. Cut the sutures and open the abdominal cavity.
4. Remove the fluid from the abdominal cavity using a pipette.
5. Carefully untie the knots of the occluder and remove it.
6. Remove the abdominal flushing catheter, as well as the temperature probe.
7. Control and note the overall condition of the kidney after experiment (e.g., surface coloring and its homogeneity).
8. Exsanguinate the animal by cutting the abdominal aorta.

3.7 Data Analysis

The MRI data acquired with the multi-echo gradient echo (MGE) protocol contain images with different T_2* weighting (determined by the echo time, TE). From these images calculate T_2* for each pixel to construct a map of the parameter T_2*.

1. Open the reconstructed MR images in *MATLAB* (💻; *see* Subheading 2.3, step 8) by loading the binary *2dseq* files of each MGE scan and import *echo time* (TE) parameters from the *method* text file.
2. Perform pixel-wise exponential curve fitting of the equation $S(t) = A + B \exp(-t/T_2^*)$ to the image intensities $S(t)$ versus time t (the 10 echo times) (💻 *MATLAB*) (*see* **Note 13**).
3. To visually inspect the time series of the registered T_2* maps display the matrix containing the fitted T_2* parameter as images using a pseudocolor scale (e.g., *jet* in MATLAB) with the limits 0–30 ms (*see* Fig. 3) (💻 *MATLAB*).
4. For benchmarking the regional T_2* effects during ischemia/reperfusion against standardized test procedures such as hypoxia and hyperoxia, calculate ΔT_2* maps by subtracting the last baseline T_2* parameter map from any of the later acquired T_2* parameter maps, e.g., at end ischemia or end reperfusion (e.g., *see* Fig. 4) (💻 *MATLAB*).
5. Open and display the baseline T_2* parameter map. Place a rectangular selection (reference rectangle) tightly around the outer

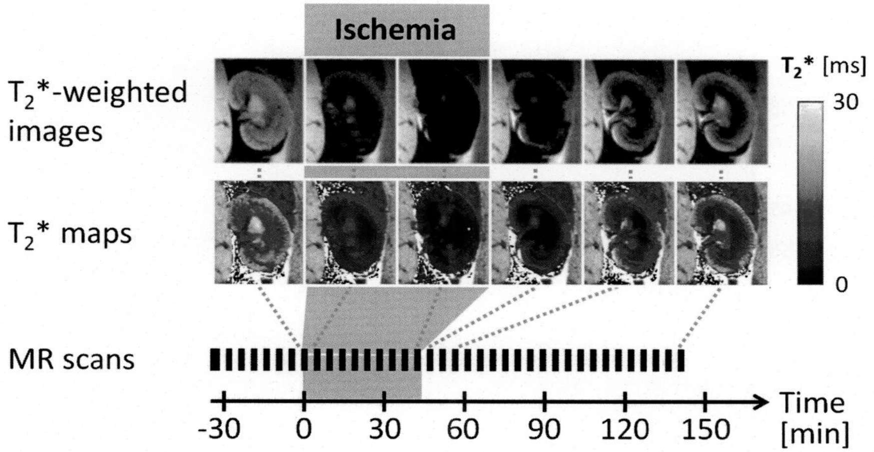

Fig. 3 T_2*-weighted MR images shown in gray scale together with derived color-coded T_2* parameter maps (*see* Subheading 3.7. 2+3). T_2* measurements (MGE scans) are repeated continuously throughout the entire experiment. Images and maps are shown for six time points: last baseline, onset of ischemia and end of ischemia after 45 min, onset of reperfusion, and 12 min and 100 min after onset of reperfusion. Adapted from ref. 10 with permission from the Public Library of Science

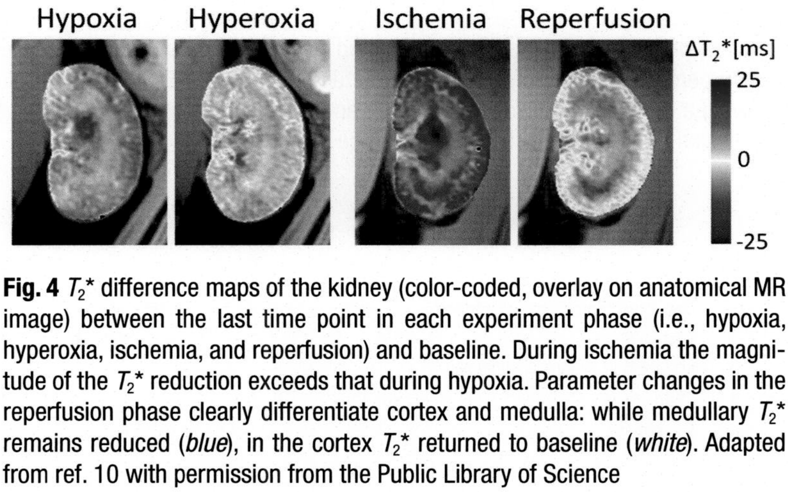

Fig. 4 T_2* difference maps of the kidney (color-coded, overlay on anatomical MR image) between the last time point in each experiment phase (i.e., hypoxia, hyperoxia, ischemia, and reperfusion) and baseline. During ischemia the magnitude of the T_2* reduction exceeds that during hypoxia. Parameter changes in the reperfusion phase clearly differentiate cortex and medulla: while medullary T_2* remains reduced (*blue*), in the cortex T_2* returned to baseline (*white*). Adapted from ref. 10 with permission from the Public Library of Science

edges of the kidney. Place nine regions of interest (ROIs) at defined positions within the rectangular reference frame (for size and positions *see* Fig. 5 and [10]) and store the mean T_2* value for each ROI (💻 *ImageJ* or *MATLAB*) (*see* **Note 14**).

6. Finally, for the detailed analysis of temporal changes in regional renal T_2* plot the mean T_2* value for each ROI versus time (e.g., *see* Fig. 6) (💻 *MATLAB*) (*see* **Note 15**).

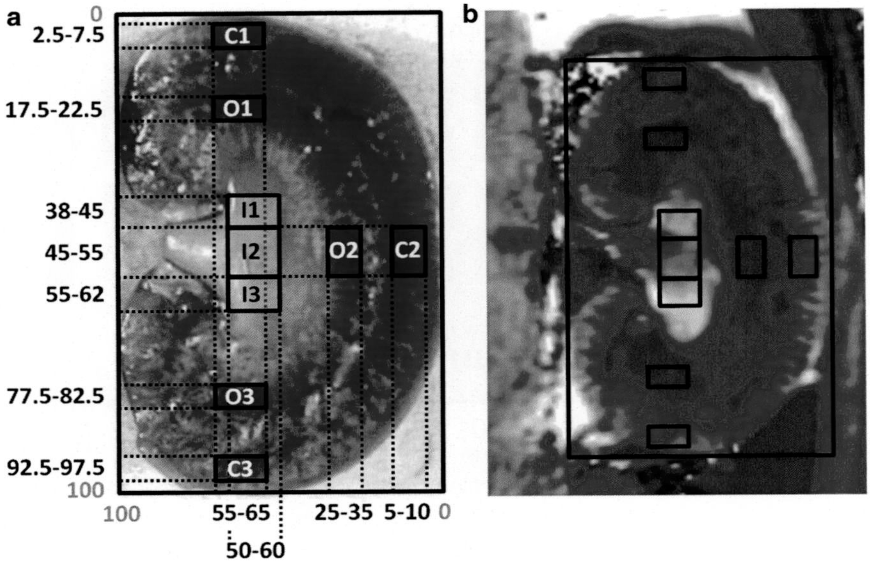

Fig. 5 Standardized segmentation model of the rat kidney that is used to semiautomatically place regions of interest (ROIs) in the cortex (C1,C2,C3), the outer medulla (O1,O2,O3), and the inner medulla (I1,I2,I3). (a) Segmentation model overlaid onto a photograph of a freshly excised coronal view of a rat kidney. During analysis the *rectangular reference frame* is manually positioned around the kidney in the coronal view. Then the ROIs are placed automatically at pre-defined relative positions within this reference frame (*number ranges* signify percentages of the reference frame dimensions). (b) Example of a color-coded T_2^* parameter map, showing a rat kidney in vivo together with an overlay of the segmentation model. Adapted from ref. 10 with permission from the Public Library of Science

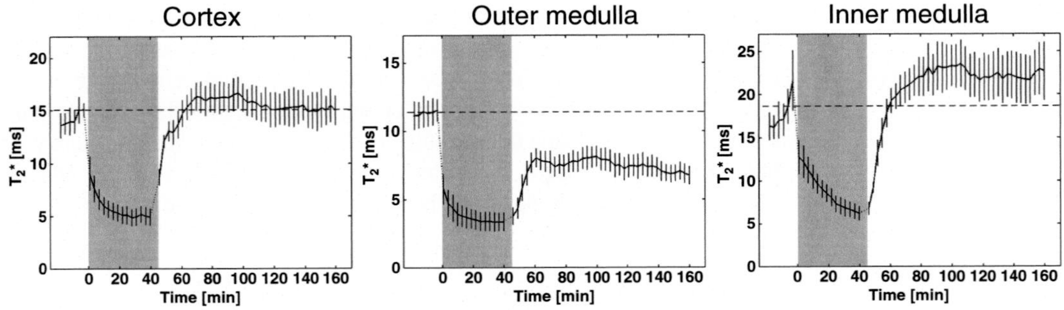

Fig. 6 Detailed analysis of temporal T_2^* changes. Plots of T_2^* (mean ± SEM averaged over six animals) versus time for an ROI in the cortex (*left*), in the outer medulla (*center*), and in the inner medulla (*right*). Ischemia (*shaded in gray*) leads to an immediate and significant T_2^* decrease ($p < 0.001$). Clear differences between cortex, outer medulla, and inner medulla can be observed, particularly after reperfusion. Adapted from ref. 10 with permission from the Public Library of Science

4 Notes

1. If the holder does not provide a respiratory (anesthetic) mask, such a mask can be easily built: take a 20 ml plastic syringe, cut off the tip approximately 15 mm from the bottom of the syringe, yielding a funnel-shaped mask. Finally deflash and smooth the cutting edges of the mask using a file.
2. The temperature of the water will be much higher than the temperature of the rubber mat and depends on the length and material of the tubing used. Hence the temperature of the bath must be adapted to the local setup.
3. Make sure to keep tubes or cable close to the animal bed so they cannot get caught anywhere on the way into the magnet!
4. Urethane supports anesthesia throughout the surgical preparation and the MRI examination (for several hours) and leaves cardiovascular and respiratory reflexes largely undisturbed.
5. An essential characteristic of this technique of occlusion is that it does not change vascular volume fraction (VVF) (vessel volume per tissue volume element) of the kidney. VVF strongly affects T_2*. Keeping VVF constant eliminates one major confounding factor and ensures that T_2* is predominantly affected by blood oxygenation.
6. (i) In order to prevent additional pressure on the vessels (and therefore development of unwanted kidney ischemia), the positioning and knotting of the occluder must be performed under careful monitoring of the overall condition of the kidney (e.g., surface coloring and its homogeneity); (ii) Depending on the objectives of the study, the occluder may also be placed around the renal vein or around the aorta above the renal arteries; this allows two additional types of occlusions to be conducted—pure arterial or pure venous. An essential disadvantage of these occlusions is a considerable change in VVF: decrease in case of arterial and increase in case of venous occlusion; (iii) Since the occluder consists of silicone and polyethylene and is positioned about 15 mm away from the kidney, it does not cause artifacts in MRI that affect the kidney; (iv) To allow control of in- and deflation of the occluder connect a second (control) occluder outside the scanner to the same syringe using a luer T-connection.
7. Special attention during transfer must be paid to the tube/cable extensions (occluder, abdominal flushing, and temperature probe). To prevent their displacement during transport, they must be fixed to the rat's tail by means of sticky tape.
8. The peak-to-peak amplitude of the respiratory trace should span about 2/3 of the vertical axis on the display. Any gross movement (for instance during repositioning the pillow) will

lead to large peaks and force the monitoring system to adapt the signal amplification, so that temporarily the signal will become much smaller on the display. Keep an eye on the magnification, which is given on the left next to the display; this will drop to a low value such as 15×— wait until it recovers back to a value around 100× before further adjusting it.

9. The registration will determine the name of the folder on the computer's hard disk in which all study data will be stored. Spaces and special characters (with a few exception such as "–" and "_") are not allowed. Check at this point in time that the coil configuration has been detected correctly. Select the *Location* of your scan protocols (directory where you stored them).
10. To repeat the pilot scan, undo it and then start it with the *Traffic Light* button while holding the shift key on the keyboard in order to force all adjustments to run again.
11. Check the length of the respiratory trigger window of the monitoring system and if necessary reduce the TR of the scan until it fits within this window.
12. If unsuccessful, repeat the occlusion and inflate the occluder with higher hydraulic pressure. Always check the inflation of the second occluder outside the MR scanner and remember to which volume the syringe was advanced.
13. To avoid curve fitting errors, you need to exclude the data points for which the signal intensity is not significantly above the noise level. The signal intensity threshold below which data points will be excluded is a crucial parameter. It must be chosen individually for each study based on the signal-to-noise ratio of the acquired MR data. It is highly recommended to develop and use a 'fit inspector' software tool, i.e. an interactive tool that displays the T_2* map and allows the user to move the mouse pointer over the map while plotting for the currect pixel the fitted curve together with the data points and indicating which data points were excluded. Using such a tool one can easily verify if a threshold is appropriate or must be modified.
14. It is important to verify that the T_2* parameter map does not show any artifacts within the automatically placed nine ROIs. For this purpose display the ROIs overlaid onto the T_2* parameter map and check carefully for susceptibility-induced signal loss/voids (typically close to adjacent fat tissue, intestine, or an air bubble accidentally introduced during surgery) or other obvious image artifacts. ROIs with artifacts must be excluded from further analysis.
15. For most purposes data of the three ROIs within each renal layer (cortex, outermedulla, or inner medulla) can be pooled into one mean T_2* value for each layer, since the intralayer differences between these ROIs were found to be much smaller than the interlayer differences [10].

Acknowledgements

This work was supported by the German Research Foundation (Deutsche Forschungsgemeinschaft, research unit: FOR 1368, grant numbers: NI 532/9-1, NI 532/9-2, SE 998/4-2). The authors wish to thank Ariane Anger, Gordana Bubalo, Kathleen Cantow, Duska Dragun, Bert Flemming, Mandy Fechner, Andrea Gerhardt, Jan Hentschel, Uwe Hoff, Mechthild Ladwig, Wolf-Hagen Schunck, Helmar Waiczies, and Sonia Waiczies for their help and support with developing the described methods.

References

1. Linas SL, Shanley PF, Whittenburg D, Berger E, Repine JE (1988) Neutrophils accentuate ischemia-reperfusion injury in isolated perfused rat kidneys. Am J Physiol 255(4 Pt 2):F728–F735
2. Willinger CC, Schramek H, Pfaller K, Pfaller W (1992) Tissue distribution of neutrophils in postischemic acute renal failure. Virchows Arch B Cell Pathol Incl Mol Pathol 62(4):237–243
3. White LE, Hassoun HT (2012) Inflammatory mechanisms of organ crosstalk during ischemic acute kidney injury. Int J Nephrol 2012: 505197
4. Whitehouse T, Stotz M, Taylor V, Stidwill R, Singer M (2006) Tissue oxygen and hemodynamics in renal medulla, cortex, and corticomedullary junction during hemorrhage-reperfusion. Am J Physiol Renal Physiol 291(3):F647–F653
5. Hoff U, Lukitsch I, Chaykovska L et al (2011) Inhibition of 20-HETE synthesis and action protects the kidney from ischemia/reperfusion injury. Kidney Int 79(1):57–65
6. Prasad PV (2006) Functional MRI of the kidney: tools for translational studies of pathophysiology of renal disease. Am J Physiol Renal Physiol 290(5):F958–F974
7. Pohlmann A, Arakelyan K, Hentschel J et al (2014) Detailing the relation between renal T2* and renal tissue pO2 using an integrated approach of parametric magnetic resonance imaging and invasive physiological measurements. Invest Radiol 49(8):547–560
8. Ogawa S (2012) Finding the BOLD effect in brain images. Neuroimage 62(2):608–609
9. Evans RG, Gardine BS, Smith DW, O'Connor PM (2008) Intrarenal oxygenation: unique challenges and the biophysical basis of homeostasis. Am J Physiol Renal Physiol 295(5):F1259–F1270
10. Pohlmann A, Hentschel J, Fechner M et al (2013) High temporal resolution parametric MRI monitoring of the initial ischemia/reperfusion phase in experimental acute kidney injury. PLoS One 8(2):e57411

Chapter 11

Assessment of Renal Hemodynamics and Oxygenation by Simultaneous Magnetic Resonance Imaging (MRI) and Quantitative Invasive Physiological Measurements

Kathleen Cantow, Karen Arakelyan, Erdmann Seeliger, Thoralf Niendorf, and Andreas Pohlmann

Abstract

In vivo assessment of renal perfusion and oxygenation under (patho)physiological conditions by means of noninvasive diagnostic imaging is conceptually appealing. Blood oxygen level-dependent (BOLD) magnetic resonance imaging (MRI) and quantitative parametric mapping of the magnetic resonance (MR) relaxation times T_2* and T_2 are thought to provide surrogates of renal tissue oxygenation. The validity and efficacy of this technique for quantitative characterization of local tissue oxygenation and its changes under different functional conditions have not been systematically examined yet and remain to be established. For this purpose, the development of an integrative multimodality approaches is essential. Here we describe an integrated hybrid approach (MR-PHYSIOL) that combines established quantitative physiological measurements with T_2* (T_2) mapping and MR-based kidney size measurements. Standardized reversible (patho)physiologically relevant interventions, such as brief periods of aortic occlusion, hypoxia, and hyperoxia, are used for detailing the relation between the MR-PHYSIOL parameters, in particular between renal T_2* and tissue oxygenation.

Key words Magnetic resonance imaging (MRI), Blood oxygenation level dependent (BOLD), Oxygenation, Perfusion, Rats, Hypoxia, Hyperoxia

1 Introduction

Kidney diseases are a global health burden with steadily increasing incidence [1, 2]. The estimated worldwide death toll of acute kidney injury (AKI) is two million per year [3]. Since the pathophysiology of AKI remains not completely understood, prevention and therapy of AKI still largely rely on empirical clinical knowledge. Animal studies suggest that AKI of various origins shares one common link in the pathophysiological chain of events, ultimately leading to AKI as well as to progression from AKI to chronic kidney diseases (CKD): imbalance between renal medullary oxygen delivery and oxygen demand [4, 5]. However, it is too early to

Tim D. Hewitson et al. (eds.), *Kidney Research: Experimental Protocols*, Methods in Molecular Biology, vol. 1397,
DOI 10.1007/978-1-4939-3353-2_11,

make ultimate statements on the role of hypoperfusion and hypoxia for these renal disorders, since in vivo characterization of renal hemodynamics and oxygenation constitutes a challenge. All established modalities available in today's experimental and translational research have shortcomings and methodological restraints [6], inter alia (1) the inability to provide sufficient spatiotemporal coverage of the kidney, (2) invasiveness, and (3) applicability only to animals. Therefore noninvasive methods providing spatiotemporal coverage are required to further elucidate the pathophysiology of AKI.

Blood oxygen level-dependent (BOLD) magnetic resonance imaging (MRI) and quantitative parametric mapping of the magnetic resonance (MR) relaxation times T_2* and T_2 are thought to provide surrogates of renal tissue oxygenation in preclinical and clinical studies [6–10]. The parametric maps of the MR relaxation parameters T_2* and T_2 offer full kidney coverage, (sub)millimeter spatial resolution, and (sub)minute temporal resolution, are noninvasive and support longitudinal studies. The validity and efficacy of this technique for quantitative characterization of local tissue oxygenation and its changes under different functional conditions have not been systematically examined yet and remain to be established. Moreover, for an unambiguous physiological interpretation of renal T_2*, a calibration with quantitative physiological measurements including renal tissue oxygenation is required. For this purpose, the development of integrative multimodality approaches is essential.

Here we describe an integrated hybrid approach (MR-PHYSIOL) that combines established quantitative physiological measurements including renal perfusion pressure (RPP), total renal blood flow (RBF), regional tissue perfusion (local blood flux), and regional tissue oxygenation (partial pressure of oxygen—pO_2) with mapping of the MR parameters T_2* and T_2 as well as MR-based kidney size measurements [6, 11]. Standardized reversible (patho)physiologically relevant interventions, such as brief periods of aortic occlusion, hypoxia, and hyperoxia, enable in vivo modulation of renal perfusion and oxygenation, and can be used for detailing the relationship between renal T_2* and tissue oxygenation.

2 Materials

2.1 Animals

Male Wistar rats with body mass 300–340 g. The spatial constraints dictated by the MR environment required the use of relatively small rats (≤350 g). The rats are allowed ad libitum food (standard diet) and must be housed at standard conditions.

2.2 Surgical Preparation

1. Anesthesia: urethane solution (Sigma-Aldrich, Steinheim, Germany; 20 % in distilled water).
2. Temperature-controlled operation table.

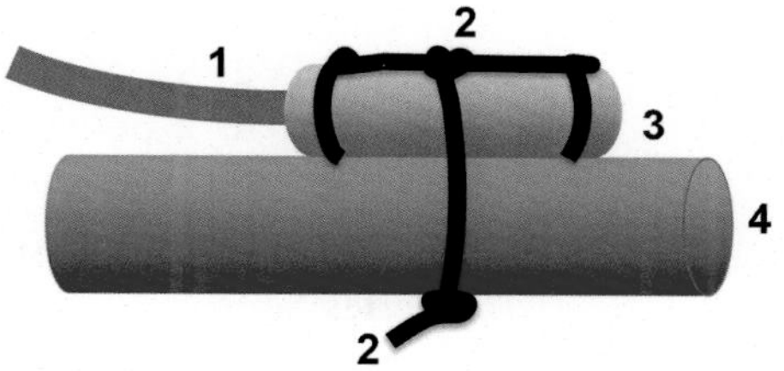

Fig. 1 Schematic view of the hydraulic occluder used for induction of aortic occlusion during MRI. *1* = Indistensible extension tube, *2* = sutures, *3* = distensible silicone tube, *4* = aorta. A water-filled syringe, connected to the indistensible extension tube, is used to create a hydraulic pressure, which leads to an inflation of the distensible tube. This causes a compression of the aorta and restricts the blood flow

3. Operation microscope (Leica MZ6; Leica Microsystems, Wetzlar, Germany) with magnification range between 6.3× and 40×.
4. Set of microsurgical instruments (including dissecting scissors, forceps, needle holders) and threads (including Vicryl polyglactin 910, 4/0; Prolene 6/0; Ethicon, Norderstedt, Germany).
5. Three occluders (*see* Fig. 1): the remotely controlled hydraulic occluder must be custom-made, as it was designed and prepared in our laboratory. The occluder consists of a distensible tube ("head" of the occluder) connected by an indispensable catheter to a syringe. The "head" of the occluder is made of a high grade silicone elastomer inner tube (Silikonkautschuk, Detakta Isolier- und Messtechnik GmbH & Co KG, Germany). For the connection between the "head" of the occluder and a regulating syringe an indistensible extension catheter (Portex Polythene Tubing; Ref 800/110/200; inner diameter 0.58 mm) was used.
6. Interferometric temperature measurement system (ACS-P4-N-62SC, Opsens, Quebec City, Canada), including a fiber-optical temperature probe (OTP-M, AccuSens, Opsens, Quebec City, Canada). If you use a GaAs crystal-based system instead be aware of the offset caused by the magnetic field (approx. 4.7 °C at 9.4 T).
7. For blood pressure measurement: (1) a pressure transducer (DT-XX, Viggo-Spectramed, Swindon, UK) connected to an amplifier (TAM-A Plugsys Transducer; Hugo Sachs Elektronik–Harvard Apparatus GmbH, Mach-Hugstetten, Germany); (2) the femoral artery catheter was designed and made in our laboratory using Portex Tubing (polythene). It must be longer than 1 m to allow placing the pressure transducer well outside the bore of the MR scanner.

8. For assessment of total renal blood flow: a perivascular flow probe (MV2PSB-MRI; Transonic Systems, Ithaca, NY, USA) connected to a perivascular flow module (TS420; Transonic Systems).
9. Combined optical Laser-Doppler-Flux and pO_2 probes (pO_2 E-Series Sensor; Oxford Optronix) for measurements of local tissue oxygenation and local tissue perfusion. The probes are attached to an OxyLite/OxyFlo™ apparatus (Oxford Optronix, Oxford, UK).
10. Laboratory power supply (e.g., Model 3200, Statron VEB, Fuerstenwalde, Germany) for creating voltage signal markers (by manual on/off switching) during the experiments. Voltage is set to 5 V and treated as a logical TTL-like signal. This will be referred to as *TTL switch* in the following.
11. For continuous logging of the signals from the probes for arterial blood pressure, renal blood flow, cortical and medullary pO_2 and flux, together with that from the TTL switch their analogue outputs must be digitized and recorded. An analogue-digital converter (DT 9800-16SE-BNC, Data Translation GmbH, Bietigheim-Bissingen, Germany) permits connection to the USB port of a PC. A dedicated data acquisition software (HAEMODYN™, Hugo Sachs Elektronik–Harvard Apparatus GmbH, March-Hugstetten, Germany) allows calibration of the probe signals and their continuous recording.
12. For fixation and stabilization of the probes and for the safe transfer of the animal to the scanner, a custom-made ***portable*** animal holder (*see* Fig. 2) must be used, in addition to the MRI

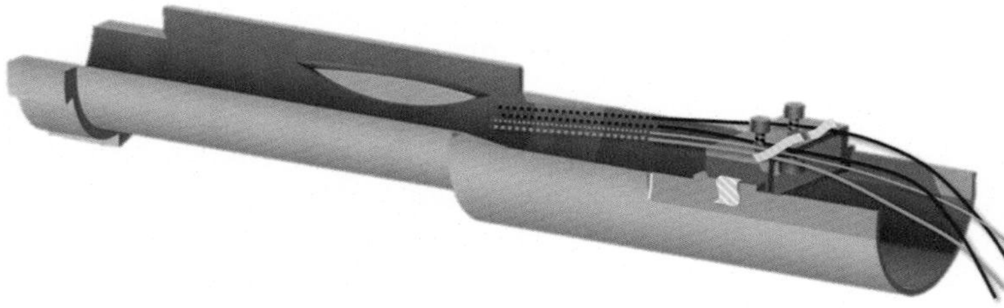

Fig. 2 Custom-made portable animal holder. A bridge-like construction (adjustable in height and distance to the animal), positioned at the end of the hind paws of the rat, enables fixation of all leads that connect the physiological probes with the equipment positioned outside the MR scanner room. The frontal part has a section with reduced outer diameter, which leaves a gap to the half-pipe-shaped animal holder of the MR system. Into this gap fits the surface RF coil, which should not be placed exactly horizontally on the holder, but be rotated by 20°–30° towards the side with the kidney of interest (here: left kidney, right side of holder in supine position of animal)

animal holder (listed in Subheading 2.3). It was designed and built in our laboratory using 3D CAD (Autodesk Inventor 2012; Autodesk, San Rafael, CA, USA) and rapid prototyping (BST 1200es; Alphacam GmbH, Schorndorf, Germany). The holder must meet the geometry of the MR setup; it has a half-pipe shape with a section of reduced diameter to allow for the four-element surface RF coil to be placed beneath. A mark on the holder indicates the center of the RF coil. A bridge-like construction, positioned at the end of the hind paws of the rat, enables fixation of all leads that connect the physiological probes with the equipment positioned outside the MR scanner room. The portable rigid animal holder in conjunction with the adjustable cable support enables safe transport of the animal to the MR scanner.

13. Patches of Dacron gauze (Woven Mesh Spacer; Merck Millipore, Billerica, MA, USA).
14. Histoacryl™ glue (Braun Surgical GmbH, Melsungen, Germany).
15. Medical sticky tape.
16. Silicone elastomer tubes (Silikonkautschuk).

2.3 Magnetic Resonance Imaging

Magnetic resonance imaging (MRI) requires access to an ultra-high-field MRI system including suitable accessories for the MR acquisition (radiofrequency antennas), positioning, anesthesia, warming, and monitoring of physiological parameters, and trained personnel for operating the MRI system.

Due to the small size of rats in comparison with humans a much higher spatial resolution is required to depict the kidney with adequate detail. This in turn demands a high signal-to-noise ratio (SNR), which must be achieved by use of tailored MR equipment.

1. MR system: a dedicated small animal MR system with a magnetic field strength of 7 T and higher is recommended. Here we describe the use of a 9.4 T 20 cm bore system (Biospec™ 94/20, Bruker Biospin) equipped with a gradient system with integrated shim set (B-GA12S2, Bruker Biospin; gradient amplitude 440 mT/m, max. slew rate 3440 T/m/s).
2. Radiofrequency (RF) coils: use RF coils (antennas for RF transmission and reception) suitable for abdominal imaging, such as a transmit/receive rat body volume coil (72 mm inner diameter, quadrature; Bruker Biospin, Ettlingen, Germany) or preferably a transmit only *rat body volume coil* (72 mm inner diameter, linear; model T10325V3, Bruker Biospin) in combination with a receive only *rat heart coil array* (curved, 2 × 2

elements; model T12814V3, Bruker Biospin). Use of the latter coil setup is assumed here, as it allows for much higher spatial resolution due to its superior SNR when compared with the transmit/receive *volume coil.*

3. Animal holder: an MRI animal holder (here model T11739, Bruker Biospin) designed for the size of the animals and the geometry of the RF coils is provided by the MR system/RF coil manufacturer (*see* **Note 1**).
4. Gases: O_2, N_2, and compressed air, as well as a gas-mixing system (FMI Föhr Medical Instruments GmbH, Seeheim-Ober Beerbach, Germany) to achieve required changes in the oxygen fraction of inspired gas mixture (FiO_2). The following gas mixtures are required during the experiment: (1) for hypoxia—10 % O_2/90 % N_2; (2) for hyperoxia—100 % O_2; (3) for normoxia—21 % O_2 (air).
5. Device for FiO_2 monitoring in gas mixtures: for example Capnomac AGM-103 (Datex GE, Chalfont St Giles, UK).
6. Device for warming the animal: use a circulating warm water-based heating system, consisting of a flexible rubber blanket with integrated tubing (part no. T10964, Bruker Biospin) connected to a conventional warm water bath (SC100-A10, ThermoFisher, Dreieich, Germany). For alternative coil setups water pipes may be built into the animal holder.
7. Monitoring of physiological parameters: for monitoring of respiration and core body temperature throughout the entire MR experiment use a small animal monitoring system (Model 1025, Small Animal Instruments, Inc., Stony Brook, NY, USA), including a rectal temperature probe and pneumatic pillow.
8. Data analysis: quantitative analysis of the data requires a personal computer and MATLAB software (R13 or higher; The Mathworks, Natick, MA, USA), ImageJ (Rasband, W.S., ImageJ, U.S. National Institutes of Health, Bethesda, Maryland, USA, http://imagej.nih.gov/ij/, 1997–2014) including the BrukerOpener plugin and NIfTi Input/Output plugin, FSL (Analysis Group, FMRIB, Oxford, UK, http://fsl.fmrib.ox.ac.uk/fsl/fslwiki/) or a similar software development environment. Analysis steps described in Subheading 3 can be performed manually by using the functions provided by the software development environment. Most of these steps benefit from (semi-)automation by creating software programs/macros—these steps are indicated by the computer symbol (💻).

3 Methods

Figure 3 illustrates the work flow, including preparations, experimental procedures, and data analyses. Steps in this work flow chart whose name starts with a number (e.g., *3.1 Preparation of MRI*) are described in the corresponding method sections with the same name.

3.1 Preparation of MRI

1. Start the *ParaVision*™ *5* software and —only before the very first experiment—create and store the following MR protocols.
   (a) *Protocol_TriPilot* (pilot scan): conventional FLASH pilot with seven slices in each direction (axial, coronal, sagittal).
   (b) *Protocol_T2axl* (axial pilot scan): RARE sequence, repetition time (TR) = 560 ms, effective echo time (TE) = 24 ms, RARE factor 4, averages = 4. Define as geometry an axial field of view (FOV) = 70×52 mm^2, matrix size (MTX) = 172×128, eight slices with a thickness of 1.0 mm and distance of 2.2 mm, and an acquisition time of approximately a minute. Respiration trigger on (per phase step), flip-back on, fat saturation on.
   (c) *Protocol_T2corsag* (coronal/sagittal pilot scan): like *Protocol_T2axl*, but only one slice in coronal orientation.
   (d) *Protocol_PRESSvoxel* (shim voxel): conventional PRESS protocol, with a voxel size of = $9 \times 12 \times 22$ mm^3.
   (e) *Protocol_MGE* (T_2* mapping): multigradient echo (MGE) sequence, TR = 50 ms, echo times = 10, first echo = 1.43 ms, echo spacing 2.14 ms, averages = 4. Define as geometry a coronal oblique image slice with an FOV = (38.2×50.3) mm^2, MTX = 169×113 zero-filled to 169×215, and a slice thickness of 1.4 mm. Respiration trigger on (per slice), fat saturation on. *See* **Note 2** for the optional *Protocol_MSME* (T_2 mapping).
   (f) *Protocol_TOF* (angiography): FLASH sequence, TR = 11 ms, TE = 3 ms, flip angle = 80°, spatial in plane resolution of 200×268 μm^2, with 15 slices of 1.0 mm thickness.
2. Switch on the gradient amplifiers of the MR system, which will also power on the automatic animal positioning system *AutoPac*™.
3. Install the transmit/receive *rat body volume coil* in the magnet bore.
4. Connect the animal holder to the animal positioning system (*AutoPac*™).
5. Install the *rat heart coil array* RF coil including its preamplifier on the animal bed.

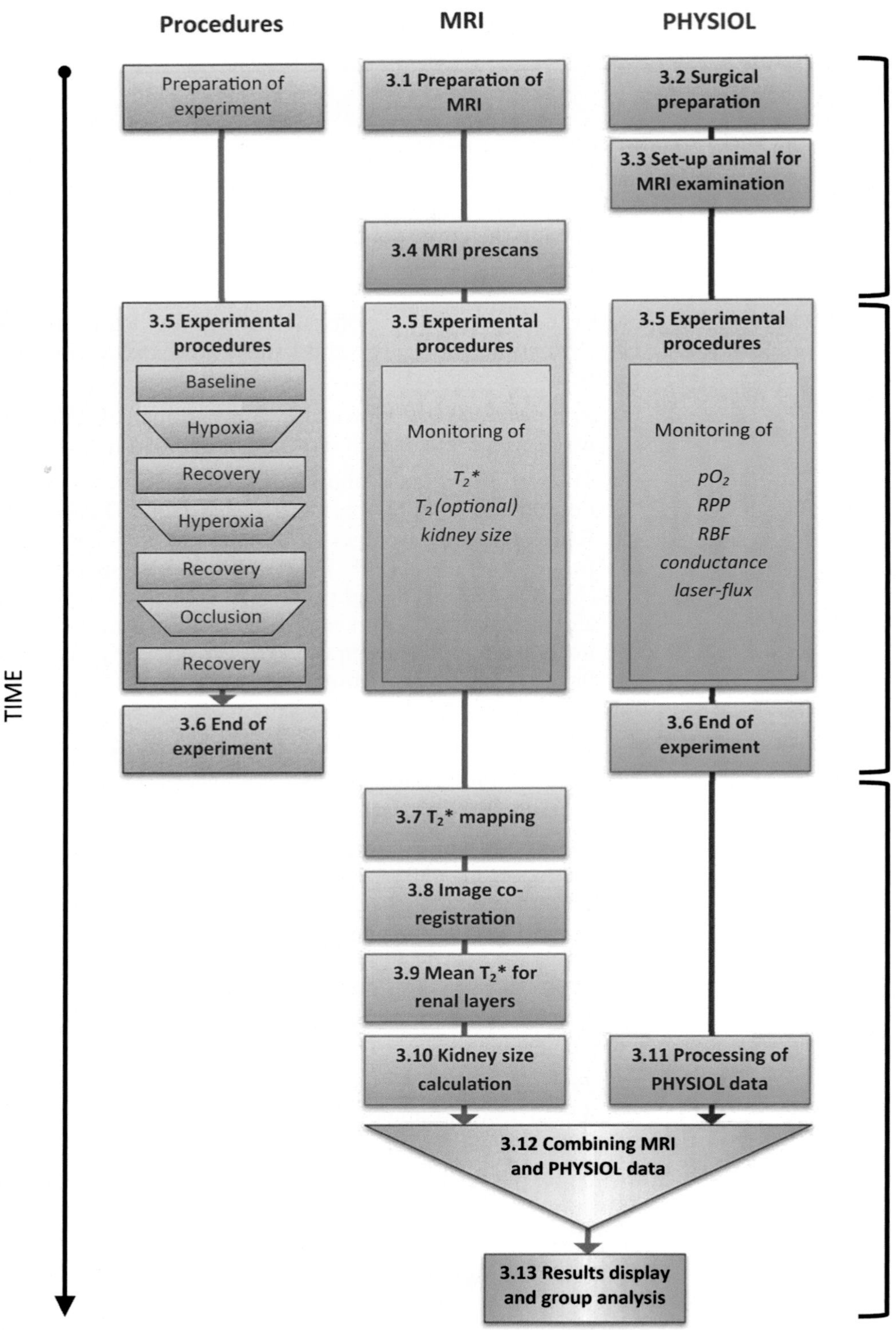

Fig. 3 Illustration of the work flow, including preparations, experimental procedures, and data analyses. Steps in this work flow chart whose name starts with a number (e.g., *3.1 Preparation of MRI*) are described in the corresponding method sections with the same name. pO_2 partial pressure of oxygen, *RPP* renal perfusion pressure, *RBF* renal blood flow

6. Attach the face mask unit (commercial or custom-made, as described in **Note 1**) to the animal holder and connect it to the inspiratory gas providing system (luer tubing).
7. Place the flexible rubber mat of the warm water-based heating system on top of the *rat heart coil array* and connect it to the warm water circulation.
8. Switch on the water bath. Adjust the temperature to approximately 45 °C (*see* **Note 3**).
9. Attach the rectal temperature probe and pneumatic pillow to the small animal monitoring system and place the probes on the animal bed, approximately at the later abdominal position of the rat.
10. Attach all tubes and cables along the length of the animal bed using tape (*see* **Note 4**).

3.2 Surgical Preparation

Surgery must be performed parallel to MRI preparation outside the MR scanner room (in a neighboring preparation room) for safety reasons.

1. Anesthetize the animal by intraperitoneal injection of urethane (20 % solution, 6 mL/kg body mass) (*see* **Note 5**).
2. After reaching the required depth of anesthesia (determined by specific physiological signs such as muscle relaxation degree, absence of the paw withdrawal reflex, absence of the swallowing reflex, whisker movements, etc.), carefully shave the coat in the abdominal area of the rat (hair clipper Aesculap Elektra II GH2, Aesculap AG, Tuttlingen, Germany).
3. Place the rat in supine position on a warmed-up (39 °C) temperature-controlled operating table and fix the paws of the animal to the table by means of sticky tapes.
4. Make an incision in the left inguinal area (approximate 12 mm) along the natural angle of the hind leg.
5. Bluntly dissect the connective tissue until the femoral artery and vein are exposed.
6. Gently separate the nerve. Do not cut or damage the nerve.
7. Using fine tip forceps separate the vein from the artery, trying to expose an approximately 7–8 mm length fragment.
8. Place three pieces of 4.0 threads under the femoral artery: the *first* piece towards the leg, the *second* towards the body, and the *third* one between them.
9. Pull the *first* thread towards the leg and tie this into a triple knot.
10. Prepare loose knots on the remaining two.
11. Pull the *second* piece of thread towards the body to stop the blood flow into the femoral artery.

12. Using fine tip scissors make an incision in the exposed segment of the femoral artery. Fill the catheter with saline.
13. Grasp the catheter with the forceps and gently push through the incision into the lumen of the femoral artery.
14. Tie the *third* knot, relax the stretched *second* thread, and push the catheter deeper ($\approx$10 mm) into the artery.
15. Rinse the catheter carefully with saline, making sure that it is patent.
16. Tie the prepared loose knot of the *second* thread.
17. Start the monitoring of arterial blood pressure.
18. Open the abdominal cavity by a midventral incision (4–5 cm). Carefully dissect the aorta directly above both renal arteries from the surrounding tissues.
19. Place the hydraulic occluder around the aorta above the renal arteries (*see* **Note 6**).
20. Using fine tip forceps carefully separate the renal artery from the vein, trying to expose an approximately 6–7 mm length fragment. Do not cut or damage the nerves.
21. Transfer the animal onto the portable animal holder such that the kidney is aligned with the mark on the holder that indicates the center of the RF coil.
22. Start the HAEMODYN™ software (*see* Subheading 2.2).
23. Place the Transonic flow probe around the renal artery and start monitoring RBF (*see* **Note 7**).
24. Remove the customary Luer-Lock connectors of the laser-flux-pO_2 probes and fix the fiber glass cores with customary silicone tubing by means of medical sticky tape. Attach tailored patches of gauze to the end of the silicone tubing (*see* Fig. 4).
25. Measure the diameter between the caudal and the cranial extremities of the left kidney with a caliper gauge. Based on this measurement the cortical laser-flux-pO_2 probe must be carefully prepared so that the distance between the insertion point and the tip exactly matches the individual kidney's diameter minus 1.5 mm.
26. Advance the cortical laser-flux-pO_2 probe meticulously from the caudal extremity of the kidney along the caudo-cranial axis (*see* Fig. 4).
27. To prevent cranio-caudal displacement the patch of gauze fixed to the silicon tubing of the probe must be glued to the capsule of kidney's ventral surface by means of the histoacryl™ glue.
28. Prepare the medullary laser-flux-pO_2 probe so that the distance between the insertion point and the tip is exactly matches 3–4 mm (*see* Fig. 4).

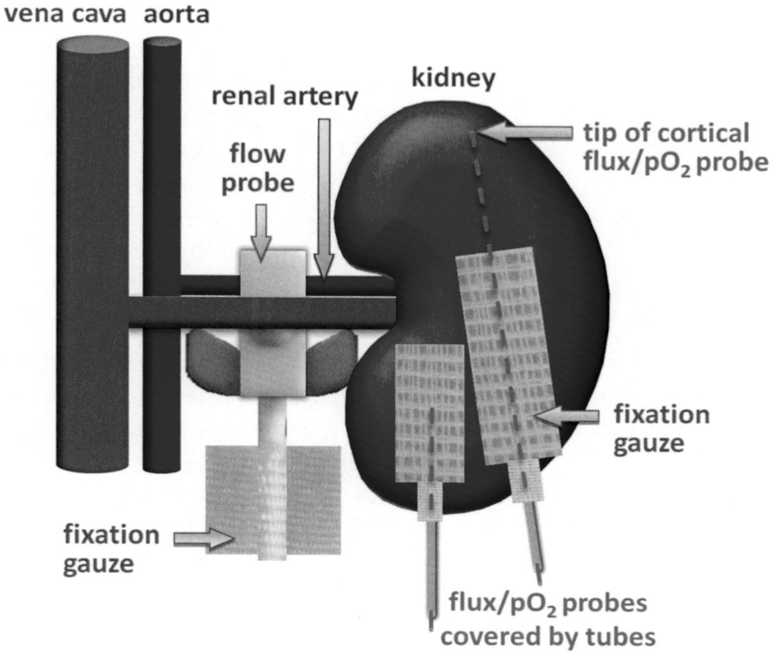

Fig. 4 Schema illustrating the position of the perivascular probe for total renal blood flow and the cortical and medullary flux/pO_2 probes together with measures taken to fix probes at their respective positions

29. Carefully advance the medullary laser-flux-pO_2 probe from the caudal extremity of the kidney along the caudo-cranial axis (*see* Fig. 4).
30. Glue the patch of gauze fixed to the silicon tubing of the probe to the capsule of kidney's ventral surface histoacryl™ glue.
31. To prevent displacement carefully fix the two probes' tubing to the retroperitoneal muscles by sutures.
32. Place the special bridge of the portable animal holder right behind the hind paws (*see* Fig. 2).
33. Connect the probes with OxyLite/OxyFlo™ apparatus and start the monitoring of tissue pO_2 and laser-Doppler-flux.
34. Place a fiber-optical temperature probe in close proximity to the kidney, in order to monitor the temperature of the kidney throughout the investigation.
35. Mark the localization of the investigated kidney's upper and lower pole on the skin of the abdomen using a pen (*see also* **step 0**).
36. Fill the abdominal cavity with warm saline (37 °C). For replenishment of abdominal saline a catheter must be placed in the abdominal cavity.

37. Fix all the extensions (temperature probe, Transonic probe, laser-flux-pO_2 probes, aortic occluder, and abdominal flushing catheter) to the bridge of the portable animal holder (cable support that prevents the disarrangement of the probes and enables a safe transport of the animal to the MR scanner).
38. Pass the extensions for aortic occluder, for abdominal flushing, as well as temperature and Transonic probe, through the caudal cutting edge of the median abdominal incision. The extensions of the laser-flux-pO_2 probes must be led through the abdominal wall using a small incision in the left inguinal region.
39. Close the abdominal cavity by continuous suture.
40. Check that the kidney (pen markings on skin) is still aligned with the mark on the portable animal holder that indicates the center of the RF coil—if necessary carefully correct the animal's position. This is essential for optimal positioning of the rat in the MR scanner (i.e., optimal position of the rat's kidney relative to the MR coil).
41. Re-start the HAEMODYN™ software (to start a new data file) and check the quality of all physiological signals, i.e., renal perfusion pressure (RPP), renal blood flow (RBF), cortical and medullary Laser-flux, cortical and medullary pO_2.

3.3 Set Up Animal for MRI Examination

1. Transfer the animal to the MR scanner room using the portable animal bed (*see* **Note 8**).
2. Position portable animal bed with the rat on the MRI animal bed.
3. Place a respiratory mask loosely around the muzzle of the spontaneously breathing rat. Open air supply to a rate of 1000 mL/min.
4. Switch on the small animal monitoring system. Insert the rectal temperature probe after cleaning it with alcohol and dipping it into Vaseline™.
5. Place the pneumatic pillow on the abdomen and cover the animal with the warming blanket. Watch the respiration trace on the monitor of the small animal monitoring system and adjust pillow position until the respiratory motion is captured well (*see* Fig. 5) (*see* **Note 9**).
6. Set the trigger options of the small animal monitoring system such that the trigger gate (indicated by white/red color horizontal bars parallel to the respiratory trace) opens for maximum 280 ms (delay 30 ms) around the expiratory peak (*see* Fig. 5).
7. Press the *Out* button of the *AutoPac*™ system to make sure the animal bed is in the reference position. Switch on the *Laser* position marker and drive animal bed until the center of the RF

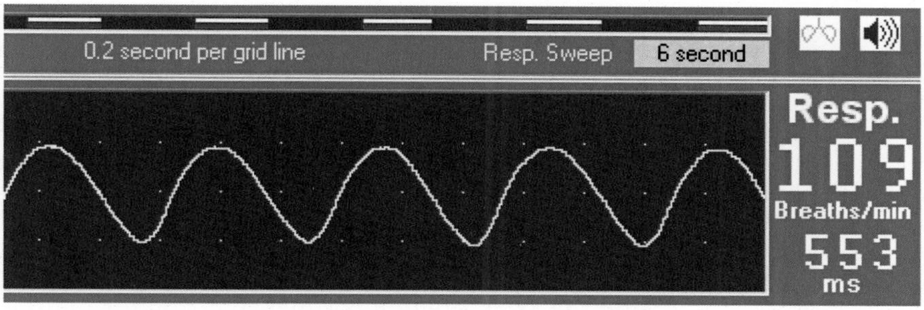

Fig. 5 Small animal monitoring system (SA Instruments, *see* Subheading 2.3). Setup of respiratory triggering for the MRI acquisition, with *begin delay* of 30 ms and *maximum width* of 280 ms for a typical respiratory rate of ~110 per minute. *Horizontal white bars* at *top* indicate trigger window

surface array coil is aligned with the Laser position. Switch off the Laser (this point is now stored as the point of interest).

8. Double-check that along the entire animal bed nothing protrudes such that it would not fit into the volume RF coil, which is installed inside the magnet.
9. Press the *Work Position* button of the *AutoPac*™ system to drive the point of interest (point on the animal bed marked by the Laser) to the iso-center of the magnet.

3.4 MRI Prescans

1. Start the *ParaVision*™ *5* software and register a new subject and study (*see* **Note 10**). For the first scan select the *Protocol_TriPilot*.
2. Tune and match the *rat body volume coil* using the *Wobble* function.
3. Start the first pilot scan and verify on the acquired images that the kidneys are in the center of the magnet (field of view) and well positioned within the signal intensity profile of the surface coil. If necessary correct the animal or/and Laser marker position and then repeat the tuning/matching and this pilot scan (*see* **Note 11**).
4. Load the *Protocol_T2axl* (axial pilot scan), edit the geometry such that the eight axial slices cover the kidney (*see* **Note 12**), and run the scan.
5. Load the *Protocol_T2corsag* (coronal/sagittal pilot scan), edit the geometry of the slice such that it is perpendicular (oblique coronal) to the axial reference image and crosses the kidney centrally along its longest dimension, and run the scan.
6. Clone the *Protocol_T2corsag* and as for the *Protocol_T2axl*, edit the geometry of the slice such that it is perpendicular (oblique sagittal) to the oblique coronal reference image and crosses the kidney centrally along its longest dimension, and run the scan.

7. Load the *Protocol_PRESSvoxel* (shim voxel) and edit the geometry (including its size) of the voxel such that it just encloses the kidney.
8. Shim on the PRESS voxel: from the *Spectrometer Control Tool> Acq> Current Adjustments* run the *Method specific adjustment for the Local Field Homogeneity* followed by *Method specific adjustment for the Basic Frequency*.
9. Load the *Protocol_MGE* (T_2* mapping) and edit the geometry (using the axial, oblique coronal, and oblique sagittal T_2 scans as references) such that the slice is strictly coronal with regard to the kidney. Run the scan and check image geometry and quality.
10. Optional (*see* **Note 2**): load the *Protocol_MSME* (T_2 mapping) and adapt the geometry such that the slice position is identical to the T_2* scan (note the difference in slice thickness). Run the scan and check image geometry and quality.
11. Load the *Protocol_TOF* (angiography) and edit the geometry such that it crosses the kidney, sagittal with regard to the kidney. Run the scan and after acquisition locate the bright signal of the large vessels within the kidney.

3.5 Experimental Procedures

1. On the PC that acquires the physiological data set a "START" marker in the HAEMODYN™ software.
2. On the MR system clone the *Protocol_MGE* five times. For the optional measurement of T_2, *see* **Note 13**.
3. Run the baseline scans with the *GOP* button of the *Spectrometer Control Tool* (not the *Traffic Lights* button of the *Scan Control Tool*). When the actual acquisition starts, as identified by the start of gradient noise (ignoring the short prescan), set a marker with the TTL switch by switching it on for ~1 s and off again.
4. Clone the *Protocol_MGE* 45 times.
5. *Start of Hypoxia*. Change the gas flowing through the respiratory mask to 10 % O_2/90 % N_2.
6. Start a *Protocol_MGE* (with GOP) immediately after the onset and another exactly 5 min after the start of hypoxia.
7. *End of Hypoxia*. Change the gas flowing through the respiratory mask back to air (21 % O_2).
8. Start a series of *Protocol_MGE* (with GOP; using the *Queued acquisition* macro) to run continuously for 20 min (recovery period).
9. *Start of Hyperoxia*. Change the gas flowing through the respiratory mask to 100 % O_2.

10. Start a *Protocol_MGE* (with GOP) immediately after the onset and another exactly 5 min after the start of hyperoxia.
11. *End of Hyperoxia*. Change the gas flowing through the respiratory mask back to air (21 % O_2).
12. Start a series of *Protocol_MGE* (with GOP; using the *Queued acquisition* macro) to run continuously for 20 min (recovery period).
13. Remove the remaining *Protocol_MGE* scans (in *ready* status) from the scan list. Clone *Protocol_TOF* once, *Protocol_MGE* once, *Protocol_TOF* once, and *Protocol_MGE* 20 times.
14. *Start of Occlusion*. Inflate the remotely controlled occluder.
15. Simultaneously start the *Protocol_TOF* and after acquisition verify the absence of the blood flow signal in the renal vessels (*see* **Note 14**).
16. Immediately afterwards start a *Protocol_MGE* (with GOP).
17. *End of Occlusion*. Rapidly deflate the occluder.
18. Simultaneously start the *Protocol_TOF* and after acquisition verify the re-occurrence of the blood flow signal in the renal vessels.
19. Start a series of *Protocol_MGE* (with GOP; using the *Queued acquisition* macro) to run continuously for 20 min (recovery period). When the acquisition of the first scan starts, as identified by the start of gradient noise (ignoring the short prescan), set another marker with the TTL switch by switching it on for ~1 s and off again.
20. After 20 min exit the *Queued acquisition* and remove excess *Protocol_MGE* scans from the scan list.
21. On the PC that acquires the physiological data set an "END" marker in the HAEMODYN software.

3.6 End of Experiment

1. Carefully remove the respiratory mask from the animal's muzzle.
2. Move the animal into preparation room and place in supine position on a warm operating table.
3. Cut the sutures and open the abdominal cavity.
4. Remove the fluid from the abdominal cavity using a pipette.
5. Control and note the overall condition of the kidney after experiment (e.g., surface coloring and its homogeneity).
6. Carefully untie the knots of the aortic occluder and remove it.
7. Remove the abdominal flushing catheter, as well as the temperature probe, the Transonic probe, and the laser-flux-pO_2 probes.
8. Exsanguinate the animal by cutting the abdominal aorta.

3.7 MRI Data Analysis: T_2* Mapping

The MRI data acquired with the multi-echo gradient echo (MGE) protocol contain images with different T_2* weighting (determined by the echo time, TE). From these images calculate T_2* for each pixel to construct a map of the parameter T_2*.

1. Open the reconstructed MR images in *MATLAB* (💻; *see* Subheading 2.3) by loading the binary *2dseq* files of each MGE scan and import *echo time* (TE) parameters from the *method* text file.
2. Perform pixel-wise exponential curve fitting of the equation $S(t) = A + B \times \exp(-t/T_2^*)$ to the image intensities $S(t)$ versus time t (the 10 echo times) (💻) (*see* **Note 15**).
3. Export T_2* parameter maps in ANALYZE/NIfTI file format (💻).

3.8 MRI Data Analysis: Image Co-registration

Subject motion during the experiment must be accounted for to permit a direct comparison between images/maps of different time points. Use automated spatial transformations of the images in order to spatially align all images to the baseline image.

1. Open the binary *2dseq* file of a baseline MGE scan in *ImageJ* (*see* Subheading 2.3) (use the *BrukerOpener* plugin) and keep only the image acquired with the shortest echo time, i.e., discard frames 2–10 (💻 *ImageJ* macro).
2. Draw a rectangular region of interest (ROI) generously around the kidney, create a mask (*Edit* > *Selection* > *Create mask*), divide by 255 to obtain a mask consisting of zeros and ones, and save in NIfTI format as *[myfilename]_mask.nii*.
3. For all MGE scans: open the binary *2dseq* file in *ImageJ*, keep only the image acquired with the shortest echo time (i.e., discard frames 2–10), and save in NIfTI format (💻 *ImageJ* macro).
4. Perform image registration for each MGE scan time series (i.e., each animal) using FSL FLIRT (*see* Subheading 2.3). Input data to FSL FLIRT are the first frame of the MGE scans and the corresponding mask. The output of FSL FLIRT consists of the registered images (needed only for visual inspection) and the spatial transformation matrices (text files that should be named *omat[frame#]*) (💻 UNIX shell script).
5. Apply this transformation matrix to the T_2* map in NIfTI format using FSL FLIRT. Repeat this step and the previous step for all MGE scans (💻 UNIX shell script).
6. To visually inspect the time series of the registered T_2* maps display the matrix containing the fitted T_2* parameters as images using a pseudo color scale (e.g., *jet* in *MATLAB*) with the limits 0–30 ms (*see* Fig. 6) (💻 *MATLAB*).

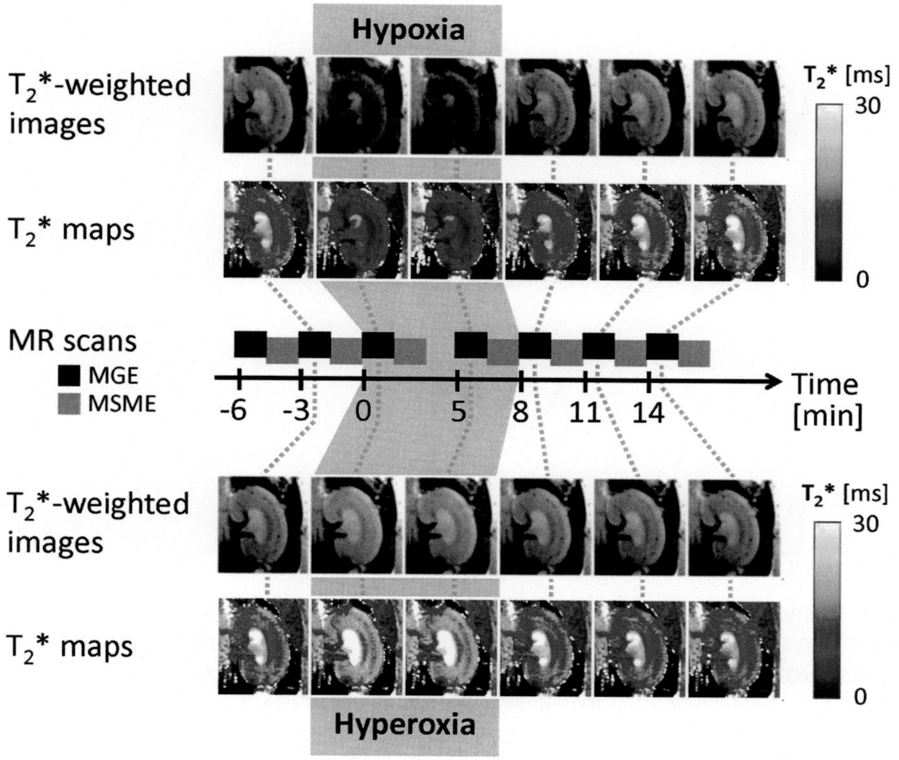

Fig. 6 Hypoxia and hyperoxia experiments. T_2*-weighted MR images (*gray scale*) and T_2* parameter maps (*color-coded*) derived from MGE acquisitions. Following baseline measurements of T_2*, hypoxia/hyperoxia was induced by changing the inhaled gas mixture from air to 10 %/90 % O_2/N_2 or 100 % O_2. T_2*/T_2 measurements were repeated directly after and exactly 5 min after onset of hypoxia/hyperoxia (to ensure equal timing of the MR scans). Subsequently inhaled gas was changed back to air and T_2*/T_2 monitored throughout their recovery phase. Reproduced from [8] with permission from the Public Library of Science

3.9 MRI Data Analysis: Mean T_2* for Renal Layers

Derive representative T_2* values for each renal layer from the T_2* maps.

1. Open and display the baseline T_2* parameter map (*see* **Note 16**) (💻 *ImageJ* macro).
2. Place a rectangular selection (reference rectangle) tightly around the outer edges of the kidney.
3. Place regions-of-interest (ROIs) at defined positions within the rectangular reference frame and store the mean T_2* value for each ROI in a text file (*see* **Note 17**). The sizes and positions of nine ROIs are defined based on renal anatomy (*see* Fig. 7 and [11]), such that they lie entirely within a renal layer and are positioned away from the image artifacts caused by the invasive pO_2/flux probes. Also store the coordinates of the reference frame in another text file to permit later calculation of the reference frame area, which serves as a surrogate of kidney size (💻 *ImageJ*).

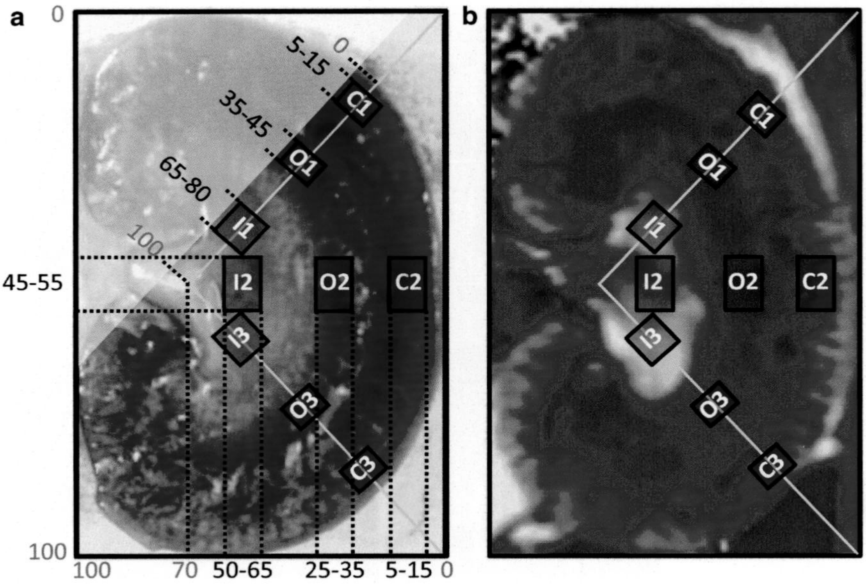

Fig. 7 Kidney segmentation model overlaid onto a photograph of a freshly excised rat kidney in coronal view (a) and superimposed to a T_2* map of a rat kidney (b). During analysis, the rectangular reference frame is manually positioned around the kidney, followed by an automated drawing of the diagonals (*yellow*). After their intersections with the kidney borders are defined manually, the ROIs (I1–I3, O1–O3, C1–C3, I: inner medulla, O: outer medulla, C: cortex) are automatically placed at predefined relative positions with regard to these references. The *numbers* shown on the horizontal and vertical axes as well as on the diagonals signify percentages of the reference frame dimensions and of the diagonals. Reproduced from [11] with permission from Wolters Kluwer Health

3.10 MRI Data Analysis: Kidney Size Calculation

1. Open the text files containing the reference rectangle coordinates (*see* Subheading 3.9, **step 3**) and the text files containing the spatial transformation matrices (*see* Subheading 3.8, **step 4**) that were calculated during image co-registration (💻 *MATLAB*).
2. Convert the matrix containing the rectangle coordinates into four points (p_1, p_2, p_3, p_4) describing the coordinates of the four corners: $p_1 = [x_1; y_1; 0; 0]$, …, $p_4 = [x_4; y_4; 0; 0]$. To apply the spatial transformations to these points, multiply them with the transformation matrices (*Mtransf*) after inverting them using the *inv* function: $p1\_transf = p1' \times inv(Mtransf)$ (💻). NB: the prime symbol (') indicates a built-in MATLAB function that denotes the transpose of a matrix.
3. Calculate the edge lengths of the rectangle as $a = \mathrm{sqrt}((p_2(1) - p_1(1))^2 + (p_2(2) - p_1(2))^2)$ and $b = \mathrm{sqrt}((p_3(1) - p_2(1))^2 + (p_3(2) - p_2(2))^2)$ for the baseline rectangle (p_1, p_2, p_3, p_4) and each spatially transformed rectangle (*p1_transf*, *p2_transf*, *p3_transf*, *p4_transf*) for images acquired through the time course of the experiment (💻).
4. Calculate the area of each rectangle as $A = a \times b$, and store the time course of this parameter as a surrogate of kidney size (💻).

3.11 Processing of PHYSIOL Data

1. Start the HAEMODYN software and open the file containing the recorded physiological signals (*.sty* file). Display the recording (*Experiment> Replay*) and scroll to the beginning of the experiment in the *Replay Detail* window. Click on the plot and set cursor (marker) "*START_RAW*". Then scroll to the end of the experiment, set the "*END_RAW*" marker, and close the window. Export the raw data to a file (will be called *RAW_DATA.txt*).
2. Open the *RAW_DATA.txt* file and reduce the sampling rate from 500 to 2 Hz by eliminating sliding window averaging. Save data in a new file called *RAW_DATA_2Hz.txt* (🖳 *MATLAB*).

3.12 Combining MRI and PHYSIOL Data

Temporally align the MRI and PHYSIOL data by identifying the starting point of the experiment within both data sets.

1. Open the file containing the recorded physiological data (*RAW_DATA_2Hz.txt*). Create a new vector called t_{PHYSIOL} with the same number of elements as the recorded signals and values starting at 0 with an equal spacing of 0.5, representing the time in ms. Plot the recorded signal of the TTL switch versus t_{PHYSIOL} (🖳 *MATLAB*).
2. Find the first TTL peak, i.e., the marker set during the experiment at the beginning of the baseline, and store the time at which it occurs as $t_{0,\mathrm{PHYSIOL}}$. Do the same for the second TTL peak to get $t_{\mathrm{end,PHYSIOL}}$ (🖳).
3. Shift the PHYSIOL time axis such that it starts at the beginning of the experiment: $t_{\mathrm{PHYSIOL}} = t_{\mathrm{PHYSIOL}} - t_{0,\mathrm{PHYSIOL}}$ (🖳).
4. Open the *acqp* file of each MRI scan (located at /opt/PV5.1/data/[user]/nmr/[registration]/[scan]), get the value of the parameter *ACQ_time* (e.g., "<15:31:33 29 Oct 2014>"), parse it into separate values for hours, minutes, seconds, etc., and calculate the start time of the scan as $t_{\mathrm{start,MRI}}(scan) = (hours \times 60 + minutes) \times 60 + seconds$ (🖳) (*see* **Note 18**).
5. Store the start time of the first baseline scan as $t_{0,\mathrm{MRI}}$ (🖳).
6. Open the *method* file of each MRI scan, find the first line of text starting with "$$" (e.g., "$$ Mon Oct 29 15:33:00 2014 CET (UT+1 h)..."), extract the time stamp (here "15:33:00"), parse it into separate values for hours, minutes, and seconds, and calculate the end time of the scan as $t_{\mathrm{end,MRI}}(scan) = (hours \times 60 + minutes) \times 60 + seconds$ (🖳).
7. Shift the MRI time axis such that it starts at the beginning of the experiment: $t_{\mathrm{start,MRI}}(scan) = t_{\mathrm{start,MRI}}(scan) - t_{0,\mathrm{MRI}}$ and $t_{\mathrm{end,MRI}}(scan) = t_{\mathrm{end,MRI}}(scan) - t_{0,\mathrm{MRI}}$ (🖳).
8. To check that the time axes are correctly synchronized confirm that the time of the "end" marker ($t_{\mathrm{end,PHYSIOL}}$) is equal to the start time of the first MR scan in the final recovery period (*see* **Note 19**).

For a direct comparison of the MR-PHYSIOL parameters, use only PHYSIOL data acquired during the relevant MRI acquisitions.

9. While the PHYSIOL parameters were measured with subsecond temporal resolution, the acquisition of the MR parameters (derived from a MR image) requires much more time (>60 s). PHYSIOL and MR parameters can only be compared based on the (low) temporal resolution of the MRI. For this purpose calculate the average value of each physiological parameter over the acquisition time of each MRI scan, i.e., from $t_{\text{start,MRI}}(scan)$ to $t_{\text{end,MRI}}(scan)$. For the Laser flux signals take instead the first 5 s after the end of the scan, i.e., from $t_{\text{end,MRI}}(scan)$ to $t_{\text{end,MRI}}(scan) + 5$ s, because during MRI acquisitions the Laser flux signal is incorrect due to interference from the MRI (💻).
10. Derive the renal conductance parameter as $conductance = RBF/RPP$ (💻).

3.13 Results Display and Group Analysis

1. Calculate for all parameters and time points the group mean and the standard error of the mean (💻 *MATLAB*).
2. For a detailed analysis of temporal changes in the MR-PHYSIOL parameters, plot the regional renal T_2^* (group mean ± SEM) for each ROI versus time, together with the corresponding PHYSIOL parameters (e.g., *see* Fig. 8) (*see* **Notes 20** and **21**) (💻).
3. A correlation analysis between parameters, in particular between MRI and PHYSIOL parameters, may reveal additional information. Use Spearman analysis (nonparametric correlation on ranks) of the individual data pairs to assess the strength of relationships that follow a monotonous function. If such a significant correlation is observed, apply additional Pearson analysis (parametric correlation) to assess the strength and parameters of linear relationships.

4 Notes

1. If the holder does not provide a respiratory (anesthetic) mask, such a mask can be easily built: take a 20 mL plastic syringe, cut off the tip approximately 15 mm from the bottom of the syringe, yielding a funnel-shaped mask. Finally deflash and smooth the cutting edges of the mask using a file.
2. Besides measurements of T_2^*, the additional measurement of T_2 may be of interest, depending on the aims of the study. Compared to T_2^*, the parameter T_2 is less sensitive to blood oxygenation (deoxyhemoglobin) but more sensitive to water content. Interleaved measurements of T_2^* (*Protocol_MGE*)

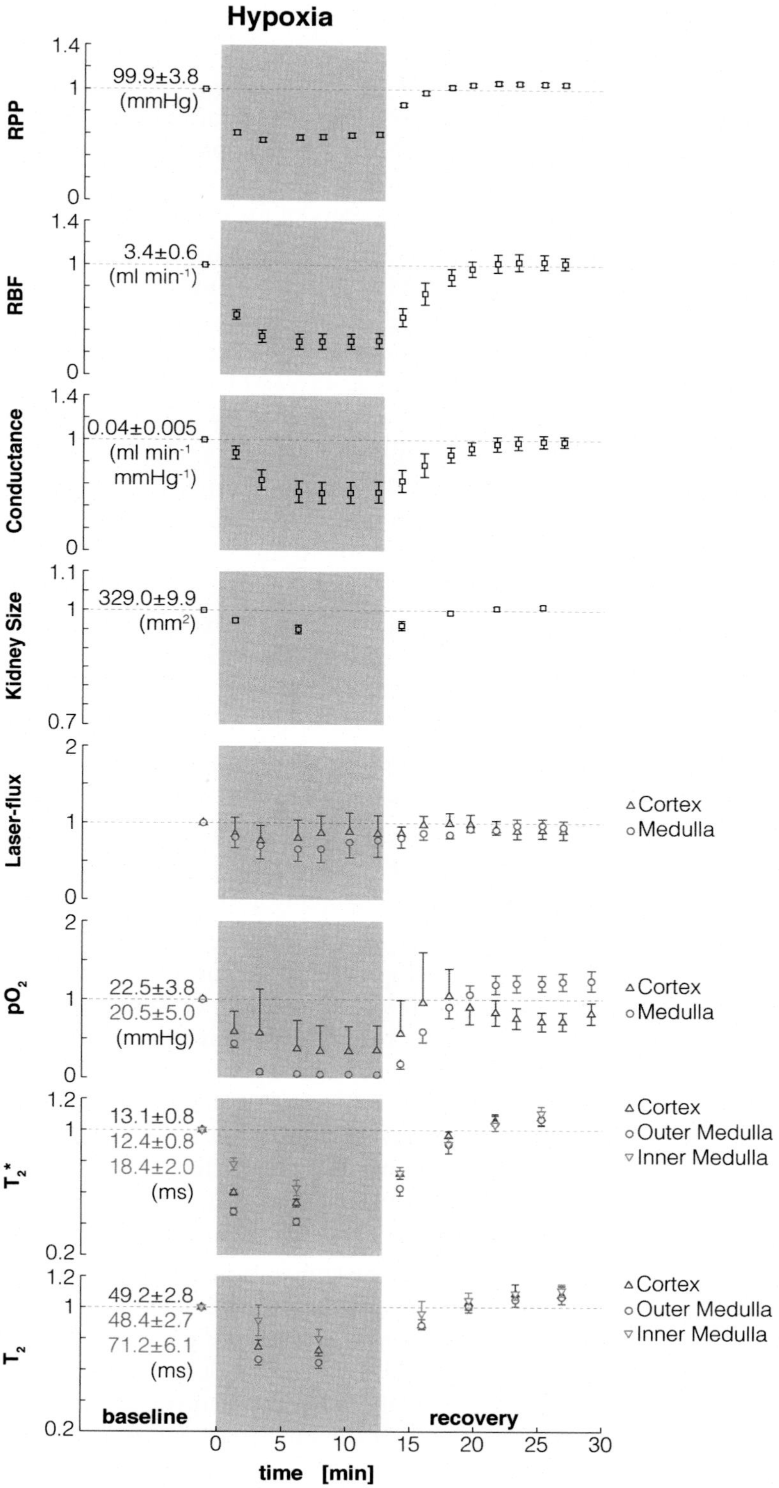

Fig. 8 Side-by-side plot of the time courses of physiological and MR parameters (group mean ± SEM) throughout baseline, hypoxia, and recovery. Relative changes in RPP, RBF, renal conductance, kidney size, cortical/medullary laser-flux, cortical/medullary pO_2, cortical/outer medullary/inner medullary T_2*, and cortical/outer medullary/inner medullary T_2. Hypoxia started at $t = 0$ and its duration is indicated by the *gray shading*. Reproduced from [11] with permission from Wolters Kluwer Health

and T_2 (*Protocol_MSME*) can be performed using the protocol below. Please note that the acquisition time for *Protocol_MGE* plus *Protocol_MSME* will be approximately double the acquisition time of *Protocol_MGE*. This will impact on the timing of the experimental protocol for baseline and interventions.

(a) *Protocol_MSME* (T_2 mapping): multi-spin echo (MGE) sequence, TR = 550 ms, echo times = 7, first echo = 10 ms, echo spacing 10 ms, averages = 1. Define as geometry a coronal oblique image slice with an FOV = (38.2 × 50.3) mm^2, matrix size = 169 × 113 zero-filled to 169 × 215, and a slice thickness of 1.5 mm. Respiration trigger on (per phase step), fat saturation on.

(b) Analysis of T_2 data: apply the same methods as described for T_2*. The range of values is different for T_2* and T_2. Use limits of 0–30 ms (T_2*) and 15–80 ms (T_2) for displaying the data in plots or color-coded maps.

3. The temperature of the water bath will be much higher than the temperature of the rubber mat and depends on the length and material of the tubing used. Hence the temperature of the bath must be adapted to the local setup.
4. Make sure to keep tubes or cable close to the animal bed so they cannot get caught anywhere on the way into the magnet!
5. Urethane supports anesthesia throughout the surgical preparation and the MRI examination (for several hours) and leaves cardiovascular and respiratory reflexes largely undisturbed.
6. (1) In order to prevent additional pressure on the aorta (and therefore development of an unwanted kidney ischemia) the positioning and knotting of the occluder to the aorta must be performed under carefully monitoring of the overall condition of the kidney (e.g., surface coloring and its homogeneity); (2) since the occluder consists of silicone and polyethylene and is positioned about 15 mm away from the kidney, it does not cause artifacts in MRI that affect the kidney; (3) to allow control of in- and deflation of the occluder connect a second (control) occluder outside the scanner using a luer T-piece.
7. (1) The signal of the conventional probe (Type 1RB; Transonic Systems) is too weak due to the long extension leads. For this setup a probe with a larger body size (reflector made of ceramics) must be used. Its reflector does not induce MR artifacts, is L-shaped, and offers no mechanism to lock the vessel which presents a significant challenge for the implantation of the probe. (2) The bulk of the intestine can cause additional pressure that can dislocate the probe and cause probe pressure on the aorta, the renal artery and vein, or the kidney itself. (3) Since Transonic measurements rely on ultrasound, an appropriate coupling into tissue is of high relevance. To this end the

abdominal cavity must be filled with saline solution (37 °C). An additional catheter was placed in the abdominal cavity to replenish saline leakage course of the experiment. (4) To avoid displacements of the probe, a gauze was attached to the cable. The gauze was fixed to the retroperitoneal muscles by means of sutures.

8. Special attention during transfer must be paid to the tube/ cable extensions (aortic occluder, abdominal flushing, and all probes).
9. The peak-to-peak amplitude of the respiratory trace should span about 2/3 of the vertical axis on the display. Any gross movement (for instance during repositioning the pillow) will lead to large peaks and force the monitoring system to adapt the signal amplification, so that temporarily the signal will become much smaller on the display. Keep an eye on the magnification, which is given on the left next to the display; this will drop to a low value such as 15×—wait until it recovers back to a value around 100× before further adjusting it.
10. The registration will determine the name of the folder on the computer's hard disk in which all study data will be stored. Spaces and special characters (with a few exception such as "–" and "_") are not allowed. Check at this point in time that the coil configuration has been detected correctly. Select the *Location* of your scan protocols (directory where you stored them).
11. To repeat the pilot scan, undo it and then start it with the *Traffic Light* button while holding the shift key on the keyboard in order to force all adjustments to run again.
12. Check the length of the respiratory trigger window of the monitoring system and if necessary reduce the TR of the scan until it fits within this window.
13. Additional measurement of T_2 (optional; *see* **Note 2**). For interleaved measurements of T_2* (*Protocol_MGE*) and T_2 (*Protocol_MSME*) replace each step involving the *Protocol_MGE* with the concatenation of *Protocol_MGE* and *Protocol_MSME*. This will require approximately double the acquisition time of *Protocol_MGE* and impact on the timing of the experimental protocol, e.g., the duration of hypoxia, hyperoxia, and occlusion must be prolonged to allow for the longer acquisition time. However, the explicitly stated time points and durations, such as the 20 min recovery time, should be maintained.
14. If unsuccessful, repeat the occlusion and inflate the occluder with higher hydraulic pressure. Always check the inflation of the second occluder outside the MR scanner and remember to which volume the syringe was advanced.
15. To avoid curve fitting errors you need to exclude the data points for which the signal intensity is not significantly above

the noise level. The signal intensity threshold below which data points will be excluded is a crucial parameter. It must be chosen individually for each study based on the signal-to-noise ratio of the acquired MR data. It is highly recommended to develop and use a 'fit inspector' software tool, i.e. an interactive tool that displays the T_2* map and allows the user to move the mouse pointer over the map while plotting for the currect pixel the fitted curve together with the data points and indicating which data points were excluded. Using such a tool one can easily verify if a threshold is appropriate or must be modified.

16. The stored parameter maps are not conventional images: the value of each pixel is a decimal number in the range 0.0–150.0 ms (rather than an image intensity with typical integer values between 0 and 65,535). When saving such a parameter map the decimal values could unintentionally by changed, for example due to automatic conversion into integer values. Make sure the stored maps contain the correct T_2* values. If parameter maps have to be stored in integer format, multiplying them with 100 beforehand can help to preserve precision of T_2* with two decimal places, e.g., storing 5.17 ms as 517 instead of as 5.
17. It is important to verify that the T_2* parameter map does not show any artifacts within the automatically placed nine ROIs. For this purpose display the ROIs overlaid onto the T_2* parameter map and check carefully for susceptibility-induced signal loss/voids (typically close to adjacent fat tissue, intestine, or an air bubble accidentally introduced during surgery) or other obvious image artifacts. ROIs with artifacts must be excluded from the further analysis.
18. Be aware that when generating a list of scans automatically from the names of the folders in the session directory /opt/PV5.1/data/[user]/nmr/[registration] the folders carry the name of the scan, i.e., usually 1, 2, 3, ..., 10, 11, ..., 20, 21, ..., but will be listed in alphabetical order as 1, 10, 11, ..., 2, 20, 21, ...—so the order of scans in the list is not chronological. If for instance scan 1 has been deleted the first scan in the list would be 10 not 2.
19. It is advised to carefully check that the PHYSIOL time axis really matches the MRI time axis to ensure PHYSIOL and MRI parameters of each scan correspond to each other. One way to do this is to confirm that the artifacts in the Laser flux signals (flux is noticeably elevated or lowered depending on the type of MRI scan) coincide with the start and end times of the MRI scans. This can be done by plotting a Laser flux signal together with a synthetic signal that changes at the start of each scan to a defined value (ID_{SCAN}) and falls back to 0 an the end of each scan. ID_{SCAN} is composed of a value indicating the

MRI method (1000: FLASH, 2000: RARE, 3000: PRESS, 4000: FieldMap, 5000: MGE, 6000: MSME, 7000: FC2D_ANGIO, or 10000: all other scan types) and the *scan number*: $ID_{SCAN} = method + scan\ number$. For instance 6023 would be the scan number 23 and of type MGE. Note that some of the scans have a short prescan (to adjust the receiver gain) which is also visible in the Laser flux signals. This procedure is also useful for fine adjustment of the temporal synchronization between MR data and PHYSIOL data (💻 *MATLAB*).

20. It is usually more practical and useful to compare relative changes in the parameters rather than absolute changes. To do this divide all parameter values by that of the last baseline value (examples *see* Fig. 8) (💻 *MATLAB*).
21. For most purposes data of the three ROIs within each renal layer (cortex, medulla, or inner medulla) can be pooled into one mean T_2* value for each layer, since the intralayer differences between these ROIs were found to be much smaller than the interlayer differences [10]. Remember to exclude ROIs in which you have detected image artifacts such as signal voids due to regional changes of magnetic susceptibility (e.g., caused by air in intestine or abdominal air bubbles unintentionally introduced during surgery) (💻 *MATLAB*).

Acknowledgements

This work was supported by the German Research Foundation (Deutsche Forschungsgemeinschaft, research unit: FOR 1368, grant numbers: NI 532/9-1, NI 532/9-2, SE 998/4-2). The authors wish to thank Ariane Anger, Bert Flemming, Andrea Gerhardt, Jan Hentschel, Uwe Hoff, Mechthild Ladwig, and Sonia Waiczies for their help and support with developing the described methods.

References

1. Eckardt KU, Coresh J, Devuyst O, Johnson RJ, Kottgen A, Levey AS, Levin A (2013) Evolving importance of kidney disease: from subspecialty to global health burden. Lancet 382:158–169
2. Leung KC, Tonelli M, James MT (2013) Chronic kidney disease following acute kidney injury-risk and outcomes. Nat Rev Nephrol 9:77–85
3. Thakar CV (2013) Perioperative acute kidney injury. Adv Chronic Kidney Dis 20:67–75
4. Evans RG, Gardiner BS, Smith DW, O'Connor PM (2008) Methods for studying the physiology of kidney oxygenation. Clin Exp Pharmacol Physiol 35(12):1405–1412
5. Seeliger E, Sendeski M, Rihal CS, Persson PB (2012) Contrast-induced kidney injury: mechanisms, risk factors, and prevention. Eur Heart J 33(16):2007–2015
6. Pohlmann A, Cantow K, Hentschel J et al (2013) Linking non-invasive parametric MRI with invasive physiological measurements (MR-PHYSIOL): towards a hybrid and integrated approach for investigation of acute kidney injury in rats. Acta Physiol (Oxf) 207(4):673–689
7. Prasad PV (2006) Functional MRI of the kidney: tools for translational studies of pathophysiology of renal disease. Am J Physiol Renal Physiol 290(5):F958–F974

8. Pohlmann A, Hentschel J, Fechner M et al (2013) High temporal resolution parametric MRI monitoring of the initial ischemia/reperfusion phase in experimental acute kidney injury. PLoS One 8(2):e57411
9. Liss P, Cox EF, Eckerbom P, Francis ST (2013) Imaging of intrarenal haemodynamics and oxygen metabolism. Clin Exp Pharmacol Physiol 40(2):158–167
10. Rossi C, Sharma P, Pazahr S, Alkadhi H, Nanz D, Boss A (2013) Blood oxygen level-dependent magnetic resonance imaging of the kidneys: influence of spatial resolution on the apparent R_2* transverse relaxation rate of renal tissue. Invest Radiol 48(9): 671–677
11. Pohlmann A, Arakelyan K, Hentschel J et al (2014) Detailing the relation between renal T_2* and renal tissue pO_2 using an integrated approach of parametric magnetic resonance imaging and invasive physiological measurements. Invest Radiol 49(8):547–560

Chapter 12

Intravital Multiphoton Imaging of the Kidney: Tubular Structure and Metabolism

David M. Small, Washington Y. Sanchez, and Glenda C. Gobe

Abstract

Multiphoton microscopy (MPM) allows the visualization of dynamic pathophysiological events in real time in live animals. Intravital imaging can be applied to investigate novel mechanisms and treatments of different forms of kidney disease as well as improve our understanding of normal kidney physiology. Using rodent models, in conjunction with endogenous fluorescence and infused exogenous fluorescent dyes, measurement can be made of renal processes such as glomerular permeability, juxtaglomerular apparatus function, interactions of the tubulointerstitium, tubulovascular interactions, vascular flow rate, and the renin-angiotensin-aldosterone system. Subcellular processes including mitochondrial dynamics, reactive oxygen species production, cytosolic ion concentrations, and death processes of apoptosis and necrosis can also be seen and measured by MPM. The current methods chapter presents an overview of MPM with a focus on techniques for intravital kidney imaging and gives examples of instances where intravital MPM has been utilized to study renal pathophysiology. Suggestions are provided for MPM methods within the confines of intravital microscopy and selected kidney structure. MPM is undoubtedly a powerful new technique for application in experimental nephrology, and we believe it will continue to create new paradigms for understanding and treating kidney disease.

Key words Intravital imaging, Multiphoton, Two-photon excitation, Fluorescence, Metabolism

1 Introduction

Multiphoton microscopy (MPM), or multiphoton excitation fluorescence microscopy, is a technology that allows quality resolution imaging of living tissues using endogenous and exogenous fluorescence. Cellular interactions, typically seen only in cell culture models, may now be investigated in living animals. This is particularly relevant in the kidney, with its complex three-dimensional structure and heterogeneous cell population, including cells of the glomeruli, tubules, interstitium, and vasculature. The kidney is ideal for imaging by MPM due to its heterogeneous three-dimensional (3D) internal structure. The renal interstitium is an extravascular space between kidney tubules that contains interstitial fibroblasts, dendritic cells, and macrophages, which rapidly react to

Tim D. Hewitson et al. (eds.), *Kidney Research: Experimental Protocols*, Methods in Molecular Biology, vol. 1397,
DOI 10.1007/978-1-4939-3353-2_12, © Springer Science+Business Media New York 2016

stress and damage of the functioning nephrons. Multiphoton tomography allows for precise 3D reconstruction of the renal tissue architecture, substantially enhancing structural and functional studies in the kidney in health and disease. This chapter presents an overview of the principles of MPM in renal pathophysiology using rodent models, describes some of the disadvantages of MPM, and provides a protocol for analyzing tubular and metabolic alterations in the kidney using intravital MPM.

1.1 Principles and Advantages of MPM

MPM enables intravital imaging at subcellular resolution at greater depths and less phototoxicity compared to conventional single-photon confocal microscopy. In single-photon microscopy, a stream of photons at a desired wavelength is used to excite a sample and capture two- and three-dimensional fluorescent images. As a result, any tissue in the beam path is nonspecifically illuminated by the laser, therefore producing out of focus emission photons that need to be eliminated using a pinhole. MPM uses lasers with a high-repetition rate to deliver pulses of near-infrared (NIR) photons to a sample for the near-simultaneous absorption of two photons by a fluorophore; hence MPM is often referred to as two-photon excitation microscopy. The simultaneous absorption of two photons by a fluorophore is equivalent to excitation at twice the energy, half the wavelength, of the incident single photon (i.e., two-photon 740 nm excitation is equivalent to one-photon 370 nm excitation). The energy of either NIR photon is too low to raise an electron to an excited state, and it is the combination of the two energies that is sufficient to excite and thus release a fluorescence emission photon from a fluorophore. The use of lower energy NIR photons increases the imaging depth range (*see* **Note 1**), eliminates out of focus nonspecific background as the focal volume (<1 μm) is the exclusive space for the two-photon process, and minimizes phototoxicity. A more detailed description of the development of MPM and general information on two-photon fluorescence is readily available [1]. MPM imaging has now been used to study the structure and function in numerous systems [2–6] including the kidney [7–10].

To date, most biological imaging has relied on labeling with conventional fluorophores or fluorescent proteins, such as green fluorescent proteins (GFP). MPM enables more specific excitation of intrinsic endogenous molecules such as the reduced form of nicotinamide adenine dinucleotide (NADH) and nicotinamide adenine dinucleotide phosphate (NADPH), flavins, serotonin, as well as second harmonic generation (SHG) of collagens, skeletal muscle, and microtubules [11]. In the kidney, endogenous fluorescence of NAD(P)H by MPM is plentiful due to the high density of mitochondria in kidney tubular epithelial cells, of which NADH is a metabolic substrate [12]. Images captured without the infusion of fluorescent dyes can rely on this signal for basic information of tubular structure and metabolism. However, glomeruli appear as dark demarcated circles due to low NAD(P)H pools. As the fluorescence excitation, emission, and fluorescence lifetime of NADH and NAD(P)H overlap, these two molecules are conventionaly referred to as NAD(P)H [13]. The spectral properties of the oxidized and reduced forms (NAD+ and NADH), as well as flavin adenine dinucleotide (FAD and $FADH_2$), can indicate dynamic metabolic processes. Other MPM modalities such as fluorescence lifetime imaging microscopy (FLIM) can provide more detailed information about the physio-chemical state of a fluorophore, which can be used to further characterize the metabolic/redox state of the cell [2, 5, 14].

Simultaneous visualization of multiple sources of fluorescence using MPM is limited by optical filter selection and detector placement. Elastin and collagen are two sources of extracellular intrinsic fluorescence that can be visualized by two-photon excitation and SHG, respectively. This is important considering the extracellular disarrangement and expansion during chronic kidney pathologies. Detection of collagen by SHG is specific for the arrangement and fibrillar composition of collagen types. Types I and III collagen are fibrillar and therefore easily detectable within the kidney capsule. However, type IV collagen, primarily located in the tubulointerstitial region that expands or proliferates following injury, has low optical activity owing to its non-fibrillar composition.

MPM is, of course, not limited to the detection of endogenous fluorescence. Many different infusible fluorophores can be used to visualize structural, cellular, and subcellular processes, as described in Table 1 (*see* **Note 2**). Rhodamine conjugates, specific to mitochondria, can reveal in vivo alterations to mitochondrial health and therefore metabolism. These include tetramethyl rhodamine methyl ester (TMRM), rhodamine-123, and rhodamine B hexyl ester. TMRM is ideal for intravital quantification since its excitation/emission does not overlap with endogenous NAD(P)H. Fluorophores for nuclei (Hoechst 33342, propidium iodide), cytosol (calcein-acetyloxymethyl), and plasma membrane (octadecyl rhodamine-B chloride R-18) offer insight into common cellular pathological events such as cell viability, apoptosis, necrosis, endocytosis and autophagy. Biochemical changes within biological structures often precede pathological events, such as calcium regulation (rhod-2 AM), pH changes (seminaphthorhodafluor-1 AM), glutathione status (monochlorobimane), and reactive oxygen species (ROS) production (dihydroethidium, dihydrorhodamine). Conjugated fluorescent dextrans of varying sizes, used singly or in combination, can provide information regarding selective permeability and vascular changes. The listed fluorophores can be visualized and quantified.

Table 1
Examples of fluorophores used for multiphoton microscopy

Probe	Characteristic	Function	2-Photon ex range
Nuclear and cell structure			
Hoechst 33342	Labels cell nuclei blue	Nuclear change (apoptosis, mitosis)	780–820 nm
Propidium iodide	Labels cell nuclei red, membrane impermeable	Necrotic change of nuclei	820–850 nm
Calcein-acetoxymethyl (AM)	Labels cytosol of tubular and endothelial cells	Cell viability	800 nm
Quinacrine	Loads preferentially into brush border and subapical vesicles	Endo- and exocytosis, and apical shedding. Renin granules in the juxtaglomerular apparatus	700–720 nm
Mitochondria and metabolism			
Tetramethyl rhodamine methyl ester (TMRM)	Mitochondrial membrane-dependent dye	Mitochondrial density and health that loads well into proximal and distal tubules, and glomeruli	800–860 nm
Rhodamine-123	Mitochondrial membrane-dependent dye	Mitochondrial density and health that loads preferentially into the proximal tubule	780–860 nm
Rhodamine B hexyl ester	Endothelial cell mitochondrial membrane-dependant dye	Labels metabolically active endothelial cells	800–860 nm
Dihydroethidium	Superoxide sensitive dye that fluoresces red and binds to DNA	Indicates levels of reactive oxygen species production	700–720 nm
Monochlorobimane	Forms a fluorescent adduct in the presence of glutathione	Indicates levels of intracellular glutathione	700–720 nm
Calcium sensors			
Rhod-2 AM	Calcium sensitive dye that is generally localized to the mitochondria	Indicates levels of intracellular calcium	850–900 nm
Fluo-4 AM	Calcium sensitive dye	Fluorescence intensity increases on calcium binding	810 nm

(continued)

Table 1 (continued)

Probe	Characteristic	Function	2-Photon ex range
Vascular and permeability			
3000–4000-Da dextran-fluorophore conjugate	Freely filtered by the glomerulus	Indicates glomerular permeability and proximal tubule endocytosis	Dependent on conjugate fluorophore
70,00–500,000-Da dextran-fluorophore conjugate	Not filtered by the kidney, but retained in the vasculature	Indicates glomerular permeability, vascular flow, and vascular permeability	Dependent on conjugate fluorophore

1.2 FLIM and Metabolism

The fluorescence lifetime is the mean time a fluorophore remains in an excited state before emitting a photon (fluorescence) and returning to the initial ground state. The fluorescence lifetime of a fluorophore, either endogenous or synthetic, depends on the type of molecule, its conformation, and the way the molecule interacts with the surrounding microenvironment [15]. FLIM constructs a spatial distribution map of fluorescence lifetimes of a sample imaged by confocal, multiphoton, endoscopic, or wide-field microscopy. Importantly, FLIM can be used to measure molecular environmental parameters, and the metabolic state of cells and tissues via endogenous fluorescence. There are two major approaches used to measure the fluorescence lifetime of a sample: time and frequency domain. Typically, frequency-domain FLIM has superior signal-to-noise ratio for complex systems with multiple fluorophores and faster acquisition times, yet requires more complex analysis compared to time-correlated FLIM. Background and principles to these two different techniques are readily available [1].

NAD(P)H lifetime measurements are widely used for metabolic and redox imaging in vitro and in vivo [16, 17]. The fluorescence lifetime of NAD(P)H is resolved as a two-component system with the short (~0.3–4 ns) and long (~2.3 ns) lifetime represented as the free and protein-bound conformations, respectively [18]. NAD+ is not excited at 740 nm whereas NAD(P)H is. The ratio of NAD(P)H:NAD(P)+ has been used previously as a measure of the redox state of the cell. Bird and colleagues demonstrated that the free-to-bound ratio of NAD(P)H, represented by the ratio of the amplitude coefficients for the short and long lifetimes (i.e. α_1/α_2), is related to the NAD(P)H/NAD(P)+ ratio and can be used as an indicator for redox changes within the cell [17]. FAD is also examined routinely by FLIM for intracellular metabolic and redox analysis. While FAD can be found in a free and protein-bound conformation, the former has a significantly higher quantum yield with a lifetime of 2.91 ns. In contrast to NAD(P)H, only the oxidized form of FAD is fluorescent and is used to measure changes in the redox state of the cell in combination with NAD(P)H [19]. The fluorescence lifetime of NAD(P)H and FAD is a viable indicator of ischemia-reperfusion injury (IR) within the liver [2] and kidney [20] in vivo. The fluorescence lifetime changes associated with necrosis [14] and apoptosis [21] have also been described and can be used to measure tissue responses to drug treatments and disease states. To date, however, studies using FLIM analysis in the kidney are limited in number (*see* **Note 3**).

1.3 Characteristics of Metabolic Changes Detected by MPM

Metabolic alterations to the kidney are readily observable via MPM. Discerning the reason for these changes requires knowledge of some basic pathophysiological processes. The oxidized and reduced redox cofactor pairs FAD/$FADH_2$ and NAD+/NADH are becoming increasingly recognized as key metabolic indicators of the state of cellular metabolism associated with health and disease.

NADH and $FADH_2$ are formed during glycolysis, fatty acid oxidation, and the citric acid cycle, and are both energy-rich molecules due to the presence of an electron pair (Fig. 1). The kidney has minimal glycolytic capacity and relies primarily on oxidative phosphorylation to produce ATP, which occurs as a result of the transfer of electrons from NADH or $FADH_2$ to O_2 by a series of electron carriers located on the inner mitochondrial membrane. NADH is a substrate for complex I of the electron transport chain (ETC) which catalyzes the oxidation to NAD+ (Fig. 1a, b). Decreased O_2 supply causes reduced complex I activity and a failure to convert NADH to NAD+ (nonfluorescent). This "backing up" of NADH is observed by MPM (excitation ~740 nm) as increased fluorescence emission compared to basal conditions (Fig. 1c, d). Oxygen deprivation reduces ATP production and ATP-dependent processes, such as actin polymerization, causing plasma membrane disruption and alterations to ETC complexes. Oxygen re-presentation to damaged components of the ETC at reperfusion is thought to result in the production of ROS, mainly superoxide (O_2^-) (Fig. 1e, f).

Fluorescence lifetime changes of NAD(P)H varying depending on the type of insult and temporal influences, and usually requires further investigation into the protein interactions of NAD(P)H such as the ratio of free-to-bound NAD(P)H (α_1/α_2). An increased ratio demonstrates either an increase in free NAD(P)H or decrease in protein-bound NAD(P)H. Binding of NAD(P)H occurs primarily on complex I of the ETC. An increased ratio is consistent with mitochondrial dysfunction and increased ROS production. FAD cannot be excluded when detecting fluorescence emission at 740 nm excitation, and further information can be provided by fluorescence lifetime differences of FAD detected by FLIM. Decreased lifetime fluorescence detected by FLIM likely indicates decreased protein-bound FAD to the succinate dehydrogenase enzyme complex of complex II of the ETC and, in addition,

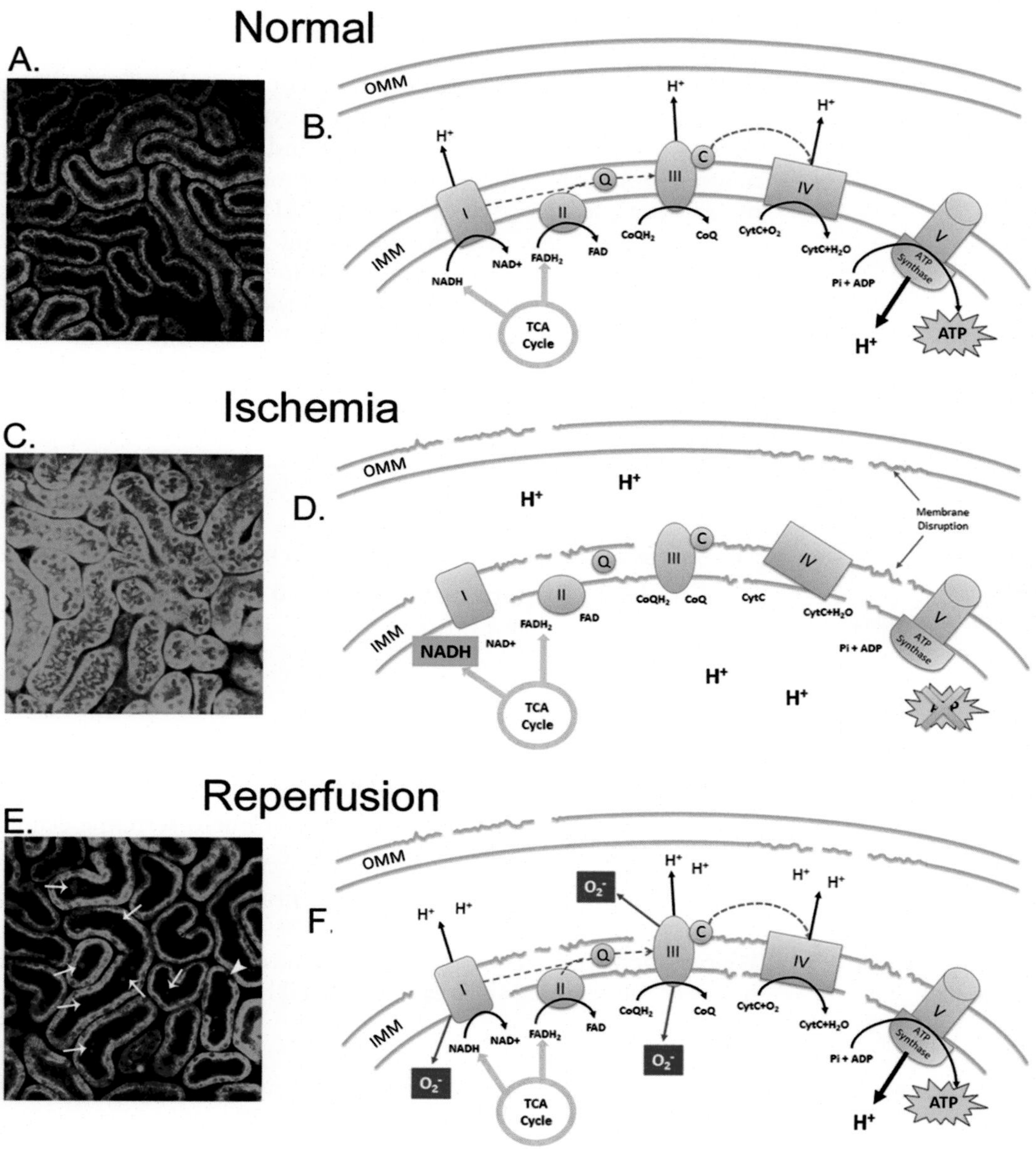

Fig. 1 Structural and metabolic alterations detected by MPM in the kidney following ischemia-reperfusion. In the normal healthy kidney cortex endogenous fluorescence at 740 nm detects NAD(P)H in the tubular epithelium indicated as *green* (**a**). The source of this fluorescence is primarily NADH which is a substrate produced by TCA cycle that acts as a proton donor for complex I of the ETC that contributes to oxidative phosphorylation and the production of ATP within mitochondria (**b**). $FADH_2$ is also a proton donor produced by TCA cycle that is converted to FAD by complex II. During ischemia, tubular swelling, reduction in interstitial space, and increased endogenous fluorescence are observed (**c**). This represents accumulation of NADH within mitochondria due to reduced electron transfer on the ETC, resulting in ATP depletion and mitochondrial membrane disruption due to reduced ATP-dependent actin polymerization (**d**). At reperfusion, tubular cell detachment (*arrow head*) and cast material formation (*arrows*) can be observed (**e**). The re-presentation of oxygen to damaged components of the ETC on the IMM results in increased O_2^- formation and mitochondrial swelling due to osmotic dysregulation (**f**). Abbreviations: *OMM* outer mitochondrial membrane, *IMM* inner mitochondrial membrane, *TCA cycle* the citric acid cycle, *NADH* nicotinamide adenine dinucleotide reduced, *NAD+* nicotinamide adenine dinucleotide oxidized, *FADH2* flavin adenine dinucleotide oxidized, *FAD* flavin adenine dinucleotide reduced, *ATP* adenosine triphosphate, *ADP* adenosine diphosphate, *Pi* inorganic phosphate, *Q* Ubiquinone, *C* cytochrome c, *CoQH2* Ubiquinol

decreased free FAD due to decreased complex II utilization and re-presentation of FAD to the citric acid cycle (Fig. 1). Both explanations underlie impaired metabolism and/or mitochondrial dysfunction. Alternately, increased FAD lifetime may indicate enhanced metabolism and increased turnover of $FADH_2$ to FAD.

The specific use of endogenous and infused fluorophores is now described, using as an example our investigation of changes in renal pathophysiology from IR injury to the kidney, and the role of mitochondrial dysfunction. Measurements of NAD(P)H and FAD were compared with other measurements of oxidative stress in progressive IR injury in the kidney.

2 Materials

2.1 Kidney Ischemia-Reperfusion Injury

1. Animals: Male 4–6 week old C57Bl6 mice.
2. Anesthetic: Mixture of ketamine (10 mg/mL) and xylazine combination (1 mg/mL).
3. Protective eye cream.
4. Small animal hair clippers.
5. Betadine antiseptic liquid.
6. Surgical instruments, sterilized.
7. Microvascular clamps (16 mm straight micro-serrefine clamps from Fine Science Tools Inc. Foster City, CA, USA).
8. 4/0 Suture (for a chronic recovery model).

2.2 Multiphoton Microscopy

We routinely worked closely with a technician specialized in MPM; however the required skills can be learned by scientists with dedication over time.

2.2.1 The MPM Room

1. The typical MPM room in any institute imaging facility is, of necessity, mostly filled with the large MPM and associated laser and computer equipment (Figs. 2 and 3). The room itself is cooled to cope with heat from the MPM and lasers, so it is necessary to carefully control any in vitro and in vivo experiments to physiological temperatures. For example, for our in vivo mouse experiments, animal body temperature is regulated with a heated pad that covered the animal on the MPM stage, and also served to weight the mouse and its kidney, and so minimize respiratory movement distorting live images (Fig. 4). A rectal probe can be used to monitor body temperature of the animals during the protracted MPM procedures.
2. The room also usually houses a Class 2 biohazard cabinet (Fig. 2) that can also be used for small animal surgery. However, the light from the biohazard cabinet interferes with the MPM

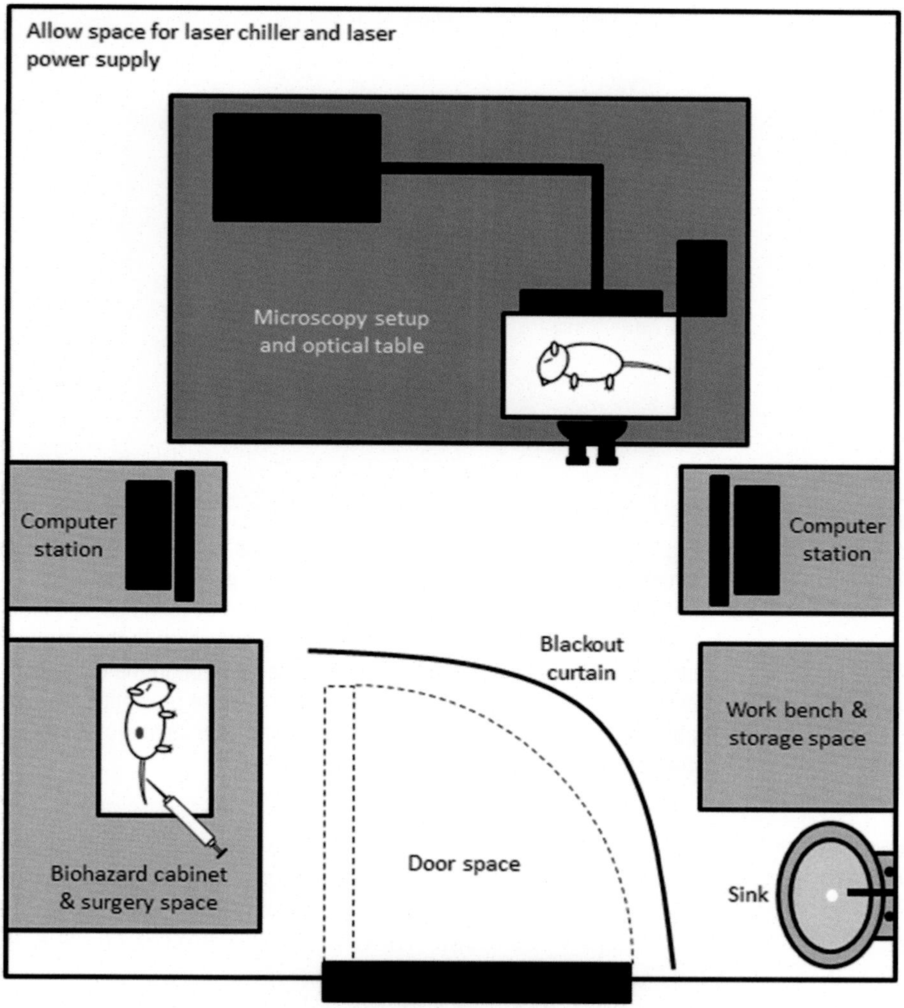

Fig. 2 Common equipment and floor plan for MPM facility. The equipment required for intravital MPM include the MPM microscope and associated computer station with automated operating software, pressurized suspension optical table, laser source, work bench, and Class 2 biohazard cabinet that can be used for small animal surgery. Interfering light needs to be considered, especially from the biohazard cabinet. The installation of a blackout curtain allows limited use of the biohazard cabinet, plus entry or exit from the room whilst the MPM is in use; however all room lights are turned off when image capture is occurring. Internal room temperature needs to be kept constant considering heat can be generated from the laser and an appropriate booster cooling system should be employed

image capture. A blackout curtain was installed to allow some use of the biohazard cabinet whilst using the MPM, but the cabinet lights must be turned off during image capture. The curtain also allows some entry or exit from the room whilst the MPM is in use.

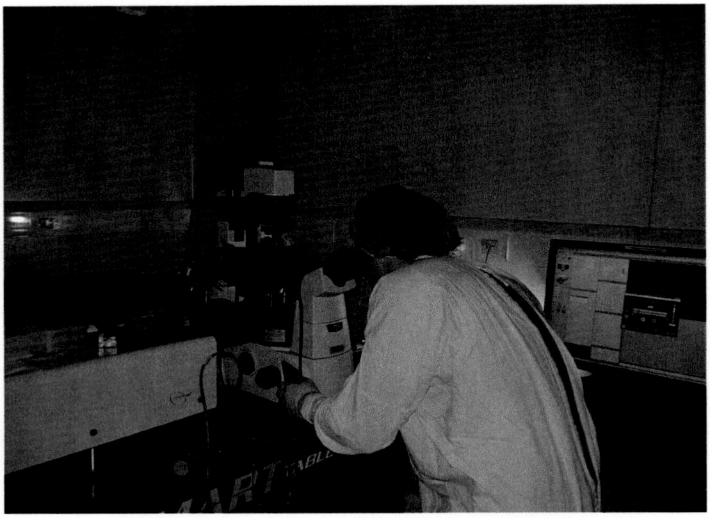

Fig. 3 Practicalities of intravital MPM operation. Kidney MPM involves a coordinated approach, especially when imaging ischemia-reperfusion injury. An approximate field of view and imaging depth can be found using a mercury laser source and the microscope eyepiece, before switching to the titanium:sapphire laser source and engaging the PMT to observe through the computer operating system. It is ideal if these two systems are in close proximity. Always apply the appropriate health and safety requirements for your institution; in this case, a lab coat, hair cover, and gloves

2.2.2 Basic Microscope for Image Capture

1. Multiphoton microscope: There are many specialized multiphoton microscopes with different lasers, objectives, and selection of direct or inverted microscopy. We use a Nikon Ti Eclipse—LaVision MPM (LaVision BioTec GmbH, Bielefeld, Germany) equipped with an ultrashort (85 fs pulse width) pulsed mode-locked 80 MHz Titanium:Sapphire MaiTai™ laser (Spectra Physics, Mountain View, CA, USA) and ImSpector Pro software. The MPM has a motorized xyz stage and is therefore capable of producing large xy and z montages. The choice of excitation wavelength is dictated by the probes and/or endogenous fluorescence being used.
2. Microscope objectives: MPM of the kidney places unique requirements on the microscope objective. Objective lenses may be designed for use with a coverslip, or in the case of "dipping objectives," designed for use without a coverslip. The microscope objective was an Olympus XLUMPFL 20XW/0.95 IR water immersion dipping lens, magnification 20×, numerical aperture 0.95, and working distance 2.0 mm. The normal curvature of the kidney may limit the region that can be imaged to a small region at the apex of this curvature. A much larger region of the kidney may be imaged when the surface is gently depressed on a coverslip.

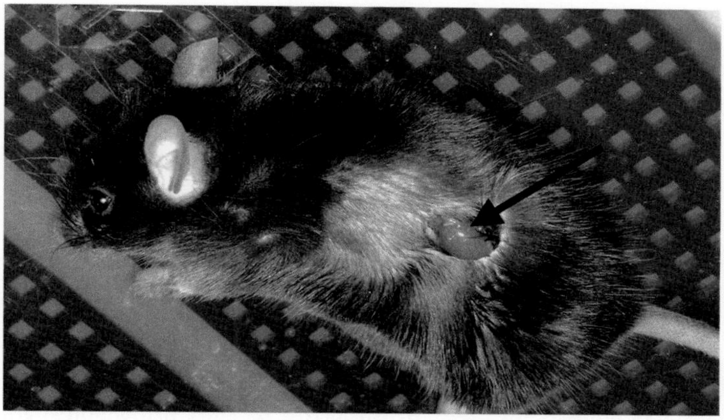

Fig. 4 Externalized mouse kidney for intravital MPM. Surgical preparation requires the externalization of the kidney whilst maintaining appropriate temperature controls. Following a left flank skin incision, the external and internal muscle layers are cut to reveal the kidney (*arrow*). A small animal heat pad can be seen beneath the mouse. The animal is then inverted and the exposed kidney placed over the MPM objective for imaging of the kidney

3. Bandpass filters: these are placed in front of the photomultiplier tubes to selectively collect emitted light into three seperate channels. We use 447–460 nm (blue), 485–550 nm (green), 575–630 nm (red).
4. Imaris ×64 7.6.0 software: To adjust the minimum and maximum exposure settings consistently throughout all acquired images for standardized quantification of fluorescence intensity.
5. ImageJ software: available as shareware (downloadable from imagej.nih.gov/ij/).

2.3 Intravital Assessment of Mitochondrial Function

1. Tetramethyl rhodamine methyl ester (TMRM) in 0.9 % saline at a concentration of 5 μg/mL.

2.4 Fluorescence Lifetime Imaging (FLIM)

1. Razor blade to section a 1 mm coronal section from the harvested kidney.
2. Glass microscope slide.
3. Glass coverslip.
4. Upright DermaInspect multiphoton microscope (JenLab GmbH, Jena, Germany) equipped with an ultrashort (85 fs pulse width) pulsed mode-locked 80 MHz Titanium:Sapphire MaiTai laser (Spectra Physics).
5. FLIM data is collected with a time-correlated single-photon counting (TCSPC) SPC-830 detector (Becker & Hickl GmbH, Berlin, Germany) integrated into the DermaInspect system.

6. Fluorescence emission is spectrally resolved between three linearly arranged photon counters through the use of three dichroic filters: 350–450 nm (NAD[P]H), 450–515 nm (NAD[P]H and FAD), and 515–610 nm (FAD).
7. SPCImage™ version 4.8 (Becker & Hickl GmbH) to analyze photon lifetime.

3 Methods

3.1 Kidney Ischemia-Reperfusion Injury

1. All animal experimentation must be carried out after appropriate ethical clearances from your institution. Our surgical procedure using MPM was for nonsurvival surgery (i.e., image acquisition is the terminal part of the experiment). Non-recovery is commonly used, because the MPM techniques can take a long period of time, during which the animal must remain anesthetized, making the procedure in itself quite stressful. Appropriate steps would need to be taken to insure rigorous sterile technique if animal recovery following image acquisition was planned. Appropriate controls have to be used for all procedures.
2. Anesthetize male 4–6 week old C57Bl6 mice with a ketamine/xylazine combination administered via intraperitoneal injection at a dose of 10 μL/g body weight.
3. In a dedicated surgery area outside of the MPM room, but close by in the imaging facility, perform bilateral kidney IR. Perform the surgery with the anesthetized animal on a 37 °C warmed operating platform.
4. Apply protective eye cream liberally to the eyes of the anesthetized animal.
5. Shave flank skin on both sides of the body and clean with Betadine antiseptic liquid.
6. Surgically expose the left kidney using a flank incision. Three layers must be incised to reach the peritoneal cavity to externalize the kidney: skin, thin outer muscle layer, and thick inner muscle layer. The incision of the inner muscle layer should be smaller than the length of the kidney, so that externalization of the kidney through the incision stabilizes the kidney without stretching the renal pedicle and reducing renal perfusion.
7. Clear the renal artery of fat before a nontraumatic vascular microclamp is placed over the vessels for 20 min.
8. During this 20-min period, the left kidney is placed within the body cavity, the mouse turned to its right side, and the procedure repeated.
9. Remove the microclamps following 20-min vessel occlusion of respective sides, and check reperfusion by noting recovery of normal color of the kidney.

10. For chronic injury to the kidney (21-days post-IR), suture the muscle and skin incisions using sterile protocols and 4-0 silk suture, and allow the animals to recover.

3.2 MPM of Kidney

1. For non-recovery intravital MPM during ischemic injury or immediately post-IR, place the anesthetized mouse on the MPM stage with the kidney for imaging just externalized through the flank incision (Fig. 4) and place over the objective bathed in sterile distilled water.
2. Place a heated jacket (38 °C) over the mouse to maintain normal body temperature. A rectal temperature probe can be used to monitor body temperature. In our experience, the heated jacket routinely maintained the animal's normal body temperature adequately.
3. Focus the objective into a field of view that visualizes detailed tubular structures and acquire serial (30–50) penetrating slice images from the renal cortex to a depth of 50–70 μm. Image capture scan speed can be set between 12.5 and 20 μs/pixel.
4. Acquire ischemic kidney images over a 20-min period from when the clamp is attached (for imaging during ischemia). To perform intravital imaging during initial reperfusion, place the animal in the biohazard cabinet and release the microclamp, then quickly replace the mouse onto the MPM stage for IR imaging. Acquire images over an approximately 40-min period from when the microclamp was removed.
5. For intravital MPM of kidneys with chronic injury, allow IR-treated mice to recover for 21 days, anesthetize the mouse, and perform the same MPM procedure used as is described above.
6. Image left and right kidneys from healthy control mice by the same procedure, except for clamping the kidneys.
7. Place the objective in a field of view that visualizes detailed tubular structures and acquire serial (30–50) penetrating slice images from the renal cortex to a depth of 50–70 μm.
8. Capture images using the Nikon Ti Eclipse—LaVision MPM described previously. Set the excitation wavelength to 740 nm for NAD(P)H fluorescence and 900 nm for collagen. Emitted light can be collected using bandpass filters: 447–460 nm (blue), 485–550 nm (green), 575–630 nm (red).
9. Obtain a minimum of three fields of view per kidney. Imaris ×64 7.6.0 is used to adjust the minimum and maximum exposure settings consistently throughout all acquired images for standardized quantification of fluorescence intensity.
10. Use ImageJ to obtain mean fluorescent intensity values for NAD(P)H (green channel). For each of the images captured per kidney, select the epithelial layer of continuous tubules and measure for mean intensity. Measure eight tubules per *z*-slice image from five selected *z*-slice images per field.

3.3 Intravital Assessment of Mitochondrial Function

1. Prepare mice as described previously (*see* Subheading 3.1).
2. Prepare TMRM fresh as a 5 μg/mL solution prepared in 0.9 % saline or PBS, immediately prior to surgery, and stored protected from light (*see* **Note 4**).
3. Prior to TMRM infusion, capture images at 740 and 800 nm to account for background emission.
4. Infuse 20 μL TMRM intravenously (tail vein) and immediately visualize via intravital MPM to determine mitochondrial health in control untreated animals and 21-days post-IR injury.
5. Infuse TMRM into the tail vein of the anesthetized mouse using a 30G needle while the mouse remains in situ on the MPM stage. This volume and concentration result in 0.1 μg/mL plasma concentration of TMRM in a 30 g mouse.
6. Capture images at 740 nm (NAD(P)H) (Fig. 1) and 800 nm (TMRM) excitation (Fig. 5) to gain a matched signal of the same region of interest. Acquire *Z*-stack images shortly after infusion at both 740 and 800 nm excitation in at least three separate fields per kidney.

3.4 Fluorescence Lifetime Imaging (FLIM)

1. As well as 2-photon MPM, FLIM can also be performed. For our work, this was carried out on mouse kidneys post-sacrifice for healthy control and for 21-day IR mice.
2. Collect kidneys following CO_2 euthanasia of mice (usually after MPM imaging) and place on ice.
3. Cut a 1 mm thick coronal section through the kidney, place on a glass slide, and overlay with a glass coverslip. Time from death to first image using FLIM capture should be no more than 30 min, but longer times have been published as satisfactory for FLIM measurements (*see* **Note 3**).

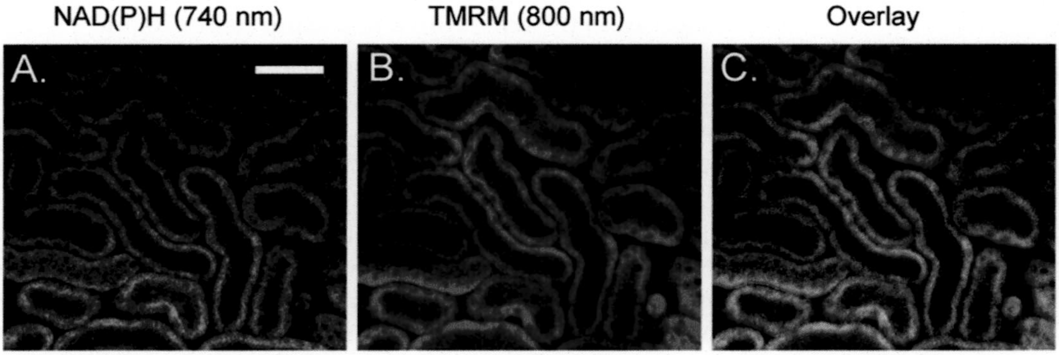

Fig. 5 In vivo mitochondrial assessment with TMRM in healthy kidney. (**a**) Endogenous NAD(P)H fluorescence detected at 740 nm. (**b**) TMRM fluorescence detected at 800 nm. (**c**) Composite image showing co-localization of NAD(P)H and TMRM within tubular epithelial cells

4. Coronal sectioning allows the visualization of the medulla. Acquire three areas each from both cortex and medulla.
5. The remaining kidney tissue was used for histology or for protein extraction and Western blot protein analysis.

3.5 Other Analyses after MPM: Histology and Western Immunoblots, and Kidney Function

Following intravital MPM, segments of kidneys can be collected for later histology and immunohistochemistry, or frozen and stored at −80 °C for protein extraction and analysis. Suitably timed collections of urine and plasma can be used to measure corresponding renal function (*see* **Note 5**).

4 Notes

1. One of the major limitations of MPM is the maximal excitation depth that can be achieved. Use of NIR wavelengths instead of ultraviolet light for fluorescence excitation has allowed reduced scattering and increased imaging penetration depth in a given sample from approximately 20 to 600 μm [22, 23]. However, in the kidney, substantial scattering of emission photons occurs because of the high refractive index and heterogeneity of kidney tissue, and the outcome is an excitation depth limited to 100–200 μm in the kidney. Attaining clear signal at 200 μm should be possible but is unusual. There is signal attenuation simply resulting from the curved shape of the kidney, with spherical aberration resulting from refractive index mismatch in the rounded structure. In addition, cytochrome c oxidase and hemoglobin, both abundant in the kidney, absorb NIR light and contribute to a reduction in the maximal excitation depth [24, 25].

   The limitation in maximal excitation depth is important in the kidney since glomerular depth from the renal capsule is routinely greater than 100 μm, particularly in rodents. In our research using C56Bl6 mice with acute and chronic IR injury in the kidney [26], it is not until there is marked tubulointerstitial atrophy in progressive chronic kidney disease that the glomeruli become situated closer to the renal capsule (Fig. 6). Thus, imaging of mouse kidney structure is usually restricted to tubules and vessels. Some examples of rodent models where glomeruli are described are: Munich-Wistar rats which possess superficial glomeruli that allow imaging of the afferent and efferent capillaries and Bowman's space with MPM, as well as the contiguous S1 segment of the proximal tubule [27]; investigations of podocyte biology in female nephrin knockout/GFP knock-in heterozygote ($Nps1^{tm1Rkl}/J$) mice using superficially located glomeruli [28]; close-to-surface glomeruli of C57Bl6 mice have been imaged using MPM in C57Bl6 mice at 3–4 weeks of age and weighing approximately 12 g, where kid-

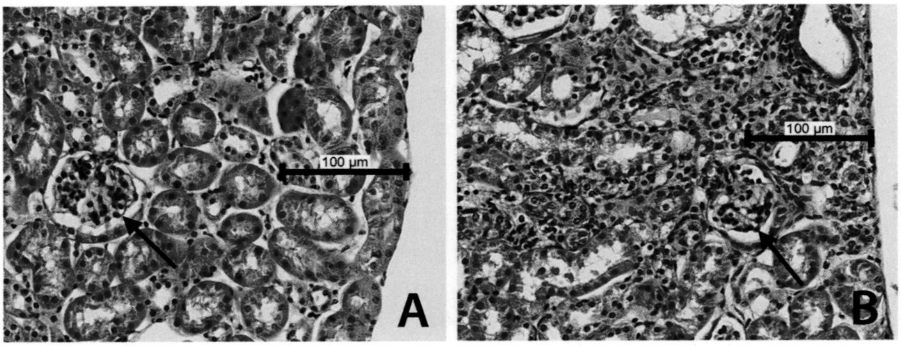

Fig. 6 Superficial glomeruli of atrophic kidney cortex. (**a**) Hematoxylin and eosin (H&E) staining of the normal healthy kidney depicts superficial glomeruli (*arrow*) that lie in the cortex at approximately 100 μm below the surface. (**b**) This kidney has undergone ischemia and long-term reperfusion (21 days). The superficial glomeruli (*example arrowed*) are now <100 μm below the kidney capsule because of considerable tubular and interstitial atrophy. The kidney capsule is on the *right of both photos*

neys are small [29]; and using the unilateral ureteric ligation model and visualizing the rim of the greatly dilated hydronephrotic kidney at 12 weeks [30]. A quantitative analysis of glomerular depths from 10 commonly used mouse strains reported that BALB/c and C57Bl6 mouse strains contained the most superficial glomeruli at both 4 and 10 weeks of age, regardless of sex [31].

2. Another important consideration of intravital MPM is the possibility of photodamage and cytotoxicity due to extended periods of excitation, and adequate controls need to be included. An important consideration when using multiple fluorophores is that they do not overlap within the excitation and emission ranges, including that of NAD(P)H.

3. FLIM techniques have been well characterized in ex vivo tissue, primarily skin, which maintains NAD(P)H endogenous fluorescence when kept at 4 °C for 5 days despite the possibility of post-death changes, such as autolysis [14]. The kidney is likely to have a shorter half-life of endogenous fluorescence given the high reliance on aerobic metabolism, although there are no previous studies that have investigated this. While the free-to-bound ratio of NAD(P)H can be used to measure redox changes within the cells, this ratio should not be confused with the widely used "redox ratio" that is calculated by measuring the ratio of FAD:NADH intensity.

4. TMRM is a cell-permeant, cationic fluorophore that is readily sequestered by active mitochondria with a polarized membrane potential. TMRM is ideal for MPM since the excitation wavelength does not overlap with NAD(P)H endogenous fluorescence excitation. This commonly occurs with other mito-

chondrial membrane potential-dependant fluorescent probes [8]. We advise that the help of one of the expert Biological Resource Facility staff from your institute is sought for tail vein injection of the TMRM because of the value of the animal at this stage for your research purposes.

5. For kidney function, blood and urine volumes are limited from mice. A 24 h urine collection can be made in the time prior to the imaging procedure, if metabolic cages are available. The urine can then be used to assess urinary creatinine (paired with plasma creatinine for creatine clearance) and protein. For plasma collections, after the imaging procedure, the mice are heparinized and euthanized. Then, blood is removed through the upper vena cava using a 23G needle. Approximately 200 μL can be collected easily. If the metabolic cages are not available for urine collection, any urine in the bladder can be collected after euthanasia and spotted onto a urine dipstick to test for proteinuria.

References

1. Small DM, Sanchez WY, Roy S et al (2014) Multiphoton fluorescence microscopy of the live kidney in health and disease. J Biomed Opt 19:020901
2. Thorling CA, Liu X, Burczynski FJ et al (2013) Intravital multiphoton microscopy can model uptake and excretion of fluorescein in hepatic ischemia-reperfusion injury. J Biomed Opt 18:101306
3. Yuryev M, Khiroug L (2012) Dynamic longitudinal investigation of individual nerve endings in the skin of anesthetized mice using in vivo two-photon microscopy. J Biomed Opt 17:046007
4. Li JL, Goh CC, Keeble JL et al (2012) Intravital multiphoton imaging of immune responses in the mouse ear skin. Nat Protoc 7:221–234
5. Sanchez WY, Obispo C, Ryan E et al (2013) Changes in the redox state and endogenous fluorescence of in vivo human skin due to intrinsic and photo-aging, measured by multiphoton tomography with fluorescence lifetime imaging. J Biomed Opt 18:061217
6. Honda M, Takeichi T, Asonuma K et al (2013) Intravital imaging of neutrophil recruitment in hepatic ischemia-reperfusion injury in mice. Transplantation 95:551–558
7. Sutton TA, Fisher CJ, Molitoris BA (2002) Microvascular endothelial injury and dysfunction during ischemic acute renal failure. Kidney Int 62:1539–1549
8. Hall AM, Rhodes GJ, Sandoval RM et al (2013) In vivo multiphoton imaging of mitochondrial structure and function during acute kidney injury. Kidney Int 83:72–83
9. Peti-Peterdi J, Sipos A (2010) A high-powered view of the filtration barrier. J Am Soc Nephrol 21:1835–1841
10. Kang JJ, Toma I, Sipos A et al (2008) The collecting duct is the major source of prorenin in diabetes. Hypertension 51:1597–1604
11. Zipfel WR, Williams RM, Christie R et al (2003) Live tissue intrinsic emission microscopy using multiphoton-excited native fluorescence and second harmonic generation. Proc Natl Acad Sci U S A 100:7075–7080
12. Hall AM, Unwin RJ, Parker N et al (2009) Multiphoton imaging reveals differences in mitochondrial function between nephron segments. J Am Soc Nephrol 20:1293–1302
13. Huang S, Heikal AA, Webb WW (2002) Two-photon fluorescence spectroscopy and microscopy of NAD(P)H and flavoprotein. Biophys J 82:2811–2825
14. Sanchez WY, Prow TW, Sanchez WH et al (2010) Analysis of the metabolic deterioration of ex vivo skin from ischemic necrosis through the imaging of intracellular NAD(P)H by multiphoton tomography and fluorescence lifetime imaging microscopy. J Biomed Opt 15:046008
15. Becker W (2012) Fluorescence lifetime imaging--techniques and applications. J Microsc 247:119–136
16. Skala MC, Riching KM, Gendron-Fitzpatrick A et al (2007) In vivo multiphoton microscopy of NADH and FAD redox states, fluorescence lifetimes, and cellular morphology in precancerous epithelia. Proc Natl Acad Sci U S A 104:19494–19499

17. Bird DK, Yan L, Vrotsos KM et al (2005) Metabolic mapping of MCF10A human breast cells via multiphoton fluorescence lifetime imaging of the coenzyme NADH. Cancer Res 65:8766–8773
18. Lakowicz JR, Szmacinski H, Nowaczyk K et al (1992) Fluorescence lifetime imaging of free and protein-bound NADH. Proc Natl Acad Sci U S A 89:1271–1275
19. Chance B, Schoener B, Oshino R et al (1979) Oxidation-reduction ratio studies of mitochondria in freeze-trapped samples. NADH and flavoprotein fluorescence signals. J Biol Chem 254:4764–4771
20. Abulrob A, Brunette E, Slinn J et al (2007) In vivo time domain optical imaging of renal ischemia-reperfusion injury: discrimination based on fluorescence lifetime. Mol Imaging 6:304–314
21. Yu JS, Guo HW, Wang CH et al (2011) Increase of reduced nicotinamide adenine dinucleotide fluorescence lifetime precedes mitochondrial dysfunction in staurosporine-induced apoptosis of HeLa cells. J Biomed Opt 16:036008
22. Helmchen F, Svoboda K, Denk W et al (1999) In vivo dendritic calcium dynamics in deep-layer cortical pyramidal neurons. Nat Neurosci 2:989–996
23. Kleinfeld D, Mitra PP, Helmchen F et al (1998) Fluctuations and stimulus-induced changes in blood flow observed in individual capillaries in layers 2 through 4 of rat neocortex. Proc Natl Acad Sci U S A 95: 15741–15746
24. Young PA, Clendenon SG, Byars JM et al (2011) The effects of spherical aberration on multiphoton fluorescence excitation microscopy. J Microsc 242:157–165
25. Young PA, Clendenon SG, Byars JM et al (2011) The effects of refractive index heterogeneity within kidney tissue on multiphoton fluorescence excitation microscopy. J Microsc 242:148–156
26. Small DM, Sanchez WH, Roy SF et al (2014) Progression to chronic kidney disease after acute kidney injury involves prolonged oxidative stress and persistent tubular metabolic alterations. Nephrology (Carlton) 19:1–104
27. Dunn KW, Sutton TA, Sandoval RM (2012) Live-animal imaging of renal function by multiphoton microscopy. Curr Protoc Cytom Chapter 14:Unit12 19
28. Khoury CC, Khayat MF, Yeo TK et al (2012) Visualizing the mouse podocyte with multiphoton microscopy. Biochem Biophys Res Commun 427:525–530
29. Devi S, Li A, Westhorpe CL et al (2013) Multiphoton imaging reveals a new leukocyte recruitment paradigm in the glomerulus. Nat Med 19:107–112
30. Steinhausen M, Snoei H, Parekh N et al (1983) Hydronephrosis: a new method to visualize vas afferens, efferens, and glomerular network. Kidney Int 23:794–806
31. Schiessl IM, Bardehle S, Castrop H (2013) Superficial nephrons in BALB/c and C57BL/6 mice facilitate in vivo multiphoton microscopy of the kidney. PLoS One 8:e52499

Part IV

Vascular Mineralization

Chapter 13

Vascular Calcification in Uremia: New-Age Concepts about an Old-Age Problem

Edward R. Smith

Abstract

A hallmark of aging, and major contributor to the increased prevalence of cardiovascular disease in patients with chronic kidney disease (CKD), is the progressive structural and functional deterioration of the arteries and concomitant accrual of mineral. Vascular calcification (VC) was long viewed as a degenerative age-related pathology that resulted from the passive deposition of mineral in the extracellular matrix; however, since the discovery of "bone-related" protein expression in calcified atherosclerotic plaques over 20 years ago, a plethora of studies have evoked the now widely accepted view that VC is a highly regulated and principally cell-mediated phenomenon that recapitulates many features of physiologic ossification. Central to this theory are changes in vascular smooth muscle cell (VSMC) phenotype and viability, thought to be driven by chronic exposure to a number of dystrophic stimuli characteristics of the uremic state. Here, dedifferentiated synthetic VSMCs are seen to spawn calcifying matrix vesicles that actively seed mineralization of the arterial matrix. This review provides an overview of the major epidemiological, histological, and molecular aspects of VC in the context of CKD, and a counterpoint to the prevailing paradigm that emphasizes the primacy of VSMC-mediated mechanisms. Particular focus is given to the import of protein and small molecule inhibitors in regulating physiologic and pathological mineralization and the emerging role of mineral nanoparticles and their interplay with proinflammatory processes.

Key words Vascular calcification, Aging, Chronic kidney disease, Osteochondrocytic transdifferentiation, Mineral nanoparticles, Inflammation

1 Chronic Kidney Disease: A Paradigm for Premature Vascular Aging

Age-associated arterial remodeling appears accelerated in patients with chronic kidney disease (CKD), who have a vastly elevated cardiovascular (CV) risk compared to the general population [1]. CV risk rises early in the progression of CKD and incrementally with the decline in glomerular filtration rate. Patients with end-stage renal disease (ESRD) initiating dialysis therapy demonstrate an age-normalized cardiovascular mortality risk that is approximately 15–30-fold higher than the general population [2]. Strikingly, even young patients on dialysis, in the absence of traditional Framingham (atherosclerotic) CV risk

Tim D. Hewitson et al. (eds.), *Kidney Research: Experimental Protocols*, Methods in Molecular Biology, vol. 1397,
DOI 10.1007/978-1-4939-3353-2_13, © Springer Science+Business Media New York 2016

factors, are at extremely high risk with CV mortality rates equivalent to octogenarians without renal failure [3].

Phenotypically, patients with ESRD often exhibit many features of normal aging: osteoporosis, insulin resistance, sarcopenia, hypogonadism, skin atrophy, impaired cognitive function, poor wound healing, and susceptibility to infection. ESRD could therefore be considered a progeroid syndrome [4]. Indeed, chronic exposure to increased oxidative stress, inflammation, and phosphate excess, all characteristic of the uremic environment, are mechanistically implicated in promoting premature cellular senescence, a state of irreversible growth arrest and fundamental phenomenon in biologic aging [5]. A hallmark of aging, and major contributor to the increase in incidence and prevalence of CV disease in patients with CKD, is the progressive structural and functional deterioration of the central arteries and concomitant accrual of insoluble mineral [6].

2 Vascular Calcification: A Hallmark of Uremic Arteriolopathy

"In extreme old age, the arteries themselves, the grand instrument of the circulation, by continual apposition of earth, become hard, as it were bony, till, having lost the power of contracting themselves they can no longer propel the blood, even through the largest channels, in consequence of which death naturally ensues"

Sermon 57-The Fall of Man

John Wesley 1703–1791

The link between vascular calcification and aging is now well established. Calcification burden has been consistently associated with age in multiple observational cohort studies of individuals with asymptomatic and established vascular disease [7]. Indeed, some authors suggest that the vascular calcification scores could be used as an index of biological age [8]. This is underscored by recent findings from the HORUS study, which examined the extent of VC in mummies across 4000 years of history, from 2000 b.c. up to the present era. Here, the number of vascular beds with evidence of vascular calcification on CT imaging was found to correlate strongly with age at the time of death [9]. Thus, far from being a recent phenomenon, calcification of the vasculature would appear to be an age-old problem of old age.

As first described by Virchow over 150 years ago, mineral deposition may occur within the intimal (*tunica intima*) or medial (*tunica media*) layer of the arteries [10]. Intimal arterial calcification is typically focal and associated with atheromatous lipid-rich flow-limiting lesions, often involving inflammatory infiltrate and generally affecting the medium-sized conduit arteries, particularly near branch points [11]. In contrast, medial arterial calcification (MAC), as first systematically studied by Mönckeberg (also known as Mönckeberg's medial sclerosis) [12], is characterized by a predominantly diffuse granular pattern of mineral deposition or, more infrequently, a plate-like circumferential distribution at some abdominal and peripheral sites [13, 14].

MAC generally occurs in the absence of immune cell infiltration or lipid deposition. For many decades MAC was considered to be a passive age-related degenerative process [12, 15] and an incidental radiological finding of little clinical significance [16]. Originally described in the context of small-to-medium sized muscular arteries of the lower limbs and extremities, MAC was subsequently observed at multiple sites throughout the arterial tree involving the large elastic vessels (aorta) and various visceral and transitional arteries including iliac, coronary, renal, temporal, and uterine, as well as vessels in which intimal (atherosclerotic) calcification is rarely seen, e.g., mammary, epigastric, and radial vessels [17]. MAC is highly prevalent (>35 %) in patients with ESRD and in those with diabetes [7].

Four stages of MAC lesion progression have been described on the basis of histological appearance [18]. In stage 1 there is evidence of fine granulations (granular media calcinosis) within VSMC (i.e., intracellular), and, more prominently, deposited in the extracellular matrix, associated with damaged or fractured elastin fibers. Stage 1 MAC is a fairly ubiquitous histological finding in aortic vessels from the elderly [13, 14]. Expansion and confluence of these granulations, often alongside the internal elastic lamina, leads to the formation of thickened solid plates of mineral that distort the medial architecture and may protrude into the intimal layer. In the absence of atherosclerotic disease, subendothelial hyperplasia is often evident (stage 2). Complete circumferential banding of mineral demarcates stage 3, while in stage 4, true bony trabeculae are observed (also known as *osseous metaplasia*) but this is and infrequent finding and rarely found to affect the large central elastic arteries.

As in bone, compositional analyses of calcified arterial lesions obtained at autopsy from patients with ESRD and in those with coronary artery disease, but normal renal function, have consistently identified hydroxyapatite (or more accurately a nonstoichiometric calcium-deficient carbonate-substituted apatite (general formula $(Ca_{10-x} (HPO_4)x (PO_4)_{6-x} (OH)_{2-x} (0<x<1))$ as the major mineral phase in affected vessels [19]. Amorphous calcium phosphate (ACP) and magnesium whitlockite $[(Ca,Mg)_3(PO_4)_2]$, in combination with hydroxyapatite, have also been variably detected [19, 20]. The absence of these other non-apatitic mineral phases in some reports (in particular, magnesium-substituted ACP) may reflect differences in tissue processing which can result in dissolution and loss of some minerals or the artifactual agglomeration of disperse structures. On detailed ultrastructural analysis, physiological and pathological calcifications do seem to differ, with some loss of hierarchical order at ectopic sites due to reduced templating by organic matrix. Pathological crystallizations generally have a planar morphology arranged in spherulitic agglomerates (calcospherites) with a crystallite size of 2–10 nm, somewhat smaller than physiologic depositions [19].

The functional implications of mineral deposition at these two sites appear largely distinct. Intimal calcification appears to relate to the overall burden of atherosclerotic disease [21] and lesion severity (stage IV-VI; according to the American Heart Association classification system). Patients with CKD tend to have more severe atherosclerotic disease than age-matched individuals without CKD, but this largely reflects qualitative differences in plaque composition (i.e., greater calcification) than greater lesion number or volume [22]. Controversially, intimal calcification may also have direct effects on plaque stability depending on the mural tension between calcified and noncalcified regions [23, 24]. In particular, regions of spotty calcifications have been predicted to have increased stress and thus more prone to rupture [25]. MAC, on the other hand, is primarily linked to alterations in vessel compliance, which may lead to changes in cardiac perfusion and afterload eventually resulting in left ventricular hypertrophy (LVH) and heart failure [26]. Loss of cushioning function and increased pulsatility may also impact on microcirculation autoregulation and perfusion [27]. However, these distinctions are probably something of an oversimplification as atherosclerotic and arteriosclerotic disease processes, particularly in adult CKD, often overlap and co-exist within the same arterial segment. For instance, MAC may potentiate atherosclerotic disease progression by restricting compensatory arterial remodeling as well as promoting thromboembolic events as a result of intimal invasion by medial VSMC [18, 28]. Thus, arteriosclerosis and atherosclerosis should not necessarily be regarded as clinically divergent pathological phenomena.

MAC is a dominant vascular pathology in patients with CKD and, like aortic stiffening, is often more severe and evident at a younger age than in the general population [29]. Intimal disease is seldom observed in young patients with CKD, but frequently co-exists with medial calcification in adult CKD, presumably due to the influence of traditional atherosclerotic risk factors such as hypertension, dyslipidemia, diabetes, adiposity, smoking, and advancing age [29]. Yet, although adult CKD patients often display a constellation of these traditional atherosclerotic CV risk factors, they fail to completely explain the grossly increased mortality rate in this population [30]. Moreover, interventions aimed at addressing some of these traditional risk factors, particularly in patients on dialysis, have yielded disappointing results [31].

While there is some evidence that the uremic state may aggravate pre-existing atherosclerotic disease (i.e., increased plaque calcification), especially in diabetic patients [32], there is little evidence of de novo induction or multiplier effects [33]. Furthermore, in some studies, the incidence of atherosclerotic disease and related CV events in matched CKD and non-CKD cohorts appears ostensibly similar [34]. However, there is consistent data to suggest that CV events are more likely to be fatal in patients with CKD than in those without CKD [35]. Thus, overall, it would seem that the principal uremia-specific effects on the cardiovascular system are mediated by the induction and potentiation of arteriosclerotic disease and ventricular dysfunction. Consistent with this hypothesis, a number of studies have now shown that patients with ESRD are more likely to die from congestive heart failure due to LVH, ventricular dysfunction, sudden death, or arrhythmia than from atherosclerotic coronary artery disease [36]. Nonetheless, uremia-induced ventricular hypertrophy and dysfunction may also predispose CKD patients to ischaemic cardiac damage due to reduced coronary reserve and capillary density [22, 37].

It should also be acknowledged that contemporary imaging techniques might not be sufficiently sensitive to reliably distinguish calcification of intimal and medial layers, particularly in smaller vessels [38]. Indeed, quantitative analyses of calcium load in arterial biopsy specimens from some patients with CKD show evidence of significant calcium accumulation, despite relatively normal calcification scores on noninvasive imaging [39]. Thus imaging techniques lack a degree of sensitivity and specificity.

Apart from uremia, the major determinants of the prevalence and severity of MAC are age and diabetic status [7]. In dialysis patients, the cumulative exposure to calcium and phosphate, as well as time on therapy, is strongly correlated with calcification scores on imaging [40]. Significantly, time-averaged serum calcium and phosphate concentrations have been correlated with actual calcium load in vessels obtained from prevalent dialysis patients at the time of transplantation and from predialysis patients about to initiate renal replacement therapy at the time of catheter insertion [39]. This is reminiscent of the seminal work of Smith and Slater in 1972, showing a striking correlation between cholesterol levels in serum and the amount of LDL in atheromatous coronary plaques [41]. However, such associative data should not be taken as evidence of causality of higher circulating calcium and phosphate concentrations on tissue mineralization. Indeed, it seems unlikely that serum calcium and phosphate concentrations are universal determinants of vascular mineral deposition given the equally high prevalence (if not greater) of MAC in patients with established type 2 diabetes mellitus, as well as in a number of rare hereditary disorders (e.g., idiopathic calcification of the newborn, pseudoxanthoma elasticum) where conventional serum biochemistry is seldom abnormal.

Progression of MAC also appears strongly related to age but is chiefly associated with the presence of pre-existing calcification [29, 42]. Indeed, this is particularly true of incident dialysis patients, where MAC has been shown to progress rapidly in those with pre-existing calcification at initiation of therapy [29]. CKD patients with low bone turnover disease appear particularly at risk of vascular calcification [43], presumably due to the inability of adynamic bone to buffer fluxes in calcium and phosphate.

Nonetheless, the presence of aortic or coronary arterial calcification (whether intimal or medial) is highly predictive of a poor CV outcome. One recent meta-analysis of almost 220,000 individuals showed that when present, vascular calcification portends to an approximately fourfold higher risk for CV mortality and 3.5-fold risk for future CV events [44]. Expectedly, calcification scores at several arterial sites (aortic, coronary, carotid) have been consistently associated with increased future risk of CV events and mortality in patients with renal failure.

There is convincing epidemiological evidence of a reciprocal relationship between loss of mineral from bone and ectopic deposition within the vascular and valvular tissue, and this would seem to be exacerbated in the uremic state [45]. Although our understanding of this apparent paradox is incomplete, it is potentially explained by the finding that certain inflammatory lipids and cytokines, which appear to inhibit normal skeletal mineralization, have also been identified as potent promoters of vascular calcification [11]. In further support of this so-called calcification paradox, there is also the suggestion that bisphosphonate therapy in dialysis patients may have beneficial effects on bone mass, while simultaneously inhibiting vascular calcification [46]. Perhaps the most persuasive evidence of a direct reciprocal relationship between vascular and skeletal mineralization comes from animal knockout models of osteoprotegerin (OPG) [47], a regulator of osteoclastogenesis and bone resorption that is also expressed by endothelial cells (EC) and VSMC [48], and α-klotho, ablation of which results in an aging-like syndrome characterized by a shortened lifespan, infertility, skin atrophy, and emphysema [49]. Both these strains of genetically modified mice develop extensive MAC with severe concomitant osteoporosis.

3 Mechanisms of Vascular Calcification: Passive Biochemistry or Recapitulation of Physiologic Ossification?

VC was long viewed as a passive age-related pathology due to the "adsorption of calcium and phosphate by degenerated tissue" [50]. Indeed, elastin, the major matrix component in aortic vascular tissue, has been shown to spontaneously calcify (particularly fragmented or fracture fibers) at physiological calcium and phosphate concentrations, both in vitro and in vivo [51]. Thus, vascular mineralization may occur passively, without an active cellular component, governed instead by biochemical equilibria, thermodynamic forces, and direct interaction of mineral with matrix components.

However, since the discovery of "bone-related" protein expression (BMP2, osteopontin, MGP) in calcified atherosclerotic plaques over 20 years ago [52], a plethora of in vitro and in vivo studies have evoked the now widely accepted view that vascular calcification is a highly regulated and principally cell-mediated phenomenon that recapitulates many features of physiologic ossification [53, 54]. According to this prevailing paradigm, vascular calcification is driven by changes in VSMC phenotype and viability. In response to persistent pathological stimuli, common in uremia and diabetes [55], VSMCs undergo maladaptive dedifferentiation to a phenotype that spawns calcifying MV, seeding mineralization of the arterial matrix, or apoptotic/necrotic cell death, yielding debris that can also nucleate mineral crystals [56, 57]. Moreover, emerging evidence suggests that other cell types, including EC and local adventitial and circulating stem cells, in addition to VSMC, may also contribute to active mineralization of the arterial wall [58].

3.1 Vascular Smooth Muscle Cell Phenotypic Plasticity

Physiologically, VSMCs primarily function to control vascular tone [59], but in response to tissue injury and specific environmental cues can differentiate from their normal contractile phenotype to a proliferative synthetic phenotype that is able to effect repair of the matrix before re-differentiating back to the quiescent contractile state [60]. Current theory holds that if pathological signals are sustained, this adaptive response can become maladaptive, and synthetic VSMC can undergo further differentiation to an osteochondrocytic phenotype that actively mineralizes the vessel wall [61, 62].

Theoretically, the uremic environment presents a perfect storm for VSMC-mediated calcification. Chronic exposure to oxidative stress, proinflammatory stimuli, and mineral stress (high phosphate/calcium flux), potentially exacerbated by concurrent therapies (e.g., excessive active vitamin D), together provide a strong impetus towards osteochondrocytic differentiation and arterial mineralization [39, 63–65]. In these differentiated VSMC, the expression of contractile proteins (e.g., α-SM actin, SM22α, and SM-MHC) is downregulated while osteogenic transcription factors (e.g., Runx2, osterix, Msx2) and mineralization-regulating proteins (e.g., alkaline phosphatase, osteocalcin, osteopontin, bone sialoprotein), usually considered to be restricted to skeleton, are upregulated [52, 56]. In particular, the expression of Runx2, a master osteogenic transcription factor and regulator of VSMC proliferative/inflammatory pathways [66, 67], although not consistently [68], appears upregulated early during phenotypic transition in response to oxidative stress and increased phosphate uptake via sodium-dependent phosphate transporters, Pit-1 and Pit-2 [69, 70]. The osteogenic phenotype of VSMC appears quite heterogeneous, however, depending on the origin of the cells (e.g., medial, adventitial or circulating progenitors) and the nature of the numerous and varied environmental stimuli

[71]. Stimulation of inflammatory, oxidative stress and mineral-related pathways results in the activation of a number of secondary signaling cascades ordinarily expressed during bone development [72]. Current evidence implicates bone morphogenetic protein (BMP), transforming growth factor (TGF)-β, Wnt and Notch signaling cascades as the major pathways in this process.

3.1.1 The Importance of VSMC Plasticity to Vascular Calcification In Vivo

The current view of VSMC plasticity is based on a large body of evidence primarily garnered from observations of cultured primary human and rodent VSMC. However, such cultured cells may not behave like all SMC in vivo as different populations show considerable genetic and phenotypic heterogeneity depending on the species [73], the embryonic origin of the cells, as well as the vessel and specific layer from which they were isolated [74]. The health and age of the animal/donor also markedly impact on phenotype and proliferative potential [75]. Indeed, of particular relevance to the current discussion, most VSMCs demonstrate an inherent tendency to mineralize in culture, since 4-week growth in standard media alone (e.g., DMEM-10 % FBS: Ca ~1.8, P ~1.0 mmol/L) results in the spontaneous formation of calcified nodules [76]. In contradistinction to these findings of isolated primary cells, VSMCs within intact aortic rings, where matrix contact and architecture (and hence the normal contractile state) are relatively persevered, are highly resistant to mineralization, even when cultured in the presence of supraphysiological concentrations of calcium and phosphate (Ca and P >2.5 mmol/L) [77].

Despite these well-known drawbacks and limitations of in vitro VSMC investigation, it would seem that several important mechanistic aspects of mineralization observed in isolated culture do appear recapitulated in some animal models of calcification. In particular, there is strong evidence suggestive of osteochondrocytic transdifferentiation of *vascular cells* in the calcified intimal plaques of high-fat-fed LDLR$^{-/-}$ and ApoE$^{-/-}$ mice [78], as well as in the arterial media of rodents with renal insufficiency induced by high phosphate/adenine diet, subtotal nephrectomy [79–81], in MGP$^{-/-}$ mice [82, 83], but interestingly not in the intima or media of fetuin-A-deficient mice [84]. As summarized in Table 1, both chondrocytic and osteoblastic features have been variously observed and it is therefore not clear whether such active mineralization processes more closely resemble endochondral or intramembranous bone formation (Table 1).

3.1.2 Technical Challenges in Identifying Phenotypically Modulated VSMC Present in Atherosclerotic Lesions In Vivo

In the context of atherosclerotic lesions the extent of the contribution of VSMC to mineralization is not clear due to difficulties in definitively identifying phenotypically modulated cells. Mature VSMCs are not terminally differentiated and demonstrate remarkable plasticity in response to varied environmental cues, switching between contractile and synthetic (proliferative/migratory) phenotypes as needed for

Table 1
Evidence for VSMC osteochondrocytic transformation in uremic and non-uremic animal models

Study	Model	Diet	Renal function	Osteochondrocytic marker(s)	Arterial mineralization	
					Location	Mechanism
Luo et al. [82]	$Mgp^{-/-}$ mice	Standard	Normal	↑Col II	Medial	Chondrocytic
Davis et al. [218]	$Ldlr^{-/-}$ mice	High fat	STX (uremic)	↑OCN	Intimal	Osteoblastic
Bobryshev [219]	$ApoE^{-/-}$ mice	Standard	Normal	↑Sox9, Col II	Intimal	Chondrocytic
Rattazi et al. [220]	$ApoE^{-/-}$ mice	Standard	Normal	↑Col II, OPN, ALP	Intimal	Chondrocytic
Al-Aly et al. [78]	$Ldlr^{-/-}$ mice	High fat	Normal	↑BMP2, Runx2, Msx2, Wnt3a/7d	Intimal/ Medial	Osteoblastic
Neven et al. [79]	Wistar rats	High P/low protein	Adenine-induced CRF	↑Sox9, Runx2, Col II	Medial	Chondrocytic
Kawata et al. [221]	Sprague–Dawley rats	High P	STX (uremic)	↑Runx2, OCN, OPN, MGP	Medial	Osteoblastic
El-Abbadi et al. [80]	DBA/2 mice	High P	STX (uremic)	↑Runx2, OPN; ↓SM22α (no Col II staining)	Medial	Osteoblastic
Koleganova et al. [222]	Sprague–Dawley rats	Standard	STX (uremic)	↑Runx2, OSN, BMP2 (MGP, OPN, Pit1: *ns* vs. sham)	Intimal/ Medial	Osteoblastic
Speer et al. [83]	$Mgp^{-/-}$ mice	Standard	Normal	↑Runx2, OPN, Col II; ↓Msx2, myocardin; Osx, Wnt3a/7d	Medial	Osteochondrocytic
				Undetected; (Sox9, BMP2: ns vs. $Mgp^{+/+}$)		
Neven et al. [79]	Wistar rats	High P/low protein	Adenine-induced CRF	↑Runx2, Sox9, Col II, aggrecan; ↓Osx, LRP6, β-catenin	Medial	Chondrocytic
Doehring et al. [223]	$Ldlr^{-/-}$ mice	High fat	Normal	↑Sox9, Col II	Intimal	Chondrocytic

CRF chronic renal failure, *STX* subtotal nephrectomy, *Pit1* Type III sodium phosphate transporter, *LRP* low-density lipoprotein receptor-related protein, *BMP2* bone morphogenetic protein, *OPN* osteopontin, *OSN* osteonectin, *OCN* osteocalcin, *Osx* osterix, *MGP* matrix Gla protein, *Col* collagen, *Runx2* runt-related transcription factor 2, *Sox9* sex determining region Y related HMG box gene 9, *Msx2* Msh homeobox 2, *Wnt* Wingless-Type MMTV Integration Site Family

vascular development during embryogenesis, and, for remodeling and repair in adulthood [85]. Dedifferentiation to a synthetic phenotype is marked by decreased (sometimes undetectable) expression of SMC-selective genes that encode components of the contractile apparatus (e.g., α-SMC, SM-MHC, SM22α) [60]. To complicate matters further, many of these contractile markers are also expressed transiently in other cell types during repair and in disease [86]. For instance, macrophages have been reported to express α-SMC and SM22α in response to TGF-β [87] and thrombin [88], while cholesterol loading of SMC results in the induction of several markers typically associated with macrophage function related to phagocytosis and antigen presentation [89]. The relative contribution of resident and myeloid progenitors to the transformed cellular pool is also uncertain. Thus, unambiguous identification of cells of SMC origin in vascular lesions is very challenging. Unfortunately, previously employed lineage-tracing methods to positively identify SMC have lacked sufficient rigor in order to define the contribution of SMC to disease progression.

Importantly, beyond issues with the in situ identification of VSMC per se, technical difficulties are also encountered in definitively ascertaining the osteochondrocytic VSMC phenotype. The identification of this putative phenotype in culture and in vivo is principally based on detecting the de novo expression of pro-osteoblastic (Runx2, osterix, Msx) or pro-chondrocytic (Runx2, Sox9) transcription factors and their downstream gene products (e.g., OPN, OCN, MGP, TNAP, Col II). However, several of these so-called bone markers lack specificity as they are synthesized by cell types besides bone cells (e.g., macrophage, endothelial cells) and are therefore not truly specific for osteogenic transformation [58, 90]. Unfortunately, relatively few studies have confirmed expression of these markers at the level of both the protein and gene. Sole reliance on changes in gene expression can be misleading due to the oversensitivity of modern-day RT-PCR. Fewer studies still, demonstrate that an often small-magnitude change in transcription factor expression translates into meaningful increases in transcriptional activity. False positive signals are also frequent with immunohistochemical analysis due to the nonspecific binding of antibodies to mineral deposits and lipidic necrotic/debris often present in atherosclerotic plaques, but perhaps more critically due to the binding of circulating "bone" proteins (e.g., MGP, OPN, OCN) to mineral phases for which they purposefully have very high affinity. Local synthesis by transformed cells cannot therefore be established by positive protein staining alone.

3.1.3 Evidence for VSMC Osteochondrogenesis in Human Vessels

With respect to studies of non-uremic human vessels obtained at autopsy or at the time of surgery (e.g., endarterectomy), "bone" protein expression has been detected both in calcified intimal and medial lesions (Table 2). In particular, while generally absent in undiseased vessels, early analyses of atherosclerotic plaques showed fairly consistent expression (gene) and staining (protein) of OPN,

Table 2
Evidence for VSMC osteochondrogenesis in human vessels

Study	Cohort	Arterial site(s)	Sample size (*n*)	Bone protein (detection method)	Localization
Bostrom et al. [52]	NS	Carotid	3	BMP2 (ISH)	Intimal SMC
Shanahan et al. [56]	NS	Coronary	18	OPN (ISH, IHC)	Intimal M, occ. medial M (no staining in normal vessels)
				MGP (ISH, IHC)	Intimal M, occ. medial M/SMC (low-level intimal and medial SMC staining in normal vessels)
Fitzpatrick et al. [224]	NS	Coronary	7	OPN (IHC)	Intimal (no staining in normal vessels)
Giachelli et al. [225]	NS	Carotid, coronary	11	OPN (IHC)	Intimal M/SMC & mineral staining (occ. intimal staining in normal vessels)
O'Brien et al. [90]	NS	Coronary	19	OPN (ISH, IHC)	Intimal M/SMC & mineral staining (no staining in normal vessels)
Shanahan et al. [56]	Diabetic	Tibial, popliteal, tibio/peroneal, femoral	9 (severe MAC)	MGP (ISH, IHC)	Highly expressed in SMC adjacent to mineral (medial SMC staining in normal vessels)
				OPN (ISH, IHC)	Intimal M, medial mineral staining (no staining in normal vessels)
				ALK, OCN, Col II (RT-PCR)	Medial (undetectable in normal vessels)
					Medial OPN BMP2, SM22a, Col I expression similar to normal vessels

(continued)

Table 2
(continued)

Study	Cohort	Arterial site(s)	Sample size (n)	Bone protein (detection method)	Localization
Tyson et al. [61]	NS	Aorta, carotid	40	Runx2 (RT-PCR) ALK, OCN (RT-PCR, ISH)	Intimal SMC & M, medial SMC; Msx2 intimal SMC
					Runx2, Sox9 undetectable by ISH; highly variable expression Sox9, MGP, Col I, Col II
					MGP expression in noncalcified vessels; MGP downregulated in some calcified samples
Bobryshev [219]	NS	Aorta, carotid	28	Col II, Sox9 (IHC)	Highly variable intimal SMC expression
Neven et al. [79]	Transplant donors	Aorta	7 (MAC in 4)	Sox9, Runx2, Col II (IHC)	Medial staining in calcified vessels (no staining in normal vessels)
CKD					
Moe et al. [226]	ESRD (predialysis/ HD)	Epigastric	39 (no MAC in 27)	OPN, ALK, OPN, Col I (IHC)	Medial staining correlated with calcification severity
					75 % vessels without MAC had staining for 1+ bone protein
Gross et al. [96]	Renal/non-renal	Coronary	46	OCN (IHC)	Intimal staining higher in renal vs. non-renal; OPN similar staining

Shroff et al. [39]	Pediatric predialysis/dialysis	Omental, epigastric, mesenteric	34 (MAC in 6 HD)	Runx2, Osx, ALK (IHC)	Medial staining in dialysis vessels; cytoplasmic transcription factor staining
					Minimal staining in normal/predialysis vessels
Yoshida et al. [93]	CKD 2–3/HD	Coronary	19 (mostly IAC)	OPN, Runx2 (IHC)	CKD 2–3: intimal (occ. medial) OPN, no Runx2; HD: intimal/medial OPN/Runx2
O'Neill et al. [94]	ESRD (predialysis/HD)	Breast	19 (MAC in 18)	OCN, Runx2, Osx, Sox9 (IHC)	No cellular staining, minimal medial OCN staining associated with mineral

NS not stated, *ESRD* end-stage renal disease, *HD* hemodialysis, *MAC* medial arterial calcification, *IAC* intimal arterial calcification, *ISH* in situ hybridization, *IHC* immunohistochemistry, *M* macrophage, *SMC* smooth muscle cell, *RT-PCR* reverse transcriptase polymerase chain reaction, *BMP2* bone morphogenetic protein, *OPN* osteopontin, *OSN* osteonectin, *MGP* matrix Gla protein, *ALK* tissue nonspecific alkaline phosphatase, *Col* collagen, *SM* smooth muscle, *OCN* osteocalcin, *Osx* osterix, *Runx2* runt-related transcription factor 2, *Sox9* sex determining region Y related HMG box gene 9, *Msx2* Msh homeobox 2 with the latter mainly localized to intimal macrophages and VSMC surrounding mineral deposits [90, 91]. Although constitutively expressed by VSMC [92], MGP expression has also been spatially related to calcified foci but appears generally downregulated in affected vessels, especially in those with severe MAC [56, 61]. Osteogenic transcription factors (Runx2, Sox9) and their downstream gene products (TNAP) have also been detected in conjunction with increased Col II expression, suggestive of a chondrocytic-like transdifferentiation and activity of VSMC, and are consistent with other histological manifestations of cartilaginous metaplasia that have previously been described [79].

What then is the evidence for VSMC-mediated VC in human CKD? Currently, data is limited to a small number of studies of medium-sized muscular arteries (epigastric, omental, mammary) acquired at transplantation and catheter insertion or of coronary vessels obtained at autopsy. In landmark studies by Shroff and colleagues, it was shown that calcium-containing deposits accumulate in the arterial wall of predialysis and dialysis vessels compared to mesenteric vessels obtained from healthy controls during elective surgery. However, in the case of predialysis vessels this deposition was associated with preserved vessel architecture, normal inhibitor concentrations, no evidence of osteochondrocytic transdifferentiation (Runx2 expression, ALP activity), MV release, and a normal number of VSMC [55]. Indeed, only in vessels from dialysis patients was there convincing evidence of VSMC loss (partly due to apoptosis—although the absolute numerical contribution was small), Runx2 staining, an increase in ALP activity, reduced levels of inhibitors (fetuin-A and MGP), and release of mineral-laden MV. Moreover, although ex vivo experiments on isolated intact vessel rings showed increased time-dependent calcium accumulation in predialysis vessels compared to controls (although of much lesser magnitude compared to the increase in dialysis vessels), there was no significant change in osteochondrocytic markers or cell viability, even in high calcium/phosphate medium (2.7 mM Ca, 2.0 mM P). Taken together these data suggest that factors specific to the dialytic milieu drive accelerated VSMC-mediated calcification in patients with ESRD, and, that other, possibly noncellular (passive), mechanisms of mineralization may predominate prior to the initiation of dialysis therapy. Critically, however, these were vessels from pediatric patients and appeared free from significant atherosclerotic disease (no intimal thickening or plaques were observed). This may be germane, as concurrent atherosclerotic disease and impairment of endothelial function, more typical of vascular disease in adult CKD, in addition to greater cumulative exposure to inappropriate environmental cues, may induce cell-mediated mineralization pathways at an earlier stage in adult CKD progression. However, arguing against this possibility, studies of coronary vessels obtained at autopsy by Yoshida et al. showed minimal intimal OPN staining (no medial staining) and absent Runx2 staining in adults with stages 2–3 CKD, with only variable intimal and medial staining present in sections from hemodialysis patients [93]. Moreover, a recent study of exclusively medial calcification in breast arteries from patients with ESRD showed no immunohistological evidence of VSMC osteochondrocytic transdifferentiation (OCN, Runx2, osterix, Sox9) or cell loss due to apoptosis in any section analyzed [94].

Based on the available evidence and aforementioned technical limitations and deficiencies, it is difficult to conclude that there is a major role for medial VSMC-mediated arterial mineralization in patients with CKD prior to the initiation of hemodialysis therapy. Even then, the occurrence would appear sporadic and inconsistent. Indeed, despite universal exposure to the pro-calcific uremic milieu only a proportion of patients receiving chronic hemodialysis have significant MAC by imaging or histology [95]. An explanation for this variable disease penetrance is not currently forthcoming. Yet it should also be stressed that with the exception of two studies on coronary vessels [93, 96], the analyses conducted to date have been on small-to-medium sized muscular arteries that are generally free from intimal/atherosclerotic disease involvement and which may therefore not be representative of changes occurring in some larger elastic vessels (e.g., aorta). This may in part also explain the apparent disparity with findings in some animal models of uremia, where studies have mainly centered on changes in VSMC phenotype in the aorta and where phenotypic switching would appear to occur relatively early in disease progression [97–99]. On the other hand, major differences in the pathogenesis of vascular disease in man and rodents are not without precedent and, specifically with respect to VC, there are examples of marked differences in phenotype resulting from mutations in key regulators of mineralization. For instance, $MGP^{-/-}$ mice have a very strong phenotype of massive medial arterial calcification that leads to vessel rupture, thrombosis, and death within several months of birth. In humans, however, inactivating mutations in MGP (Keutel syndrome) manifest with diffuse cartilage calcification, short distal phalanges, and pulmonary disease, but otherwise a very mild cardiovascular phenotype in which arterial calcification is surprisingly infrequent [100]. Conversely, inactivating mutations in *ENNP1*, encoding the pyrophosphate synthesizing ectoenzyme nucleotide pyrophosphatase/phosphodiesterase, results in the devastating (but variably penetrant) syndrome, generalized arterial calcification of infancy (GACI), which is characterized by severe systemic MAC that develops in utero, leading to widespread neointimal expansion and vessel occlusion, multiorgan failure, and non-atherosclerotic-associated myocardial infarction within a few months of birth [101]. Ablation of the homologous gene, *Npps*, in mice, however, results in ossification of the spinal ligaments and periarticular calcification but with relatively minor arterial involvement and only a modestly shortened lifespan [102, 103].

3.1.4 On the Primacy of Changes in VSMC Phenotype to Vascular Calcification

Notwithstanding these issues, even if VSMC osteochondrocytic transformation does occur in vivo, it is not certain whether such changes in phenotype are a primary initiating event driving subsequent calcification or an adaptive secondary response to increased cellular mineral exposure. A similar argument can be made for the synthesis of osteopontin by macrophage resident in calcifying lesions in fetuin-A$^{-/-}$ and MGP$^{-/-}$ mice [104]. From a teleological perspective the latter appears logical as it may represent an attempt by the cell to efficiently organize and immobilize mineral in a less toxic form outside of the cell, and in so doing, limit the spread of potentially noxious nanocrystalline particles. Although several studies in MGP$^{-/-}$ [83] and high-phosphate-fed uremic mice [98] suggest that VSMC transformation may precede arterial calcification, mineral deposition was only assessed by histochemical staining in these studies and not by actual measurement of calcium content, which may not be sufficiently sensitive to detect early microcalcifications. Furthermore, other studies in identical models have failed to replicate such findings [105], and such a temporal relationship has yet to be unequivocally demonstrated in either animals or man. These mechanistic considerations may hold important therapeutic implications, as targeting VSMC transdifferentiation processes prematurely may disrupt appropriate adaptive responses and be harmful.

3.2 Vascular Smooth Muscle Cell Matrix Vesicle Release

Ghadially and colleagues first described finding electron-dense granular and membranous bodies in the matrix of human rabbit articular cartilage in 1965 [106]. Originally termed "matrical lipidic debris," these authors and others considered such particles to be "extruded cellular processes" of chondrocytes and by-products of normal cell remodeling and turnover [106, 107]. At some sites (e.g., the calcifying zone of cartilage) this material was seen to calcify. Subsequently, detailed EM studies by Anderson and Bonucci suggested that these extracellular double membrane-bound vesicles, termed matrix vesicles (MV), were closely associated with the very early stages of bone and cartilage mineralization [108, 109]. MV represent one of several types of crudely classified extracellular vesicle that have been isolated from the "cell dust," which in addition to MV encompasses vesicular inclusions such as exosomes, ectosomes/microparticles, and apoptotic bodies that differ in size, density, morphology, and lipid/protein composition and apparent function, although there appears to be a considerable overlap. As defined, MV are a heterogeneous population of typically ovoid-shaped membrane-bound extracellular bodies (30–300 nm) released from the plasma membrane of cells involved in physiological mineralization such as osteoblasts, chondrocytes, and odontoblasts [110, 111]. Depending on their cargo, MV are capable of nucleating hydroxyapatite crystals on their exterior and interior surfaces. Mineralization-competent MV appear enriched for annexins [2, 5, 6] and acidic phospholipids (e.g., phosphatidylserine) which facilitate calcium binding, ion influx into the vesicle lumen, and formation of the nucleation core [112]. In osteoblast and chondrocyte-derived MV, alkaline phosphate has also been variously detected, which, by unknown mechanisms, is thought to drive the generation and uptake of phosphate from PPi [113]. In the absence of inhibitors (e.g., fetuin-A and MGP), intravesicular mineral accrual drives the formation and expansion of initially amorphous calcium phosphate particles before undergoing ripening to form crystalline hydroxyapatite [63, 114]. On release from the MV membrane (also by unknown mechanisms) and exposure to the extracellular milieu, these crystals may form a nidus or template for further mineral crystal formation via homologous nucleation [115]. MV appear to bind avidly to certain ECM components such as glycosaminoglycans and collagen, with MV-associated mineral crystals consequently seeding these structures in situ [116].

The existence of such specific "organelles of calcification" continues to be questioned by some authors, who cite the fact that mineral-containing vesicles indistinguishable from MV are also frequently present in noncalcifying regions of articular cartilage and noncalcifying elastic cartilages and, further, that a distinct and uniform population of vesicles of constant morphology and size has not been observed. Indeed, lipidic debris from dead and dying cells has long been known to serve as a nidus for mineralization. Isolated vesicles are also reportedly unable to actively transport calcium [117]. Despite these often vociferous protestations [118], MV are now widely regarded as having an important role in the initiation of skeletal mineralization [110] and likely represent at least one plausible pathway by which mineral is delivered to the mineralization front in calcifying tissues.

Emerging evidence also suggests a role for calcifying MV in early pathological mineralization. MV have been observed in atherosclerotic plaques [119], MAC [114], as well as in calcifying aortic value disease [120], where they may initiate mineralization in a manner similar to that proposed for bone and the dentition. In vitro studies show that high extracellular calcium concentrations induce the release of mineral-loaded MV from VSMC [63], while macrophages have also recently been shown to release MV in response to mineral and inflammatory stressors like TNF-α [119].

Elevated calcium levels can also induce apoptotic death of VSMC leading to the extrusion of intracellular calcium stores, further augmenting local tissue concentrations, and the release of mineral crystal-nucleating apoptotic bodies [57, 63]. Calcium also plays a role in triggering MMP2-mediated degradation of elastin and in the exposure of annexin 6-phosphatidylserine nucleation complexes [63, 121]. Perhaps more importantly, chronic exposure to high calcium levels and the resultant cell attrition/mineral deposition is postulated to deplete the vascular tissues from essential calcification inhibitors [63, 114]. Thus the high calcium and phosphate concentrations frequently encountered in patients with advanced CKD (assumed to reflect a net positive total body balance of calcium and phosphate) appear to act synergistically, driving disparate processes that together lead to cell-mediated mineralization of the vasculature [122].

4 Biomineralization: Loss of Inhibition

Beyond these cell-mediated mechanisms, it is instructive to consider the physicochemical pathways driving all mineralization processes that are harnessed biologically to create ordered deposition in the body. Crystallization, the transition of matter from the solvated phase into a crystal lattice, is a complex physiochemical process that is governed by various thermodynamic and kinetic considerations [123]. These relate to the degree of solute supersaturation relative to a given solid phase, which in turn is dependent on the composition and size of the initial solid-phase nuclei, the presence of active heterogeneous nucleators or templates that facilitate nucleation, and the presence (or absence) of crystal growth inhibitors/modifiers [124]. All these factors impinge on the molecular "energy landscape," which determines the rate and pathway of transformation of a metastable, high free energy solution phase to a lower free energy crystalline phase.

Hydroxyapatite is the most stable calcium phosphate phase (having a lower free energy compared to ionic and precursor states) and its formation although kinetically unfavorable is strongly thermodynamically favored [124]. Thus, over time, hydroxyapatite will eventually seed from fluids containing physiological concentrations of calcium and phosphate, if crystal nucleators are present. Consequently, multiple local and systemic inhibitory systems have evolved to regulate mineralization and restrict hydroxyapatite nucleation, growth, and deposition to physiological sites (bone, cartilage growth plate, dentin, and cementum), as well as cargo or "chaperone" systems to facilitate clearance of mineral nuclei from extraosseous locales. Indeed, it is worth noting that Gersh and McLean demonstrated more than 70 years ago that the loading of plasma with supraphysiological amounts of calcium and phosphate does not result in spontaneous precipitation of mineral in the circulation due to the formation of soluble colloidal protein-bound complexes of calcium phosphate that are cleared rapidly by the reticuloendothelial system [125–127].

The importance of these inhibitory networks is strongly reinforced by the fact that genetic ablation of protein inhibitors of mineral crystal nucleation (e.g., matrix Gla protein, MGP) or disruption of enzyme systems that generate potent biochemical inhibitors of mineral crystal growth (e.g., pyrophosphate, PPi) in mice and in humans (see below) can result in a phenotype of spontaneous severe calcification of the vasculature and other soft tissues [82, 128, 129]. Critically, this occurs in the absence of disturbances in systemic mineral biochemistry (i.e., normal serum calcium and phosphate concentrations). Indeed, the spatial restriction of mineral deposition to physiological sites of mineralization appears to relate to the co-expression of (1) enzyme systems that degrade endogenous inhibitors of mineral crystal growth (e.g., TNAP, tissue nonspecific alkaline phosphatase); (2) an active self-assembling macromolecular scaffold (e.g., type I collagen) that provides the necessary spatial constraints and local mineral supersaturation for crystal formation; and (3) noncollagenous acidic phosphoproteins (e.g., dentin matrix protein 1, DMP-1; matrix Gla protein, MGP) and phospholipids that can direct mineral crystal nucleation, aggregation, and the hierarchical assembly of oriented crystalline hydroxyapatite platelets within the organic matrix [130, 131]. Such co-expression appears unique to terminally differentiated mineralization-competent osteoblasts, odontoblasts, and chondrocytes [132].

These seminal works have led to the "loss-of-inhibition" model of mineralization whereby deposition of mineral only occurs when these inhibitory systems are downregulated, either physiologically during skeletal ossification or pathologically leading to ectopic tissue calcification.

The corollary of these findings is that, for ectopic calcification to occur, endogenous inhibitory systems must be overwhelmed or compromised and a nidus for mineral crystal deposition must be present (e.g., cell debris, apoptotic bodies, damaged/fragmented elastin fibers, matrix vesicles). Although elevations in calcium or phosphate may clearly promote ectopic calcification (by virtue of increased supersaturation), they are neither required nor alone sufficient to induce ectopic mineralization at concentrations observed in vivo.

4.1 Local and Systemic Inhibitory Networks

VSMCs are normally highly resistant to mineralization and only when inhibitory mechanisms become compromised does calcification occur. Amongst the best characterized inhibitors of calcification are fetuin-A, a systemically acting glycoprotein secreted by the liver [84], and MGP and PPi, both produced locally by osteoblasts and VSMC [82, 133]. Reynolds et al. demonstrated that fetuin-A is internalized by synthetic VSMC exposed to high calcium and phosphate concentrations, wherein it regulates several of the key steps that lead to VSMC calcification: blocking caspase-mediated apoptotic signaling cascades; promoting phagocytosis of calcifying cells and apoptotic bodies; and loading into matrix vesicles, preventing mineral crystal growth [114, 134]. Interestingly, fetuin-A also appears to be loaded into matrix vesicles elaborated from chondrocytes during physiologic chondrogenesis [135]. Ablation of the fetuin-A gene results in a phenotype of extensive soft-tissue calcification and altered bone growth [136]. Like fetuin-A, MGP is also loaded into matrix vesicles wherein it inhibits mineral crystal nucleation [63, 114]. Additionally, MGP serves as soluble decoy receptor for osteoinductive BMP2 [137, 138], a potent activator of SMAD (Mothers Against Decapentaplegic homolog proteins)-dependent Runx2, Msx2, and p21 signaling cascades, as well as other members of the TGF-β superfamily. Critically, MGP activity requires vitamin K-dependent γ-carboxylation [139]; thus nutritional vitamin K deficiency or inhibition of active vitamin recycling with warfarin is associated with impaired functionality of this protein [140]. MGP knockout mice develop severe vascular calcification rapidly after birth [82]. PPi, on the other hand, is a crystal poison, directly binding to mineral crystals blocking further growth [141]. PPi is degraded by tissue nonspecific alkaline phosphatase (TNAP) [142], which appears consistently upregulated during early VSMC mineralization in a number of in vitro and in vivo calcification models [142–145]. PPi is synthesized via ATP-dependent pathways catalyzed by ENPP1 [146]. As described above, inactivating mutations in ENNP1 result in the devastating phenotype of idiopathic infantile arterial calcification or GACI [101].

5 Mineral Nanocrystals: True Mediators of Mineral Stress?

The elevation in plasma phosphate concentrations often seen in patients with advanced CKD (eGFR <30 ml/min/1.73 m^2) [147, 148] is widely considered to be the principal driver of accelerated vascular calcification in this setting [122, 149–153]. However, while there is in vitro evidence that phosphate may be directly damaging to certain cell types (e.g., EC, VSMC) at levels equivalent to plasma concentrations found in some ESRD patients (>2 mmol/L) [65, 154], emerging data suggests that the toxic effects of phosphate may be mediated by calcium phosphate nanocrystals rather than soluble ionic phosphate itself [68, 141, 155–157].

Standard culture media are relatively simple buffered salt solutions without the complex biochemistry and colligative properties of serum. Importantly, many culture media contain much higher ionized Ca^{2+} concentrations than found in human serum (e.g., DMEM ~1.8 mmol/L vs. 1.2 mmol/L) and, although supplementation with FBS does supply some calcium-binding/mineral crystal inhibitors (e.g., fetuin-A), mineral precipitation is strongly kinetically favored compared to precipitation in human serum. Thus, even relatively modest elevations in calcium or phosphate may lead to spontaneous mineral nanocrystal formation in culture media.

This was recently highlighted by Sage et al. who found that mineral nanocrystals were formed within 72 h in cell-free culture medium (10 % FCS-DMEM) containing 2 mmol/L phosphate (and 1.8 mmol/L free calcium) incubated at 37 °C [68]. Nanocrystal formation was abrogated by the addition of PPi, a crystal poison, and grossly augmented by the omission of FCS from the medium, indicating that components in serum have a marked effect on nanocrystal formation [68].

Nanocrystals of basic calcium phosphate (BCP) consisting of hydroxyapatite or octacalcium phosphate have been reported to strongly stimulate a number of cell types, in vitro [68, 156]. The rise in intracellular calcium concentration that accompanies lysosomal dissolution of internalized BCP crystals, or calcium influx following direct or receptor-mediated crystal-cell interactions, induces the synthesis of various intracellular and secreted mediators, alters membrane permeability and reduces cell viability. Of particular importance, BCP crystals have been shown to induce the production of osteogenic proteins by VSMC (e.g., BMP2 and osteopontin), independent of Runx2, as well as inducing cell death at high levels without any increase in culture phosphate concentration [68, 155]. Critically, exposure of VSMC to high phosphate medium (2 mmol/L) after removal of nanocrystals by centrifugation does not appear to generate the same osteogenic response [68]. Crystal nucleation inhibitors, PPi and phosphonoformic acid, also reportedly block high phosphate-induced VSMC mineralization [156]. Interestingly, there is evidence that calcium phosphate nanocrystals may regulate the osteogenic response in osteoblasts. Khoshniat et al. reported that cell-crystal interactions mediated by caveolae-containing lipid rafts led to early response kinase (ERK) 1/2-stimulated mineralization-associated gene expression [158]. Signaling via this pathway was abolished by inhibition of crystal formation with phosphocitrate, despite high phosphate/calcium culture conditions [158].

BCP-induced apoptosis of VSMC may lead to atherosclerotic plaque destabilization and rupture, while in the medial layer, enhanced VSMC mineralization may potentiate the loss of large vessel compliance [23]. In the macrophage, BCP crystals induce secretion of tumor necrosis factor (TNF)-α, interleukin (IL)-1β, and IL-18, via Toll-like receptor 4, protein kinase C-α/mitogen-activated protein kinase and nucleotide-binding domain, leucine-rich-repeat-containing family pyrin domain-containing 3 (NLRP3) inflammasome-dependent pathways [159, 160]. The inflammasomes are cytosolic multicomponent proteolytic complexes, which once primed, activate caspase 1-mediated processing of pre-formed pro-IL1β and pro-IL-18 cytokines to their mature secreted forms [161]. Considerable interest now surrounds the role of the NLRP3 inflammasome activation in response to a diverse array of metabolic damage-associated signals (DAMPs) and its role in various chronic metabolic disorders (e.g., atherosclerosis, type 2 diabetes, obesity, and the gouty arthritis) [162]. In particular, NLRP3 priming and activation appears to be central to the proinflammatory response to crystalline and aggregated particulate matter (e.g., cholesterol, MSU and BCP crystals, and islet amyloid polypeptide). BCP-induced inflammation contributes to the cartilage deterioration, subchondral remodeling, and synovitis seen in severe osteoarthritis and other destructive arthropathies [163]. Indeed intra-articular deposition of BCP crystals is strongly associated with the severity of cartilage degeneration [164]. In fibroblasts, on the other hand, exposure to BCP results in matrix metalloproteinase production and a proliferative response [165, 166]. Consequently, BCP crystals may play an important, but cell-specific, role in the genesis of a number of pathologies.

Crystal size appears to be a key determinant of proinflammatory and apoptotic potency [167]. Crystals isolated from aortic and carotid plaques range in size from 50 nm to 8 μm [155, 168]; however, studies in both the macrophage and VSMC suggest that only crystals less than 1-2 μm in diameter are strongly stimulatory [155, 169]. Interestingly, in studies of synthetic hydroxyapatite crystals released from prosthetic implants, Laquerriere et al. reported that needle-shaped particles were most potently inflammatory [170]; thus crystal shape may also be important. Indeed, phagolysosomal overloading with crystalline particles of acute-angle geometries, as seen with silica dust, cholesterol crystals, asbestos, and carbon nanotubes, may result in phagolysosomal rupture and subseuqent activation of the NRLP3 inflammasome [171, 172].

Agglomeration of mineral nanocrystals has also been reported to induce ROS generation and proinflammatory cascades and affect membrane integrity and cell viability of human and murine macrophages in vitro [173]. The pathophysiological significance of nanoparticulate agglomerates is uncertain, however, as these effects have only been observed in serum-free or low serum media. Indeed, there is evidence that serum components may stabilize smaller particles and reduce agglomeration potential [173].

In aggregate, it is likely that both "passive" and "active" components contribute to vascular mineralization, each with potentially greater or lesser importance, depending on the vessel or pathological context. Moreover, the distinction between these elements appears overly simplistic. For instance, matrix components (e.g., elastin, MGP, PPi) not only are synthesized by cells but also have independent effects on cell function and phenotype, distinct from their effects on mineral crystal nucleation and growth [174, 175]. Thus, it would be more accurate to consider VC as a hybrid of interacting biochemical and cellular processes. However, what is far less clear is whether changes in VSMC phenotype (osteochondrocytic transdifferentiation) or viability represents the initial event driving subsequent VC, as supported by some studies [83, 98], or, as suggested by others [157], whether such cellular transformations and attrition are mainly a secondary adaptive/maladaptive response to mineral deposition in the surrounding matrix that results from compromised inhibitory mechanisms and the thermodynamic tendency for calcium and phosphate to spontaneously precipitate.

6 Inflammation and Vascular Calcification: Cause and Effect?

Aging is associated with the development of a chronic low-grade proinflammatory phenotype, referred to as "inflammaging" [176]. This is characterized by amplified inflammatory cytokine production and dysregulation of the immune system, which results in increased susceptibility to infection but also has important effects on the function of other tissues like the vasculature [177]. Multiple lines of evidence suggest that proinflammatory cascades are amplified in CKD. Indeed, age-associated central arterial stiffening has been independently associated with serum C-reactive protein (CRP) levels and other proinflammatory cytokines in patients with non-dialysis-dependent and dialysis-dependent CKD, as well as in the non-CKD elderly population [178, 179]. Interestingly, there is also strong evidence of increased arterial stiffness in patients with chronic inflammatory conditions such as inflammatory bowel disease, rheumatoid arthritis, and systemic lupus erythematosus, who are also at higher CV risk [180–184]. In this setting, treatment with anti-inflammatory agents (e.g., TNFα blockade) is associated with reduced arterial stiffness and inflammation (assessed by ^{18}F-fluorodeoxyglucose (FDG)-PET) [183, 185]. Importantly, these chronic inflammatory disorders are also associated with increased prevalence and severity of vascular calcification [186–188], and variously with lower levels of systemic mineralization inhibitors such as fetuin-A [189, 190].

Like many chronic diseases, CKD is considered a proinflammatory state with increased circulating concentrations of cytokines such as TNFα, and in particular, IL-6 [191], a key marker of the senescence-associated secretory phenotype. Indeed, in hemodialysis patients, serum IL-6 levels are strongly correlated with oxidative stress, telomere attrition, and all-cause mortality [192]. Elevated serum C-reactive protein levels have been associated with reduced telomere length in mononuclear cells from hemodialysis patients [193]. Moreover, changes in the stem cell pool usually observed with aging have also been noted in younger uremic patients [194], typified by the depletion of endothelial progenitors (CD34+), bone marrow-derived stromal cells, and expansion of the proinflammatory and pro-atherogenic $CD14^+/CD16^+$ monocytic pool [195, 196]. Together these findings implicate accelerated age-related changes in immune cell function, as an important component of the premature aging phenotype seen in CKD [4]. Indeed, cells of the immune system are subject to enormous proliferative demand, especially in advanced CKD, and are therefore particularly sensitive to telomere attrition and telomere-initiated senescence [197]. Furthermore, the secretory products of these senescent cells (e.g., BMP2, IL-6) may be critical in reinforcing proinflammatory, osteogenic and growth arrest signaling in the vasculature [198].

In support of the direct effect of proinflammatory cytokines on calcification propensity, in vitro studies have demonstrated the ability of these factors to induce osteochondrocytic differentiation and resultant mineralization of VSMC through convergent intracellular signaling pathways. TNFα, mainly released by activated macrophages but also variably by VSMC, upregulates expression of osteoblast specific transcription factor (Osf2) and enhances expression and activity of ALP via cAMP-dependent pathways [199]. TNFα also induces VSMC apoptosis via cAMP-mediated inhibition of AMP-activated protein kinase (AMPK)-dependent Gas6/Akt signaling and increases Msx2 expression via the nuclear factor κ-light chain enhancer of activated B-cells (NF-κB) pathway, also leading to enhanced ALP activity [200]. Systemic and locally synthesized IL-6 can also enhance ALP activity and promote osteochondrocytic differentiation via activation of Janus kinases and signal transducers and activators of transcription (STAT) signaling cascades [201]. Furthermore, Awan et al. recently reported that administration of a monoclonal antibody against Il-1β attenuated the development of aortic calcifications in high-fat-fed adult LDLR$^{-/-}$ mice compared to placebo [202].

Elegant molecular imaging studies have provided compelling evidence of the close temporal and spatial relationship of inflammatory and osteogenic vascular processes in vivo. Using intravital microscopy and sensitive fluorogenic probes, Aikawa et al. were able to demonstrate the co-localization and increasing intensity of mineralization (bisphosphonate-conjugated agent) and inflammatory activity/macrophage burden (NIRF-conjugated iron nanoparticles) in intimal plaques of carotid arteries isolated from ApoE$^{-/-}$ mice [203]. The same group subsequently extended these findings in uremic ApoE$^{-/-}$ mice, but also showed a reciprocal relationship between bone mineral density (osteoporosis), inflammatory activity, and mineral accrual within vascular and valvular lesions [204]. Bone mineral loss in uremic mice appeared accelerated compared to nonuremic animals. More recently, Abdelbaky et al. have reported on the longitudinal association between focal inflammation determined by ^{18}F-FDG uptake and calcification in the thoracic aorta in patients with serial PET/CT analyses [205]. Here, these authors showed evidence that FDG uptake preceded calcium deposition at the same arterial sites, emphasizing the primacy of inflammatory changes in driving subsequent mineralization of atherosclerotic plaques.

Increased reactive oxygen species (ROS) production via NADH oxidases, xanthine oxidase, uncoupled nitric oxide synthase, and various other mitochondrial enzyme systems is strongly associated with tissue inflammation. ROS are also potent activators of VSMC osteogenesis via phosphatidylinositol (PI)3 kinase/Akt pathways, stimulating increased Runx2, ALP, and osteocalcin expression and concomitant suppression of contractile markers [64, 72]. Accumulation of proinflammatory advanced glycation end products and advanced oxidation protein products has also been demonstrated to directly induce VSMC mineralization under Runx2 transcriptional control via p38 MAPK and ERK-dependent pathways, respectively [206, 207]. Likewise, oxidized LDL (oxLDL) can enhance phosphate-induced calcification via increased MAPK/ERK pathways resulting in the activation of downstream mediators regulated by osterix [207]. Critically, oxLDL has also been reported to upregulate TNFα and BMP2 expression, further amplifying pro-osteogenic VSMC signaling cascades [207, 208].

Arterial aging is also associated with increased proinflammatory signaling induced by a number of vasoactive stressors, in particular angiotensin II and endothelin-1 [209]. Chronic stimulation of these pathways leads to a shift in the pattern of transcription factors and downstream effectors activated in vascular cells, enhancing proinflammatory remodeling via NF-κB and v-ets erythroblastosis virus E26 oncogene homolog 1 (Ets-1) while downregulating protective anti-oxidative regulators like nuclear factor (erythroid-derived2)–like2 (Nrf2) [210–212]. Collectively, these changes in transcriptional regulation result in augmented synthesis of monocyte chemoattractant protein (MCP)-1, TGF-β1, ROS, and MMP2 activation, driving ECM degradation, intimal invasion, fibrosis and thickening, mineralization, and ultimately vascular cell senescence [213].

Finally, and with particular relevance to the uremic state, inflammation may downregulate the synthesis of important systemic calcification inhibitors like fetuin-A. Lebreton et al. found that serum fetuin-A concentrations were suppressed in individuals with acute infectious disease and thus classified fetuin-A as a negative acute-phase reactant [214]. Proinflammatory cytokines suppress hepatic fetuin-A expression by inducing a switch to short isoforms of CCAAT/enhancer binding protein (C/EBP) that bind the CRE in the gene promoter and which block transcription [215]. Significant changes in expression appear evident at subclinical levels of inflammation (CRP <5 mg/L) as observed in patients with CKD and on HD [216, 217].

7 Summary

There is compelling evidence of accelerated age-related arterial remodeling in CKD patients with an exacerbation of structural and functional changes seen with normal aging. This vascular phenotype is characterized by increased calcification and arterial stiffening and is associated with greatly increased risk of adverse cardiovascular outcomes. With respect to the mechanism(s) of vascular calcification, the prevailing dogma dictates that chronic exposure of the vasculature to dysregulated mineral metabolism and an increasingly proinflammatory environment, all characteristic of the uremic milieu, drives phenotypic changes in key vascular cell types leading to active cell mineralization, senescence, and attrition. Recent studies highlight the importance of proinflammatory mineral nanocrystals, rather than soluble mineral ions, in mediating many of these cellular effects, potentiating the vicious cycle of vascular inflammation and mineralization. However, the relative contribution of active and passive mineralization processes in different vascular territories and disease settings has been questioned and remains controversial. Moreover, while it is likely that active cell-mediated mineralization and phenotypic transitions do contribute significantly to vascular mineral burden in some situations, it is unclear whether these changes precede subsequent mineral deposition or occur as a response to preexisting calcification in human disease. Nonetheless, that patients with chronic inflammatory disease but normal renal function also exhibit increased vascular calcification, arterial stiffness, and cardiovascular mortality compared to the non-inflamed suggests that proinflammatory cascades per se may be an important driver of mineral stress and premature vascular aging and thus represents a promising target for therapeutic intervention.

Acknowledgements

Apologies to all authors whose important contributions could not be cited due to space restrictions.

References

1. Go AS, Chertow GM, Fan D et al (2004) Chronic kidney disease and the risks of death, cardiovascular events, and hospitalization. N Engl J Med 351(13):1296–1305
2. de Jager DJ, Grootendorst DC, Jager KJ et al (2009) Cardiovascular and noncardiovascular mortality among patients starting dialysis. JAMA 302(16):1782–1789
3. Foley RN, Parfrey PS (1998) Cardiovascular disease and mortality in ESRD. J Nephrol 11(5):239–245
4. Stenvinkel P, Larsson TE (2013) Chronic kidney disease: a clinical model of premature aging. Am J Kidney Dis 62(2):339–351
5. Yang H, Fogo AB (2010) Cell senescence in the aging kidney. J Am Soc Nephrol 21(9): 1436–1439
6. Lakatta EG (2013) The reality of aging viewed from the arterial wall. Artery Res 7(2):73–80
7. Lehto S, Niskanen L, Suhonen M et al (1996) Medial artery calcification. A neglected harbinger of cardiovascular complications in non-insulin-dependent diabetes mellitus. Arterioscler Thromb Vasc Biol 16(8):978–983
8. Shaw LJ, Raggi P, Berman DS et al (2006) Coronary artery calcium as a measure of biologic age. Atherosclerosis 188(1):112–119
9. Thompson RC, Allam AH, Lombardi GP et al (2013) Atherosclerosis across 4000 years of human history: the Horus study of four ancient populations. Lancet 381(9873):1211–1222
10. Virchow R (1863) Cellular pathology as based upon physiological and pathological histology (trans: Frank Chance). John Churchill, London
11. Demer LL, Tintut Y (2008) Vascular calcification: pathobiology of a multifaceted disease. Circulation 117(22):2938–2948
12. Monckeberg JG (1903) Über die reine Mediaverkalkung der Extremitä tenarterien und ihr Verhalten zur Arteriosklerose. Virchows Arch Pathol Anat 171:141–167
13. Mohr W, Gorz E (2001) Granular media calcinosis of the aorta. Structural findings, historical review and pathogenetic significance. Z Kardiol 90(12):916–928
14. Mohr W, Gorz E (2002) Morphogenesis of media calcinosis in Monckeberg disease. Light microscopy, scanning electron microscopy

and roentgen microanalysis findings. Z Kardiol 91(7):557–567
15. Klotz O (1905) Studies upon calcareous degeneration: I. The process of pathological calcification. J Exp Med 7(6):633–674
16. Silbert S, Lippmann HI, Gordon E (1953) Monckeberg's arteriosclerosis. J Am Med Assoc 151(14):1176–1179
17. Elliott RJ, McGrath LT (1994) Calcification of the human thoracic aorta during aging. Calcif Tissue Int 54(4):268–273
18. Janzen J, Vuong PN (2001) Arterial calcifications: morphological aspects and their pathological implications. Z Kardiol 90(Suppl 3):6–11
19. Schlieper G, Aretz A, Verberckmoes SC et al (2010) Ultrastructural analysis of vascular calcifications in uremia. J Am Soc Nephrol 21(4):689–696
20. Verberckmoes SC, Persy V, Behets GJ et al (2007) Uremia-related vascular calcification: more than apatite deposition. Kidney Int 71(4):298–303
21. Schmermund A, Mohlenkamp S, Erbel R (2003) Coronary artery calcium and its relationship to coronary artery disease. Cardiol Clin 21(4):521–534
22. Drueke TB, Massy ZA (2010) Atherosclerosis in CKD: differences from the general population. Nat Rev Nephrol 6(12):723–735
23. Vengrenyuk Y, Carlier S, Xanthos S et al (2006) A hypothesis for vulnerable plaque rupture due to stress-induced debonding around cellular microcalcifications in thin fibrous caps. Proc Natl Acad Sci U S A 103(40):14678–14683
24. Lin TC, Tintut Y, Lyman A et al (2006) Mechanical response of a calcified plaque model to fluid shear force. Ann Biomed Eng 34(10):1535–1541
25. Aikawa M, Libby P (2004) The vulnerable atherosclerotic plaque: pathogenesis and therapeutic approach. Cardiovasc Pathol 13(3):125–138
26. London GM, Pannier B, Marchais SJ (2013) Vascular calcifications, arterial aging and arterial remodeling in ESRD. Blood Purif 35(1-3):16–21
27. Mitchell GF (2008) Effects of central arterial aging on the structure and function of the peripheral vasculature: implications for end-organ damage. J Appl Physiol (1985) 105(5):1652–1660
28. Lanzer P, Boehm M, Sorribas V et al (2014) Medial vascular calcification revisited: review and perspectives. Eur Heart J 35(23):1515–1525
29. Goodman WG, Goldin J, Kuizon BD et al (2000) Coronary-artery calcification in young adults with end-stage renal disease who are undergoing dialysis. N Engl J Med 342(20): 1478–1483
30. Longenecker JC, Coresh J, Powe NR et al (2002) Traditional cardiovascular disease risk factors in dialysis patients compared with the general population: the CHOICE Study. J Am Soc Nephrol 13(7):1918–1927
31. Wanner C, Krane V, Marz W et al (2005) Atorvastatin in patients with type 2 diabetes mellitus undergoing hemodialysis. N Engl J Med 353(3):238–248
32. Ritz E, Strumpf C, Katz F et al (1985) Hypertension and cardiovascular risk factors in hemodialyzed diabetic patients. Hypertension 7(6 Pt 2):II118–II124
33. Parfrey PS, Foley RN, Harnett JD et al (1996) Outcome and risk factors of ischemic heart disease in chronic uremia. Kidney Int 49(5): 1428–1434
34. Weiner DE, Tabatabai S, Tighiouart H et al (2006) Cardiovascular outcomes and all-cause mortality: exploring the interaction between CKD and cardiovascular disease. Am J Kidney Dis 48(3):392–401
35. Meisinger C, Doring A, Lowel H et al (2006) Chronic kidney disease and risk of incident myocardial infarction and all-cause and cardiovascular disease mortality in middle-aged men and women from the general population. Eur Heart J 27(10):1245–1250
36. Parfrey PS (2000) Cardiac disease in dialysis patients: diagnosis, burden of disease, prognosis, risk factors and management. Nephrol Dial Transplant 15(Suppl 5):58–68
37. Middleton RJ, Parfrey PS, Foley RN (2001) Left ventricular hypertrophy in the renal patient. J Am Soc Nephrol 12(5):1079–1084
38. Lau WL, Ix JH (2013) Clinical detection, risk factors, and cardiovascular consequences of medial arterial calcification: a pattern of vascular injury associated with aberrant mineral metabolism. Semin Nephrol 33(2):93–105
39. Shroff RC, McNair R, Figg N et al (2008) Dialysis accelerates medial vascular calcification in part by triggering smooth muscle cell apoptosis. Circulation 118(17):1748–1757
40. Raggi P, Boulay A, Chasan-Taber S et al (2002) Cardiac calcification in adult hemodialysis patients. A link between end-stage renal disease and cardiovascular disease? J Am Coll Cardiol 39(4):695–701
41. Smith E, Slater R (1972) Relationship between low-density lipoprotein in aortic intima and serum-lipid levels. Lancet 299(7748):463–469
42. Moe SM, O'Neill KD, Fineberg N et al (2003) Assessment of vascular calcification in ESRD patients using spiral CT. Nephrol Dial Transplant 18(6):1152–1158
43. London GM, Marchais SJ, Guerin AP et al (2008) Association of bone activity, calcium

load, aortic stiffness, and calcifications in ESRD. J Am Soc Nephrol 19(9):1827–1835
44. Rennenberg RJ, Kessels AG, Schurgers LJ et al (2009) Vascular calcifications as a marker of increased cardiovascular risk: a meta-analysis. Vasc Health Risk Manag 5(1):185–197
45. Raggi P, Bellasi A, Ferramosca E et al (2007) Pulse wave velocity is inversely related to vertebral bone density in hemodialysis patients. Hypertension 49(6):1278–1284
46. Toussaint ND, Lau KK, Strauss BJ et al (2010) Effect of alendronate on vascular calcification in CKD stages 3 and 4: a pilot randomized controlled trial. Am J Kidney Dis 56(1):57–68
47. Bucay N, Sarosi I, Dunstan CR et al (1998) osteoprotegerin-deficient mice develop early onset osteoporosis and arterial calcification. Genes Dev 12(9):1260–1268
48. D'Amelio P, Isaia G, Isaia GC (2009) The osteoprotegerin/RANK/RANKL system: a bone key to vascular disease. J Endocrinol Invest 32(4 Suppl):6–9
49. Kuro-o M, Matsumura Y, Aizawa H et al (1997) Mutation of the mouse klotho gene leads to a syndrome resembling ageing. Nature 390(6655):45–51
50. Boyd W (1932) A text-book of pathology, 1st edn. Lea & Febiger, Philadelphia
51. Vyavahare N, Ogle M, Schoen FJ et al (1999) Elastin calcification and its prevention with aluminum chloride pretreatment. Am J Pathol 155(3):973–982
52. Bostrom K, Watson KE, Horn S et al (1993) Bone morphogenetic protein expression in human atherosclerotic lesions. J Clin Invest 91(4):1800–1809
53. Sallam T, Cheng H, Demer LL et al (2013) Regulatory circuits controlling vascular cell calcification. Cell Mol Life Sci 70(17):3187–3197
54. Sage AP, Tintut Y, Demer LL (2010) Regulatory mechanisms in vascular calcification. Nat Rev Cardiol 7(9):528–536
55. Shroff RC, McNair R, Skepper JN et al (2010) Chronic mineral dysregulation promotes vascular smooth muscle cell adaptation and extracellular matrix calcification. J Am Soc Nephrol 21(1):103–112
56. Shanahan CM, Cary NR, Salisbury JR et al (1999) Medial localization of mineralization-regulating proteins in association with Monckeberg's sclerosis: evidence for smooth muscle cell-mediated vascular calcification. Circulation 100(21):2168–2176
57. Proudfoot D, Skepper JN, Hegyi L et al (2000) Apoptosis regulates human vascular calcification in vitro: evidence for initiation of vascular calcification by apoptotic bodies. Circ Res 87(11):1055–1062
58. Yao Y, Jumabay M, Ly A et al (2013) A role for the endothelium in vascular calcification. Circ Res 113(5):495–504
59. Qiu H, Zhu Y, Sun Z et al (2010) Short communication: vascular smooth muscle cell stiffness as a mechanism for increased aortic stiffness with aging. Circ Res 107(5):615–619
60. Gomez D, Owens GK (2012) Smooth muscle cell phenotypic switching in atherosclerosis. Cardiovasc Res 95(2):156–164
61. Tyson KL, Reynolds JL, McNair R et al (2003) Osteo/chondrocytic transcription factors and their target genes exhibit distinct patterns of expression in human arterial calcification. Arterioscler Thromb Vasc Biol 23(3):489–494
62. Bostrom KI, Rajamannan NM, Towler DA (2011) The regulation of valvular and vascular sclerosis by osteogenic morphogens. Circ Res 109(5):564–577
63. Kapustin AN, Davies JD, Reynolds JL et al (2011) Calcium regulates key components of vascular smooth muscle cell-derived matrix vesicles to enhance mineralization. Circ Res 109(1):e1–e12
64. Byon CH, Javed A, Dai Q et al (2008) Oxidative stress induces vascular calcification through modulation of the osteogenic transcription factor Runx2 by AKT signaling. J Biol Chem 283(22):15319–15327
65. Jono S, McKee MD, Murry CE et al (2000) Phosphate regulation of vascular smooth muscle cell calcification. Circ Res 87(7):E10–E17
66. Speer MY, Li X, Hiremath PG et al (2010) Runx2/Cbfa1, but not loss of myocardin, is required for smooth muscle cell lineage reprogramming toward osteochondrogenesis. J Cell Biochem 110(4):935–947
67. Sun Y, Byon CH, Yuan K et al (2012) Smooth muscle cell-specific runx2 deficiency inhibits vascular calcification. Circ Res 111(5):543–552
68. Sage AP, Lu J, Tintut Y et al (2011) Hyperphosphatemia-induced nanocrystals upregulate the expression of bone morphogenetic protein-2 and osteopontin genes in mouse smooth muscle cells in vitro. Kidney Int 79(4):414–422
69. Li X, Yang HY, Giachelli CM (2006) Role of the sodium-dependent phosphate cotransporter, Pit-1, in vascular smooth muscle cell calcification. Circ Res 98(7):905–912
70. Crouthamel MH, Lau WL, Leaf EM et al (2013) Sodium-dependent phosphate cotransporters and phosphate-induced calcification of vascular smooth muscle cells: redundant roles for PiT-1 and PiT-2. Arterioscler Thromb Vasc Biol 33(11):2625–2632
71. Nguyen AT, Gomez D, Bell RD et al (2013) Smooth muscle cell plasticity: fact or fiction? Circ Res 112(1):17–22

72. Liu Y, Shanahan CM (2011) Signalling pathways and vascular calcification. Front Biosci 16:1302–1314
73. Shanahan CM, Weissberg PL, Metcalfe JC (1993) Isolation of gene markers of differentiated and proliferating vascular smooth muscle cells. Circ Res 73(1):193–204
74. Gittenberger-de Groot AC, DeRuiter MC, Bergwerff M et al (1999) Smooth muscle cell origin and its relation to heterogeneity in development and disease. Arterioscler Thromb Vasc Biol 19(7):1589–1594
75. Proudfoot D, Shanahan C (2012) Human vascular smooth muscle cell culture. Methods Mol Biol 806:251–263
76. Proudfoot D, Skepper JN, Shanahan CM et al (1998) Calcification of human vascular cells in vitro is correlated with high levels of matrix Gla protein and low levels of osteopontin expression. Arterioscler Thromb Vasc Biol 18(3):379–388
77. Lomashvili KA, O'Neill WC (2009) On vascular calcification prevention with phosphonoformate and bisphosphonates. Kidney Int 75(12):1356
78. Al-Aly Z, Shao JS, Lai CF et al (2007) Aortic Msx2-Wnt calcification cascade is regulated by TNF-alpha-dependent signals in diabetic Ldlr-/- mice. Arterioscler Thromb Vasc Biol 27(12):2589–2596
79. Neven E, Dauwe S, De Broe ME et al (2007) Endochondral bone formation is involved in media calcification in rats and in men. Kidney Int 72(5):574–581
80. El-Abbadi MM, Pai AS, Leaf EM et al (2009) Phosphate feeding induces arterial medial calcification in uremic mice: role of serum phosphorus, fibroblast growth factor-23, and osteopontin. Kidney Int 75(12):1297–1307
81. Davies MR, Lund RJ, Hruska KA (2003) BMP-7 is an efficacious treatment of vascular calcification in a murine model of atherosclerosis and chronic renal failure. J Am Soc Nephrol 14(6):1559–1567
82. Luo G, Ducy P, McKee MD et al (1997) Spontaneous calcification of arteries and cartilage in mice lacking matrix GLA protein. Nature 386(6620):78–81
83. Speer MY, Yang HY, Brabb T et al (2009) Smooth muscle cells give rise to osteochondrogenic precursors and chondrocytes in calcifying arteries. Circ Res 104(6):733–741
84. Jahnen-Dechent W, Heiss A, Schafer C et al (2011) Fetuin-A regulation of calcified matrix metabolism. Circ Res 108(12):1494–1509
85. Owens GK, Kumar MS, Wamhoff BR (2004) Molecular regulation of vascular smooth muscle cell differentiation in development and disease. Physiol Rev 84(3):767–801
86. Li L, Miano JM, Cserjesi P et al (1996) SM22 alpha, a marker of adult smooth muscle, is expressed in multiple myogenic lineages during embryogenesis. Circ Res 78(2):188–195
87. Stewart HJ, Guildford AL, Lawrence-Watt DJ et al (2009) Substrate-induced phenotypical change of monocytes/macrophages into myofibroblast-like cells: a new insight into the mechanism of in-stent restenosis. J Biomed Mater Res A 90(2):465–471
88. Martin K, Weiss S, Metharom P et al (2009) Thrombin stimulates smooth muscle cell differentiation from peripheral blood mononuclear cells via protease-activated receptor-1, RhoA, and myocardin. Circ Res 105(3):214–218
89. Rong JX, Shapiro M, Trogan E et al (2003) Transdifferentiation of mouse aortic smooth muscle cells to a macrophage-like state after cholesterol loading. Proc Natl Acad Sci U S A 100(23):13531–13536
90. O'Brien ER, Garvin MR, Stewart DK et al (1994) Osteopontin is synthesized by macro phage, smooth muscle, and endothelial cells in primary and restenotic human coronary atherosclerotic plaques. Arterioscler Thromb 14(10):1648–1656
91. Shanahan CM, Cary NR, Metcalfe JC et al (1994) High expression of genes for calcification-regulating proteins in human atherosclerotic plaques. J Clin Invest 93(6):2393–2402
92. Wallin R, Cain D, Sane DC (1999) Matrix Gla protein synthesis and gamma-carboxylation in the aortic vessel wall and proliferating vascular smooth muscle cells--a cell system which resembles the system in bone cells. Thromb Haemost 82(6):1764–1767
93. Yoshida H, Yokoyama K, Yaginuma T et al (2011) Difference in coronary artery intima and media calcification in autopsied patients with chronic kidney disease. Clin Nephrol 75(1):1–7
94. O'Neill WC, Adams AL (2013) Breast arterial calcification in chronic kidney disease: absence of smooth muscle apoptosis and osteogenic transdifferentiation. Kidney Int 85(3): 668–676
95. Moe SM, O'Neill KD, Duan D et al (2002) Medial artery calcification in ESRD patients is associated with deposition of bone matrix proteins. Kidney Int 61(2):638–647
96. Gross ML, Meyer HP, Ziebart H et al (2007) Calcification of coronary intima and media: immunohistochemistry, backscatter imaging, and x-ray analysis in renal and nonrenal patients. Clin J Am Soc Nephrol 2(1):121–134
97. Fang Y, Ginsberg C, Sugatani T et al (2014) Early chronic kidney disease-mineral bone disorder stimulates vascular calcification. Kidney Int 85(1):142–150

98. Pai A, Leaf EM, El-Abbadi M et al (2011) Elastin degradation and vascular smooth muscle cell phenotype change precede cell loss and arterial medial calcification in a uremic mouse model of chronic kidney disease. Am J Pathol 178(2):764–773
99. Mathew S, Tustison KS, Sugatani T et al (2008) The mechanism of phosphorus as a cardiovascular risk factor in CKD. J Am Soc Nephrol 19(6):1092–1105
100. Keutel J, Jorgensen G, Gabriel P (1971) A new autosomal-recessive hereditary syndrome. Multiple peripheral pulmonary stenosis, brachytelephalangia, inner-ear deafness, ossification or calcification of cartilages. Dtsch Med Wochenschr 96(43):1676–1681, passim
101. Rutsch F, Ruf N, Vaingankar S et al (2003) Mutations in ENPP1 are associated with 'idiopathic' infantile arterial calcification. Nat Genet 34(4):379–381
102. Okawa A, Nakamura I, Goto S et al (1998) Mutation in Npps in a mouse model of ossification of the posterior longitudinal ligament of the spine. Nat Genet 19(3):271–273
103. Sakamoto M, Hosoda Y, Kojimahara K et al (1994) Arthritis and ankylosis in twy mice with hereditary multiple osteochondral lesions: with special reference to calcium deposition. Pathol Int 44(6):420–427
104. Jahnen-Dechent W, Schafer C, Ketteler M et al (2008) Mineral chaperones: a role for fetuin-A and osteopontin in the inhibition and regression of pathologic calcification. J Mol Med 86(4):379–389
105. Khavandgar Z, Roman H, Li J et al (2014) Elastin haploinsufficiency impedes the progression of arterial calcification in MGP-deficient mice. J Bone Miner Res 29(2):327–337
106. Ghadially FN, Meachim G, Collins DH (1965) Extra-cellular lipid in the matrix of human articular cartilage. Ann Rheum Dis 24:136–146
107. Holtrop ME (1972) The ultrastructure of the epiphyseal plate. II The hypertrophic chondrocyte. Calcif Tissue Res 9(2):140–151
108. Anderson HC (1967) Electron microscopic studies of induced cartilage development and calcification. J Cell Biol 35(1):81–101
109. Bonucci E (1967) Fine structure of early cartilage calcification. J Ultrastruct Res 20(1):33–50
110. Anderson HC (2003) Matrix vesicles and calcification. Curr Rheumatol Rep 5(3):222–226
111. van der Pol E, Boing AN, Harrison P et al (2012) Classification, functions, and clinical relevance of extracellular vesicles. Pharmacol Rev 64(3):676–705
112. Genge BR, Wu LN, Wuthier RE (1989) Identification of phospholipid-dependent calcium-binding proteins as constituents of matrix vesicles. J Biol Chem 264(18): 10917–10921
113. Anderson HC, Sipe JB, Hessle L et al (2004) Impaired calcification around matrix vesicles of growth plate and bone in alkaline phosphatase-deficient mice. Am J Pathol 164(3):841–847
114. Reynolds JL, Joannides AJ, Skepper JN et al (2004) Human vascular smooth muscle cells undergo vesicle-mediated calcification in response to changes in extracellular calcium and phosphate concentrations: a potential mechanism for accelerated vascular calcification in ESRD. J Am Soc Nephrol 15(11):2857–2867
115. Anderson HC (1988) Mechanisms of pathologic calcification. Rheum Dis Clin North Am 14(2):303–319
116. Wu LN, Genge BR, Wuthier RE (1991) Association between proteoglycans and matrix vesicles in the extracellular matrix of growth plate cartilage. J Biol Chem 266(2): 1187–1194
117. Felix R, Fleisch H (1976) The role of matrix vesicles in calcification. Calcif Tissue Res 21(Suppl):344–348
118. Ghadially FN (2001) As you like it, part 3: a critique and historical review of calcification as seen with the electron microscope. Ultrastruct Pathol 25(3):243–267
119. New SE, Goettsch C, Aikawa M et al (2013) Macrophage-derived matrix vesicles: an alternative novel mechanism for microcalcification in atherosclerotic plaques. Circ Res 113(1): 72–77
120. Kim KM (1976) Calcification of matrix vesicles in human aortic valve and aortic media. Fed Proc 35(2):156–162
121. Kumata C, Mizobuchi M, Ogata H et al (2011) Involvement of matrix metalloproteinase-2 in the development of medial layer vascular calcification in uremic rats. Ther Apher Dial 15(Suppl 1):18–22
122. Shanahan CM, Crouthamel MH, Kapustin A et al (2011) Arterial calcification in chronic kidney disease: key roles for calcium and phosphate. Circ Res 109(6):697–711
123. De Yoreo JJ, Vekilov PG (2003) Principles of crystal nucleation and growth. Rev Mineral Geochem 54(1):57–93
124. Wang L, Nancollas GH (2008) Calcium orthophosphates: crystallization and dissolution. Chem Rev 108(11):4628–4669
125. Gersh I (1938) The fate of colloidal calcium phosphate in the dog. Am J Physiol 121(3): 589–594
126. Gersh I (1938) Histochemical studies on the fate of colloidal calcium phosphate in the rat. Anat Rec 70(3):331–349

127. McLean FC, Hinrichs MA (1938) The formation and behavior of colloidal calcium phosphate in the blood. Am J Physiol 121(3):580–588
128. Ho AM, Johnson MD, Kingsley DM (2000) Role of the mouse ank gene in control of tissue calcification and arthritis. Science 289(5477):265–270
129. Rutsch F, Vaingankar S, Johnson K et al (2001) PC-1 nucleoside triphosphate pyrophosphohydrolase deficiency in idiopathic infantile arterial calcification. Am J Pathol 158(2):543–554
130. Kirsch T (2012) Biomineralization--an active or passive process? Connect Tissue Res 53(6):438–445
131. Murshed M, Harmey D, Millan JL et al (2005) Unique coexpression in osteoblasts of broadly expressed genes accounts for the spatial restriction of ECM mineralization to bone. Genes Dev 19(9):1093–1104
132. Murshed M, McKee MD (2010) Molecular determinants of extracellular matrix mineralization in bone and blood vessels. Curr Opin Nephrol Hypertens 19(4):359–365
133. Prosdocimo DA, Wyler SC, Romani AM et al (2010) Regulation of vascular smooth muscle cell calcification by extracellular pyrophosphate homeostasis: synergistic modulation by cyclic AMP and hyperphosphatemia. Am J Physiol Cell Physiol 298(3):C702–C713
134. Reynolds JL, Skepper JN, McNair R et al (2005) Multifunctional roles for serum protein fetuin-a in inhibition of human vascular smooth muscle cell calcification. J Am Soc Nephrol 16(10):2920–2930
135. Wallin R, Schurgers LJ, Loeser RF (2010) Biosynthesis of the vitamin K-dependent matrix Gla protein (MGP) in chondrocytes: a fetuin-MGP protein complex is assembled in vesicles shed from normal but not from osteoarthritic chondrocytes. Osteoarthritis Cartilage 18(8):1096–1103
136. Schafer C, Heiss A, Schwarz A et al (2003) The serum protein alpha 2-Heremans-Schmid glycoprotein/fetuin-A is a systemically acting inhibitor of ectopic calcification. J Clin Invest 112(3):357–366
137. Yao Y, Bennett BJ, Wang X et al (2010) Inhibition of bone morphogenetic proteins protects against atherosclerosis and vascular calcification. Circ Res 107(4):485–494
138. Zebboudj AF, Shin V, Bostrom K (2003) Matrix GLA protein and BMP-2 regulate osteoinduction in calcifying vascular cells. J Cell Biochem 90(4):756–765
139. Schurgers LJ, Cranenburg EC, Vermeer C (2008) Matrix Gla-protein: the calcification inhibitor in need of vitamin K. Thromb Haemost 100(4):593–603
140. Schurgers LJ, Uitto J, Reutelingsperger CP (2013) Vitamin K-dependent carboxylation of matrix Gla-protein: a crucial switch to control ectopic mineralization. Trends Mol Med 19(4):217–226
141. Villa-Bellosta R, Sorribas V (2011) Calcium phosphate deposition with normal phosphate concentration. -Role of pyrophosphate. Circ J 75(11):2705–2710
142. O'Neill WC (2006) Pyrophosphate, alkaline phosphatase, and vascular calcification. Circ Res 99(2), e2
143. Schoppet M, Shanahan CM (2008) Role for alkaline phosphatase as an inducer of vascular calcification in renal failure? Kidney Int 73(9): 989–991
144. Lomashvili KA, Garg P, Narisawa S et al (2008) Upregulation of alkaline phosphatase and pyrophosphate hydrolysis: potential mechanism for uremic vascular calcification. Kidney Int 73(9):1024–1030
145. Price PA, Toroian D, Chan WS (2009) Tissue-nonspecific alkaline phosphatase is required for the calcification of collagen in serum: a possible mechanism for biomineralization. J Biol Chem 284(7):4594–4604
146. Prosdocimo DA, Douglas DC, Romani AM et al (2009) Autocrine ATP release coupled to extracellular pyrophosphate accumulation in vascular smooth muscle cells. Am J Physiol Cell Physiol 296(4):C828–C839
147. Block GA (2000) Prevalence and clinical consequences of elevated Ca x P product in hemodialysis patients. Clin Nephrol 54(4): 318–324
148. Block GA, Klassen PS, Lazarus JM et al (2004) Mineral metabolism, mortality, and morbidity in maintenance hemodialysis. J Am Soc Nephrol 15(8):2208–2218
149. Shroff R (2013) Phosphate is a vascular toxin. Pediatr Nephrol 28(4):583–593
150. Giachelli CM (2009) The emerging role of phosphate in vascular calcification. Kidney Int 75(9):890–897
151. Lau WL, Pai A, Moe SM et al (2011) Direct effects of phosphate on vascular cell function. Adv Chronic Kidney Dis 18(2):105–112
152. Ketteler M, Wolf M, Hahn K et al (2013) Phosphate: a novel cardiovascular risk factor. Eur Heart J 34(15):1099–1101
153. Hruska K, Mathew S, Lund R et al (2011) Cardiovascular risk factors in chronic kidney disease: does phosphate qualify? Kidney Int Suppl 121:S9–S13
154. Di Marco GS, Hausberg M, Hillebrand U et al (2008) Increased inorganic phosphate induces human endothelial cell apoptosis in vitro. Am J Physiol Renal Physiol 294(6):F1381–F1387

155. Ewence AE, Bootman M, Roderick HL et al (2008) Calcium phosphate crystals induce cell death in human vascular smooth muscle cells: a potential mechanism in atherosclerotic plaque destabilization. Circ Res 103(5):e28–e34
156. Villa-Bellosta R, Sorribas V (2009) Phosphonoformic acid prevents vascular smooth muscle cell calcification by inhibiting calcium-phosphate deposition. Arterioscler Thromb Vasc Biol 29(5):761–766
157. Villa-Bellosta R, Millan A, Sorribas V (2011) Role of calcium-phosphate deposition in vascular smooth muscle cell calcification. Am J Physiol Cell Physiol 300(1):C210–C220
158. Khoshniat S, Bourgine A, Julien M et al (2011) Phosphate-dependent stimulation of MGP and OPN expression in osteoblasts via the ERK1/2 pathway is modulated by calcium. Bone 48(4):894–902
159. Jin C, Frayssinet P, Pelker R et al (2011) NLRP3 inflammasome plays a critical role in the pathogenesis of hydroxyapatite-associated arthropathy. Proc Natl Acad Sci U S A 108(36):14867–14872
160. Pazar B, Ea HK, Narayan S et al (2011) Basic calcium phosphate crystals induce monocyte/macrophage IL-1beta secretion through the NLRP3 inflammasome in vitro. J Immunol 186(4):2495–2502
161. Latz E, Xiao TS, Stutz A (2013) Activation and regulation of the inflammasomes. Nat Rev Immunol 13(6):397–411
162. Robbins GR, Wen H, Ting JP (2014) Inflammasomes and metabolic disorders: old genes in modern diseases. Mol Cell 54(2): 297–308
163. Ea HK, Liote F (2009) Advances in understanding calcium-containing crystal disease. Curr Opin Rheumatol 21(2):150–157
164. McCarthy GM (2009) Inspirational calcification: how rheumatology research directs investigation in vascular biology. Curr Opin Rheumatol 21(1):47–49
165. Halverson PB, Greene A, Cheung HS (1998) Intracellular calcium responses to basic calcium phosphate crystals in fibroblasts. Osteoarthritis Cartilage 6(5):324–329
166. McCarthy GM, Augustine JA, Baldwin AS et al (1998) Molecular mechanism of basic calcium phosphate crystal-induced activation of human fibroblasts. Role of nuclear factor kappab, activator protein 1, and protein kinase c. J Biol Chem 273(52):35161–35169
167. Jiang W, Kim BY, Rutka JT et al (2008) Nanoparticle-mediated cellular response is size-dependent. Nat Nanotechnol 3(3):145–150
168. Schmid K, McSharry WO, Pameijer CH et al (1980) Chemical and physicochemical studies on the mineral deposits of the human atherosclerotic aorta. Atherosclerosis 37(2): 199–210
169. Nadra I, Boccaccini AR, Philippidis P et al (2008) Effect of particle size on hydroxyapatite crystal-induced tumor necrosis factor alpha secretion by macrophages. Atherosclerosis 196(1):98–105
170. Laquerriere P, Grandjean-Laquerriere A, Jallot E et al (2003) Importance of hydroxyapatite particles characteristics on cytokines production by human monocytes in vitro. Biomaterials 24(16):2739–2747
171. Hornung V, Bauernfeind F, Halle A et al (2008) Silica crystals and aluminum salts activate the NALP3 inflammasome through phagosomal destabilization. Nat Immunol 9(8):847–856
172. Duewell P, Kono H, Rayner KJ et al (2010) NLRP3 inflammasomes are required for atherogenesis and activated by cholesterol crystals. Nature 464(7293):1357–1361
173. Peng HH, Wu CY, Young D et al (2013) Physicochemical and biological properties of biomimetic mineralo-protein nanoparticles formed spontaneously in biological fluids. Small 9(13):2297–2307
174. Kim HJ, Delaney JD, Kirsch T (2010) The role of pyrophosphate/phosphate homeostasis in terminal differentiation and apoptosis of growth plate chondrocytes. Bone 47(3):657–665
175. Simionescu A, Philips K, Vyavahare N (2005) Elastin-derived peptides and TGF-beta1 induce osteogenic responses in smooth muscle cells. Biochem Biophys Res Commun 334(2):524–532
176. Franceschi C, Bonafe M, Valensin S et al (2000) Inflamm-aging. An evolutionary perspective on immunosenescence. Ann N Y Acad Sci 908:244–254
177. Collerton J, Martin-Ruiz C, Davies K et al (2012) Frailty and the role of inflammation, immunosenescence and cellular ageing in the very old: cross-sectional findings from the Newcastle 85+ Study. Mech Ageing Dev 133(6):456–466
178. Schumacher W, Cockcroft J, Timpson NJ et al (2009) Association between C-reactive protein genotype, circulating levels, and aortic pulse wave velocity. Hypertension 53(2):150–157
179. Lilitkarntakul P, Dhaun N, Melville V et al (2011) Blood pressure and not uraemia is the major determinant of arterial stiffness and endothelial dysfunction in patients with chronic kidney disease and minimal co-morbidity. Atherosclerosis 216(1):217–225
180. Zanoli L, Cannavo M, Rastelli S et al (2012) Arterial stiffness is increased in patients with inflammatory bowel disease. J Hypertens 30(9):1775–1781

181. Wallberg-Jonsson S, Caidahl K, Klintland N et al (2008) Increased arterial stiffness and indication of endothelial dysfunction in long-standing rheumatoid arthritis. Scand J Rheumatol 37(1):1–5
182. Pieringer H, Schumacher S, Stuby U et al (2009) Augmentation index and large-artery remodeling in patients with longstanding rheumatoid arthritis compared with healthy controls. Semin Arthritis Rheum 39(3): 163–169
183. Pieringer H, Pichler M (2011) Cardiovascular morbidity and mortality in patients with rheumatoid arthritis: vascular alterations and possible clinical implications. QJM 104(1): 13–26
184. Greene ER, Lanphere KR, Sharrar J et al (2009) Arterial distensibility in systemic lupus erythematosus. Conf Proc IEEE Eng Med Biol Soc 2009:1109–1112
185. Maki-Petaja KM, Elkhawad M, Cheriyan J et al (2012) Anti-tumor necrosis factor-alpha therapy reduces aortic inflammation and stiffness in patients with rheumatoid arthritis. Circulation 126(21):2473–2480
186. Wang S, Yiu KH, Mok MY et al (2009) Prevalence and extent of calcification over aorta, coronary and carotid arteries in patients with rheumatoid arthritis. J Intern Med 266(5):445–452
187. Yiu KH, Wang S, Mok MY et al (2011) Relationship between cardiac valvular and arterial calcification in patients with rheumatoid arthritis and systemic lupus erythematosus. J Rheumatol 38(4):621–627
188. Molad Y, Levin-Iaina N, Vaturi M et al (2006) Heart valve calcification in young patients with systemic lupus erythematosus: a window to premature atherosclerotic vascular morbidity and a risk factor for all-cause mortality. Atherosclerosis 185(2):406–412
189. Sato H, Kazama JJ, Wada Y et al (2007) Decreased levels of circulating alpha2-Heremans-Schmid glycoprotein/Fetuin-A (AHSG) in patients with rheumatoid arthritis. Intern Med 46(20):1685–1691
190. Marhaug G, Shah V, Shroff R et al (2008) Age-dependent inhibition of ectopic calcification: a possible role for fetuin-A and osteopontin in patients with juvenile dermatomyositis with calcinosis. Rheumatology 47(7):1031–1037
191. Stenvinkel P, Ketteler M, Johnson RJ et al (2005) IL-10, IL-6, and TNF-alpha: central factors in the altered cytokine network of uremia--the good, the bad, and the ugly. Kidney Int 67(4):1216–1233
192. Carrero JJ, Stenvinkel P, Fellstrom B et al (2008) Telomere attrition is associated with inflammation, low fetuin-A levels and high mortality in prevalent haemodialysis patients. J Intern Med 263(3):302–312
193. Ramirez R, Carracedo J, Soriano S et al (2005) Stress-induced premature senescence in mononuclear cells from patients on long-term hemodialysis. Am J Kidney Dis 45(2):353–359
194. Seidler S, Zimmermann HW, Bartneck M et al (2010) Age-dependent alterations of monocyte subsets and monocyte-related chemokine pathways in healthy adults. BMC Immunol 11:30
195. Noh H, Yu MR, Kim HJ et al (2012) Uremia induces functional incompetence of bone marrow-derived stromal cells. Nephrol Dial Transplant 27(1):218–225
196. Ramirez R, Carracedo J, Merino A et al (2011) CD14+CD16+ monocytes from chronic kidney disease patients exhibit increased adhesion ability to endothelial cells. Contrib Nephrol 171:57–61
197. Andrews NP, Fujii H, Goronzy JJ et al (2010) Telomeres and immunological diseases of aging. Gerontology 56(4):390–403
198. Rodier F, Coppe JP, Patil CK et al (2009) Persistent DNA damage signalling triggers senescence-associated inflammatory cytokine secretion. Nat Cell Biol 11(8):973–979
199. Tintut Y, Patel J, Parhami F et al (2000) Tumor necrosis factor-alpha promotes in vitro calcification of vascular cells via the cAMP pathway. Circulation 102(21): 2636–2642
200. Son BK, Akishita M, Iijima K et al (2008) Adiponectin antagonizes stimulatory effect of tumor necrosis factor-alpha on vascular smooth muscle cell calcification: regulation of growth arrest-specific gene 6-mediated survival pathway by adenosine 5′-monophosphate-activated protein kinase. Endocrinology 149(4):1646–1653
201. Abedin M, Lim J, Tang TB et al (2006) N-3 fatty acids inhibit vascular calcification via the p38-mitogen-activated protein kinase and peroxisome proliferator-activated receptor-gamma pathways. Circ Res 98(6):727–729
202. Awan Z, Denis M, Roubtsova A et al. (2015) Reducing vascular calcification by anti-IL-1beta monoclonal antibody in a mouse model of familial hypercholesterolemia. Angiology doi: 10.1177/0003319715583205
203. Aikawa E, Nahrendorf M, Figueiredo JL et al (2007) Osteogenesis associates with inflammation in early-stage atherosclerosis evaluated by molecular imaging in vivo. Circulation 116(24):2841–2850
204. Hjortnaes J, Butcher J, Figueiredo JL et al (2010) Arterial and aortic valve calcification inversely correlates with osteoporotic bone remodelling: a role for inflammation. Eur Heart J 31(16):1975–1984

205. Abdelbaky A, Corsini E, Figueroa AL et al (2013) Focal arterial inflammation precedes subsequent calcification in the same location: a longitudinal FDG-PET/CT study. Circ Cardiovasc Imaging 6(5):747–754
206. Tanikawa T, Okada Y, Tanikawa R et al (2009) Advanced glycation end products induce calcification of vascular smooth muscle cells through RAGE/p38 MAPK. J Vasc Res 46(6):572–580
207. Bear M, Butcher M, Shaughnessy SG (2008) Oxidized low-density lipoprotein acts synergistically with beta-glycerophosphate to induce osteoblast differentiation in primary cultures of vascular smooth muscle cells. J Cell Biochem 105(1):185–193
208. Goettsch C, Rauner M, Hamann C et al (2011) Nuclear factor of activated T cells mediates oxidised LDL-induced calcification of vascular smooth muscle cells. Diabetologia 54(10):2690–2701
209. Wang M, Jiang L, Monticone RE et al (2014) Proinflammation: the key to arterial aging. Trends Endocrinol Metab 25(2):72–79
210. Wang M, Zhang J, Telljohann R et al (2012) Chronic matrix metalloproteinase inhibition retards age-associated arterial proinflammation and increase in blood pressure. Hypertension 60(2):459–466
211. Ungvari Z, Bailey-Downs L, Sosnowska D et al (2011) Vascular oxidative stress in aging: a homeostatic failure due to dysregulation of NRF2-mediated antioxidant response. Am J Physiol Heart Circ Physiol 301(2): H363–H372
212. Csiszar A, Wang M, Lakatta EG et al (2008) Inflammation and endothelial dysfunction during aging: role of NF-kappaB. J Appl Physiol 105(4):1333–1341
213. Wang M, Wang HH, Lakatta EG (2013) Milk fat globule epidermal growth factor VIII signaling in arterial wall remodeling. Curr Vasc Pharmacol 11(5):768–776
214. Lebreton JP, Joisel F, Raoult JP et al (1979) Serum concentration of human alpha 2 HS glycoprotein during the inflammatory process: evidence that alpha 2 HS glycoprotein is a negative acute-phase reactant. J Clin Invest 64(4):1118–1129
215. Gangneux C, Daveau M, Hiron M et al (2003) The inflammation-induced downregulation of plasma Fetuin-A (alpha2HS-Glycoprotein) in liver results from the loss of interaction between long C/EBP isoforms at two neighbouring binding sites. Nucleic Acids Res 31(20):5957–5970
216. Cozzolino M, Gallieni M, Brancaccio D (2009) Inflammation and vascular calcification in chronic kidney disease: the role of Fetuin-A. Cytokine 45(2):70–71
217. Memoli B, Salerno S, Procino A et al (2010) A translational approach to micro-inflammation in end-stage renal disease: molecular effects of low levels of interleukin-6. Clin Sci (Lond) 119(4):163–174
218. Davies MR, Lund RJ, Hruska KA (2003) BMP-7 is an efficacious treatment of vascular calcification in a murine model of atherosclerosis and chronic renal failure. J Am Soc Nephrol 14(6):1559–1567
219. Bobryshev YV (2005) Transdifferentiation of smooth muscle cells into chondrocytes in atherosclerotic arteries in situ: implications for diffuse intimal calcification. J Pathol 205 (5):641–650
220. Rattazzi M, Bennett BJ, Bea F et al (2005) Calcification of advanced atherosclerotic lesions in the innominate arteries of ApoE-deficient mice: potential role of chondrocyte-like cells. Arterioscler Thromb Vasc Biol 25 (7):1420–1425
221. Kawata T, Nagano N, Obi M et al (2008) Cinacalcet suppresses calcification of the aorta and heart in uremic rats. Kidney Int 74 (10):1270–1277
222. Koleganova N, Piecha G, Ritz E et al (2009) A calcimimetic (R-568), but not calcitriol, prevents vascular remodeling in uremia. Kidney Int 75(1):60–71
223. Doehring LC, Heeger C, Aherrahrou Z et al (2010) Myeloid CD34+CD13+ precursor cells transdifferentiate into chondrocyte-like cells in atherosclerotic intimal calcification. Am J Pathol 177(1):473–480
224. Fitzpatrick LA, Severson A, Edwards WD et al (1994) Diffuse calcification in human coronary arteries. Association of osteopontin with atherosclerosis. J Clin Invest 94(4): 1597–1604
225. Giachelli CM, Bae N, Almeida M et al (1993) Osteopontin is elevated during neointima formation in rat arteries and is a novel component of human atherosclerotic plaques. J Clin Invest 92(4):1686–1696
226. Moe SM, O'Neill KD, Duan D et al (2002) Medial artery calcification in ESRD patients is associated with deposition of bone matrix proteins. Kidney Int 61(2):638–647

Chapter 14

An In Vitro Murine Model of Vascular Smooth Muscle Cell Mineralization

Kristen J. Kelynack and Stephen G. Holt

Abstract

Vascular calcification (VC) is seen ubiquitously in aging blood vessels and prematurely in disease states like renal failure. It is thought to be driven by a number of systemic and local factors that lead to extra-osseous deposition of mineral in the vascular wall and valves as a common endpoint. The response of resident vascular smooth muscle cell to these dystrophic signals appears to be important in this process. Whilst *in vivo* models allow the observation of global changes in a pro-calcific environment, identifying the specific cells and mechanisms involved has been largely garnered from in vitro experiments, which provide added benefits in terms of reproducibility, cost, and convenience. Here we describe a 7–21 day cell culture model of calcification developed using immortalized murine vascular smooth muscle cells (MOVAS-1). This model provides a method by which vascular smooth muscle cell involvement and manipulation within a mineralizing domain can be studied.

Key words Smooth muscle cell, Vascular calcification, Calciprotein particles, In vitro

1 Introduction

Vascular calcification (VC) is a feature of aging and a number of diseases including chronic kidney disease (CKD), where it is an important predictor of adverse outcomes and mortality in patients [1]. The process is characterized by the deposition of insoluble mineral in cardiovascular tissues, especially the arterial vasculature, both in intimal plaque and the tunica media [2, 3].

The underlying molecular mechanisms of VC are thought to resemble those of physiological skeletal bone formation. Whilst in health, mineralization is tightly regulated, in disease, vascular calcification is thought to be driven by mineral imbalance, bone morphogenetic proteins, some pro-atherosclerotic lipids as well as loss of negative regulators of calcification [4, 5].

Our current understanding suggests that vascular calcification is not purely due to passive precipitation of calcium phosphate but

Tim D. Hewitson et al. (eds.), *Kidney Research: Experimental Protocols*, Methods in Molecular Biology, vol. 1397,
DOI 10.1007/978-1-4939-3353-2_14, © Springer Science+Business Media New York 2016

may be cell driven to some extent [6]. Several cell types have been implicated in extra-osseous calcification [7], but the vascular smooth muscle cell (VSMC) appears to play a pivotal role. In response to pro-calcific triggers, many of which remain poorly understood, the VSMC develops phenotypic changes characteristic of osteoblast or chondrocyte-like cells. Thus, a better understanding of these VSMC pro-osteoblastic traits and their regulatory mechanisms may be beneficial in the prevention of calcification. Given that research in animal models indicates that VC is multifaceted and complex, the study of contributing pathways in simpler settings provides many advantages.

In vitro models of VC enable the study of mechanisms influencing this process in isolation. While cultivation of primary cell cultures is a useful method for studying VSMC behavior in vitro, the need to reduce inter-assay variation, convenience, and cost has driven research towards the use of cell lines for kinetic and mechanistic investigation.

One such model employs the SV40 large T antigen-immortalized murine VSMC line MOVAS-1. First described by Afroze et al. the cell line has been used in the study of vascular circadian rhythms [8], cell cycling [9] and lipid loading [10].

A variety of methods can be used to induce procalcific and osteochondrocytic behavior in these cells, as they have an inherent propensity to calcify. Here we describe several *in vitro* techniques for studying calcification in MOVAS-1. Whilst more sensitive techniques can be employed to detect changes in genotype that occur early in culture, the formation of calcium containing nodules is not observed prior to 14 days of cultivation. After this time, staining techniques can be used to demonstrate presence of nodules in the cell monolayer. This model enables the study of mechanisms driving calcification as well as potential agonists and antagonists of the process in a controlled and reproducible environment.

2 Materials

All aqueous solutions are prepared in double deionized water (ddH_2O) unless otherwise stated.

2.1 Cell Culture

2.1.1 Cell Culture

1. MOVAS-1 (ATCC, Manassas, VA, USA).
2. Tissue culture flasks (75 or 150 cm^2).
3. 6-well tissue culture plates (34.5 mm diameter).
4. Sterile 50 mL plastic tubes.
5. Sterile glass or plastic 10 mL pipettes.

6. Pipetting aid suitable for use in cell culture.
7. DMEM + 10 % fetal calf serum (FCS) (*see* **Note 1**).
8. Sterile Tris buffered saline (TBS): 50 mM Tris–HCl, 150 mM NaCl, pH 7.4.
9. Sterile trypsin–EDTA (0.25 % trypsin in 0.02 % EDTA).
10. Laminar flow hood.
11. Phase-contrast microscope suitable for viewing cells in culture flasks.
12. Trypan blue solution (0.4 % w/v in ddH_2O).
13. Hemocytometer or automated cell counter.

2.2 Additives to Promote Calcification

2.2.1 Calcium and Phosphate Salt Solutions

1. 20 mM calcium chloride ($CaCl_2$) in TBS, pH 7.4.
2. 14 mM sodium phosphate ($NaHPO_4$) in TBS, pH 7.4.
3. Sterile TBS.
4. 0.22 μM syringe microfilter.

2.2.2 Calciprotein Particles (CPP)

1. 50 mL sterile tubes.
2. Sterile TBS.
3. FCS or human serum (as a source of fetuin-A and other plasma proteins).
4. 20 mM $CaCl_2$ in TBS, pH 7.4 (Ca).
5. 14 mM $NaHPO_4$ in TBS pH 7.4 (Pi).
6. High-speed refrigerated centrifuge.

2.2.3 β-Glycerophosphate

1. β-glycerophosphate (Cell culture grade, Sigma-Aldrich), in ddH_2O at a concentration of 10–25 mM.
2. 50 μg/mL ascorbic acid solution in ddH_2O.
3. Aliquot solutions and store at −20 °C prior to use in cell culture.

2.2.4 Naked Apatite Crystals

1. Synthetic hydroxyapatite nanocrystals, <200 nm 10 % w/v in ddH_2O (Sigma-Aldrich).
2. Spin filter columns, 300 kDa molecular weight cutoff (Sartorius AG, Dandenong South, Victoria, Australia).
3. Sterile TBS.
4. High-speed refrigerated centrifuge.

2.3 Histological Methods for Detection of Calcification

2.3.1 Alizarin Red Staining

1. 10 % Neutral buffered formalin.
2. Transfer pipettes.
3. TBS.
4. 2 % Alizarin Red (w/v) in ddH_2O, pH 4, adjust using 0.5 % ammonium hydroxide and filter to remove particulates. Stable when stored in dark for 1 month.

2.3.2 von Kossa Staining of Nodules

1. 4 % paraformaldehyde (w/v) in TBS.
2. 1 % silver nitrate (w/v) in ddH_2O.
3. 5 % sodium thiosulfate (w/v) in ddH_2O.
4. Harris hematoxylin (*see* **Note 2**).
5. Ultraviolet light source.
6. Aqueous mounting medium.

2.4 Quantifying Mineralization

2.4.1 Calcium and Phosphate Assays

1. 0.6 M hydrochloric acid in ddH_2O.
2. TBS (cold).
3. Cell scrapers.
4. 96-well flat bottom plates.
5. RIPA® Buffer (cold) supplemented with protease inhibitor cocktail (Sigma).
6. Calcium colorimetric assay (Sigma-Aldrich).
7. Phosphate Quantichrom® colorimetric kit (BioAssay Systems, Hayward, CA, USA).
8. Micro BCA® colorimetric protein kit (Thermo Scientific, Waltham, MA, USA).
9. Microplate reader.

3 Methods

3.1 MOVAS-1 Cell Culture and Characterization

3.1.1 Culturing and Maintaining Cells

1. Immortalized murine vascular smooth muscle cell line (MOVAS-1) are maintained in 75–150 cm^2 sterile vented cell culture flasks in DMEM + 10 % FCS supplemented with glutamine and antibiotics (*see* **Note 1**).
2. Cells are incubated at 37 °C in a humidified atmosphere of 95 % air–5 % CO_2.
3. Media is replaced with 10–15 mL fresh media on alternate days (Fig. 1; *see* **Notes 1** and **3**).

3.1.2 Seeding Cells for Experiments

1. Using a sterile pipette, remove media from flask.
2. Wash MOVAS-1 monolayer twice with 5 mL cold, sterile TBS.
3. Add 5 mL Trypsin/EDTA solution per 75 cm^2 flask.

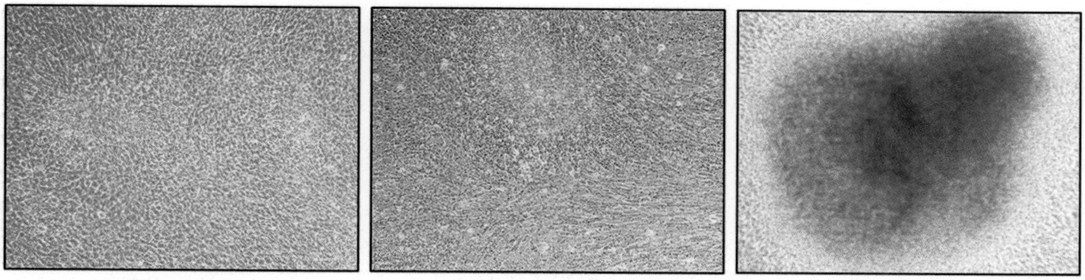

Fig. 1 Phase-contrast images of nodule formation in cultured MOVAS-1 monolayers over time

4. Replace flask lid and return flask to incubator for approximately 5 min to allow dislodging of monolayer from plastic surface. Monitor progress after 2.5 min. Gentle tapping on the outside of the flask may aid cells lifting.
5. Stop trypsin action by adding 5 mL media.
6. Mix by gentle aspiration.
7. Remove cell solution and place into a clean, sterile 50 mL tube.
8. Sediment cells by centrifugation for 5 min at $800 \times g$ at 4 °C.
9. Remove media and replace with 1–5 mL fresh media.
10. Resuspend pellet by gentle aspiration.
11. Perform cell count.
12. Seed cells at desired density in 6-well plates (*see* **Notes 3** and **4**). Add 2 mL media to each well.
13. Grow to confluence in media changing media on alternate days (*see* **Note 5**).
14. Replace with DMEM + 10 % FCS containing treatments (*see* **Note 6**).
15. Change media every 2–3 days for the duration of the experiment.

3.2 Calcifying Growth Conditions

3.2.1 Spontaneous Calcification

MOVAS-1 are known to calcify following long-term culture (approximately 30 days) in serum containing media. This process can be accelerated by addition of procalcific reagents to the growth media.

3.2.2 Calcium Phosphate Induced Calcification

Calcification can be induced by incubating cells in the presence of moderate doses of either calcium or phosphate or both in serum-containing media.

Under sterile conditions:

1. Double filter $CaCl_2$ and $NaHPO_4$ solutions separately through sterile 0.22 μM microfilters.
2. Prepare desired treatment doses of Ca and Pi using 1 M stock solutions diluted in media.
3. A range of doses should be used, including a combination of Ca and Pi. For example, 2, 3.6 and 4.8 mM Ca; 1.5, 3 and 5 mM Pi; 2 mM Ca and 2 mM Pi; 2.8 mM Ca and 3 mM Pi.
4. Culture cells for 7–21 days using the method detailed (*see* Subheading 3.1.1) for nodule formation.

3.2.3 Calciprotein Particle (CPP) Preparation

1. Under sterile conditions combine the following with end-over-end mixing after each addition:

   5.0 mL sterile FCS (*see* **Note 7**).

   10.0 mL 20 mM $CaCl_2$ in TBS, pH 7.4.

   10.0 mL 14 mM $NaHPO_4$ in TBS, pH 7.4.

   15.0 mL sterile TBS.

   in a sterile 50 mL tube.

2. Incubate with continual slow mixing on rotator for 12 h at room temperature.
3. Aliquot into sterile 2 mL tubes.
4. Pellet particles by centrifugation at 24,000 ×*g* for 2 h at 4 °C.
5. Remove supernatant and wash pellet twice with ice-cold TBS.
6. Resuspend the pellet in 200 μL warmed (37 °C) TBS (*see* **Note 8**).
7. Pool supernatant mixture into a single tube.
8. Spin at 1,000 ×*g* for 10 min at room temperature to pellet large aggregates and collect supernatant (CPP stock).
9. Test concentration of CPP mix using calcium assay and adjust to 1 mg/mL with sterile TBS.
10. Aliquot CPP suspension into sterile 1.5 mL tubes and store at 4 °C (1–2 days) or at −80 °C for long-term storage.
11. Aliquots should be thawed and diluted in media immediately prior to incubation with cells (*see* **Note 9**).

3.2.4 β-Glycerophosphate

β-glycerophosphate serves as a free phosphate ion donor (cleavable by alkaline phosphatase) to cells in culture. Ascorbic acid is thought to enhance both the collagen-producing and proliferative capabilities of the cells.

1. Add β-glycerophosphate solution and ascorbic acid solution in media to confluent cells (*see* **Note 10**).
2. Change media every 2–3 days, for 21 days.

3.2.5 Naked Apatite Particles

Hydroxyapatite particles formed by the premixing of calcium and phosphate salt solutions are known to induce calcification.

1. Filter hydroxyapatite particles through a Vivaspin® column by filling column to its maximum volume.
2. Centrifuge at 10,000 × *g* for 20 min at room temperature.
3. Discard the filtrate and resuspend the retentate with a maximum volume of TBS.
4. Centrifuge at 10,000 × *g* for 20 min at room temperature.
5. Discard the filtrate and recover apatite in TBS solution from the concentrator pocket of the column.
6. Test the concentration of apatite crystals by performing a calcium assay (*see* Subheading 3.4.2).
7. Add to media and mix well (*see* **Note 11**). Treat confluent cell monolayers.

3.3 Qualitative Methods for Assessing Calcification

Alizarin Red is used to stain calcium deposits. The dye forms a complex with calcium during the chelation process and appears as a red salt in stained sections and cell monolayers.

3.3.1 Alizarin Red Staining for Calcium Deposition

1. Aspirate media from cell monolayers.
2. Gently wash monolayers twice with Dulbecco's PBS (Ca^{2+}/Mg^{2+} free) twice.
3. Aspirate wash and fix cells with enough neutral buffered formalin (10 %) to completely cover the cell monolayer at room temperature for at least 30 min.
4. Remove fixative and wash with ddH_2O.
5. Carefully aspirate and cover cell monolayer with Alizarin Red staining solution. Incubate at room temperature for 45 min (*see* **Note 12**).
6. Remove excess stain and wash gently four times with ddH_2O.
7. Add PBS to each well.
8. View under a light microscope. Calcium deposits stain red (Fig. 2a).

3.3.2 von Kossa Staining of Calcium-Containing Nodules

The von Kossa staining protocol is used to identify calcium phosphate deposits localized to cells or within tissue sections. Silver nitrate solution is used to deposit silver in areas of concentrated calcium phosphate. This is based on a reaction between silver nitrate and phosphate ions. The resulting silver phosphate salt is then degraded to silver under UV light (or strong white light).

1. Remove media.
2. Rinse monolayers twice with ice-cold TBS.
3. Fix cells in 4 % paraformaldehyde in TBS for 5 min at room temperature.

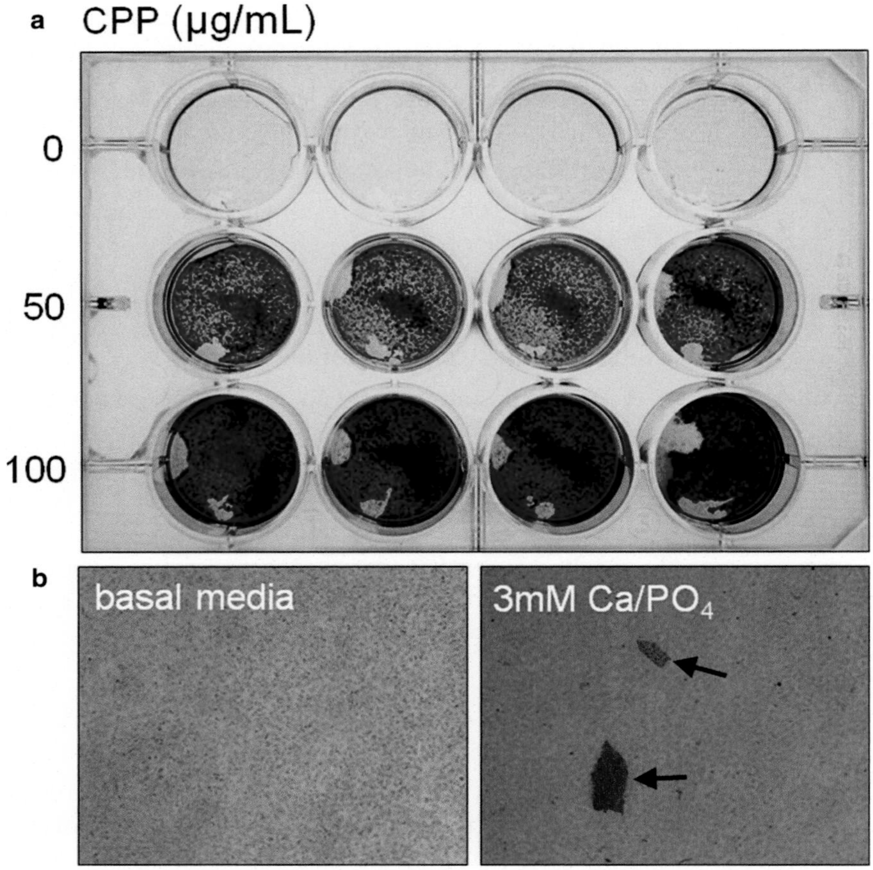

Fig. 2 (**a**) Alizarin Red staining of MOVAS-1 monolayers treated with basal media or CPP (at the stated dose) for 7 days. (**b**) Appearance of mineral nodules (*arrows*) in MOVAS-1 monolayers cultured in Ca/P supplemented media for 21 days and visualized with von Kossa stain

4. Remove fixative and air-dry (approximately 30 min).
5. Incubate cell monolayers in 1 % sodium nitrate for 30–90 min at room temperature under ultraviolet light (*see* **Note 13**).
6. Remove sodium nitrate and wash cells with two changes of ddH_2O.
7. Remove unreacted silver particles by adding 5 % sodium thiosulfate for 5 min at room temperature.
8. Counterstain using Harris hematoxylin for 30 s (*see* **Note 2**).
9. Remove and rinse twice with tap water.
10. Mount by placing a small amount of aqueous mounting media on a coverslip. Invert.
11. Place over cells in well.
12. View under a light microscope. Calcium-containing nodules appear as black deposits (Fig. 2b).

3.4 Quantitative Methods of Assessing Calcification

3.4.1 Harvesting Samples for Estimation of Calcium Concentration

1. Wash cell monolayers with TBS.
2. Aspirate wash and dissolve mineral within cell monolayers by incubating with 0.6 M hydrochloric acid (6–24 h) at 4 °C (*see* **Note 14**).
3. Collect extracts and centrifuge at 10,000 × *g* for 20 min at 4 °C to pellet debris.
4. Collect supernatants in clean 1.5 mL tubes and store at −20 °C until analysis.

3.4.2 Estimating Calcium Concentration by Colorimetric Assay

This assay measures the chromogenic complex formed by calcium ions binding with *o*-cresophthalein.

1. Prepare calcium standards by diluting 10 μL of calcium standard solution (500 mM) in 990 μL of ddH_2O to prepare the first standard of 5 mM. Mix by aspiration.
2. From this, add 0, 2, 4, 6, 8, and 10 μL into the wells of a 96-well plate. Bring the volume to 50 μL with ddH_2O. Your standards will have the concentrations 0 (blank), 0.4, 0.8, 1.2, 1.6, and 2.0 μg/mL Ca (*see* **Note 15**).
3. Aliquot 50 μL each sample into 96-well plate. Prepare in duplicate.
4. Add 90 μL of chromogenic reagent to each well. Mix gently.
5. Add 60 μL of Calcium Assay reagent to each well. Mix gently.
6. Protect plate from light by wrapping whole plate in foil. Incubate the reaction for 5–10 min at room temperature.
7. Read absorbance of wells at 575 nM (*see* **Note 16**).
8. Plot concentration (*x* axis) versus average absorbance (*y* axis) to obtain a standard curve.
9. Read average sample absorbances from the standard curve to obtain concentrations in μg/mL (*see* **Note 17**).
10. Calcium concentrations should be normalized to total protein. Neutralize and solubilize calcium extracts with 10 % TBS supplemented with 1 % SDS. Determine total protein concentration using micro BCA assay. Exemplar data showing dose-dependent CPP-induced calcium deposition on MOVAS-1 cell monolayers after 7 day treatment compared to basal media or media supplemented with equivalent doses of free calcium ($CaCl_2$) are shown in Fig. 3.

3.4.3 Estimation of Phosphate Concentration

This commercial phosphate kit is based on the reaction of malachite green and molybdate with ionic phosphate.

1. Wash cell monolayers twice using ice-cold TBS.
2. Lyse cells in 200 μL ice-cold RIPA® buffer.

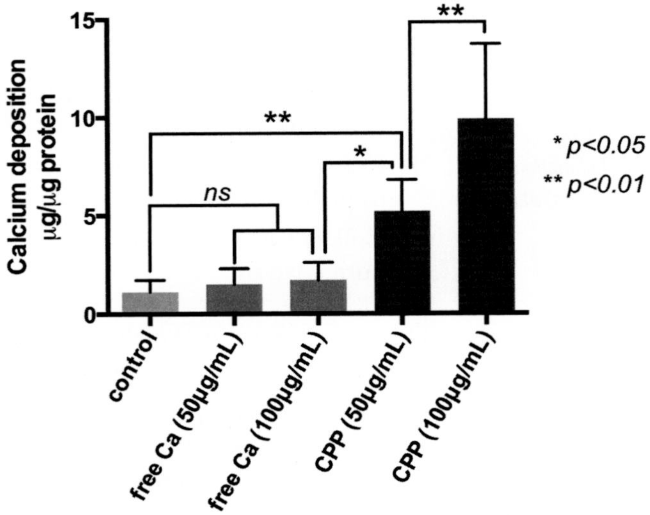

Fig. 3 Calcium concentration (normalized to total protein) of MOVAS-1 monolayers incubated over 7 days ± CPP or equivalent does of free calcium compared to treatment with basal media (control). Pairwise comparisons made with Tukey's multiple comparison test

3. Scrape cells from the plastic using a cell scraper.
4. Collect the cell suspension in sterile 1.5 mL tubes.
5. Incubate on ice for 45 min. Gently vortex suspensions periodically.
6. Centrifuge at 10,000 × *g* for 20 min at 4 °C to pellet debris.
7. Place supernatants in clean 1.5 mL tubes (can be stored at −80 °C prior to analysis).
8. Prepare standards by performing serial dilutions of phosphate (Pi, 30 μM) standard in ddH_2O. The working linear range of the assay is 0.3–50 μM, so standards and samples should be kept within this range. Dilute with ddH_2O as necessary.
9. In a 96-well plate, add 50 μL standards and samples into each well. Each standard and sample should be tested in duplicate. However, triplicate is optimal.
10. Add 100 μL working reagent (provided in kit) to each well. Tap plate gently to mix.
11. Seal and incubate plate at room temperature for 30 min.
12. Read absorbance at 620 nM (600–660 nM).
13. Generate a standard curve by plotting concentration (*x* axis) versus average absorbance (*y* axis) of the standards.

14. Read the average sample absorption from the standard curve. Alternatively, curve plotting software may be used to devise an equation to estimate phosphate concentration within samples.
15. Phosphate concentrations should be normalized to total protein using the micro BCA assay.

4 Notes

1. DMEM + 10 % FCS is prepared by combining 10 mL HEPES Buffer (1 M), 4 mL glutamine (200 mM), 8 mL penicillin–streptomycin (5000 U/mL), 40 mL fetal calf serum (FCS), and 348 mL 1× DMEM and passing through a 0.22 μM filter. Store at 4 °C in a sterile bottle for up to 2 weeks. Warm to 37 °C immediately prior to use.
2. Harris hematoxylin is used to add definition to the cell monolayer for the purpose of orientation under the microscope. Other stains may be used, e.g., Nuclear Fast Red.
3. Cells should be maintained at 1×10^4 to 1×10^5 cells/cm^2. The doubling time of MOVAS-1 cultures is approximately 15 h. This should be factored in when choosing initial seeding density and timing of your experiment.
4. Cell counts can be performed either using a hemocytometer or automated cell counter. A Bio-Rad TC20® Automated Cell counter is used routinely in our lab.
5. Media can be changed every 2–3 days without compromising experiments.
6. DMEM + 10 % FCS is used as the base media into which treatments are placed immediately prior to addition to cells.
7. Alternatively cell culture-grade human serum (Sigma) may be used in the preparation of CPP.
8. The pellet does not resuspend easily. Mix vigorously by aspiration. Warm TBS is used to aid resuspension.
9. Sonicate particulate suspension and media well by aspiration immediately prior to treating cells. Prepare fresh on the day of use and discard unused treatment media. A detailed methodology for CPP preparation is provided elsewhere by Smith [11].
10. Current literature indicates calcification can be induced in MOVAS-1 and vascular smooth muscle cells within the concentration range provided.
11. For best results, sonicate hydroxyapatite in media for 1 min prior to addition to cell cultures.
12. The pH of Alizarin Red solution affects the overall staining quality. We recommend using 0.5 % ammonium hydroxide to adjust pH when preparing 2 % aqueous (w/v) solution. Store at room temperature. Discard unused portion after 1 month.

13. Light from a 60 W (or higher) bulb can be used instead of UV light. However, slides may need a longer incubation time. If the silver is removed during the wash step following incubation, the light source is not strong enough.
14. When demineralizing monolayers, agitation such as use of a plate mixer is recommended.
15. The linear range of this assay is 0.4–2.0 µg/mL. Samples may need to be diluted to achieve readings within the working range of the assay. A new standard curve needs to be set up with each assay run.
16. Read samples within 30 min as the calcium chromophore is prone to fading over time.
17. Alternatively, curve-fitting software may be employed to generate equation and concentrations can be calculated using this method.

Acknowledgements

Development of these methods was supported by a Jacquot Research Establishment Award to SGH. The authors are grateful to A/Prof. Grant Drummond for supplying MOVAS-1.

References

1. Blacher J, Guerin AP, Pannier B et al (2001) Arterial calcifications, arterial stiffness, and cardiovascular risk in end-stage renal disease. Hypertension 38:938–942
2. Schwarz U, Buzello M, Ritz E et al (2000) Morphology of coronary atherosclerotic lesions in patients with end-stage renal failure. Nephrol Dial Transplant 15:218–223
3. Schlieper G, Aretz A, Verberckmoes SC et al (2010) Ultrastructural analysis of vascular calcifications in uremia. J Am Soc Nephrol 21(4): 689–696
4. Sage AP, Tintut Y, Demer LL (2010) Regulatory mechanisms in atherosclerotic calcification. Nat Rev Cardiol 7:528–536
5. Jia G, Stormont RM, Gangahar DM et al (2012) Role of matrix Gla protein in angiotensin II-induced exacerbation of vascular calcification. Am J Physiol Heart Circ Physiol 303:H523–H532
6. Wu M, Rementer C, Giachelli CM (2013) Vascular calcification: an update on mechanisms and challenges in treatment. Calcif Tissue Int 93:365–373
7. Johnson RC, Leopold JA, Loscalzo J (2006) Vascular calcification: pathobiological mechanisms and clinical implications. Circ Res 99:1044–1059
8. Chalmers JA, Martino TA, Tata N et al (2008) Vascular circadian rhythms in a mouse vascular smooth muscle cell line (MOVAS-1). Am J Physiol Regul Integr Comp Physiol 295:R1529–R1538
9. Afroze T, Yang LL, Wang C et al (2003) Calcineurin-independent regulation of plasma membrane Ca^{2+} ATPase-4 in the vascular smooth muscle cell cycle. Am J Physiol Cell Physiol 285:C88–C95
10. Rivera J, Walduck AK, Thomas RR et al (2013) Accumulation of serum lipids by vascular smooth muscle cells involves a macropinocytosis-like uptake pathway and is associated with the downregulation of the ATP-binding cassette transporter A1. Naunyn Schmiedebergs Arch Pharmacol 386:1081–1093
11. Smith ER (2015) Isolation, characterization of calciprotein particles in biological fluids. Methods Mol Biol (in press)

Chapter 15

The Isolation and Quantitation of Fetuin-A-Containing Calciprotein Particles from Biological Fluids

Edward R. Smith

Abstract

Multiple overlapping systemic and local inhibitory networks have evolved to prevent the unwanted deposition of mineral at ectopic sites. Fetuin-A is a liver-derived glycoprotein abundant in plasma that binds and stabilizes nascent mineral ion nuclei to form soluble colloidal high molecular weight complexes, called calciprotein particles (CPP). The binding of fetuin-A to mineral retards crystal ripening and precipitation from the aqueous phase, thereby facilitating the regulated clearance of mineral debris from the extracellular fluid. However, persistent disturbances in this humoral homeostatic system, as frequently seen in patients with Chronic Kidney Disease, may lead to the accumulation and aggregation of these nanoparticles in extraosseous tissues like the vasculature, driving inflammatory cascades, aberrant tissue remodeling, and functional impairment. Consistent with this conceptual framework, higher circulating CPP levels are associated with reduced renal function, increments in systemic inflammatory markers, derangements in bone morphogenetic cytokines, higher vascular calcification scores, aortic stiffening and an increased risk of death. This chapter describes optimized sample collection and preparative procedures for the isolation and enrichment of CPP from biological fluids. Methods for CPP quantitation are critically reviewed and detailed.

Key words Fetuin-A, Calciprotein particles, Ultracentrifugation, Gel filtration chromatography, Immunoisolation, Flow cytometry

1 Introduction

Even under physiological conditions, plasma is considered a metastable aqueous solution of calcium and phosphate, sustaining mineral precipitation once crystals are nucleated. Multiple systemic and local inhibitory networks have evolved to constrain mineral deposition to physiologic calcified tissues such as bone, cartilage, teeth (dentin, cementum, enamel) and inner ear otoconia. For ectopic calcification to occur, these inhibitory mechanisms must be compromised. Fetuin-A (also known as α_2 -Heremans Schmid glycoprotein; AHSG) is a liver-derived, partially phosphorylated, glycoprotein that circulates in plasma at high concentration and has multiple roles in the regulation of matrix mineralization at

Tim D. Hewitson et al. (eds.), *Kidney Research: Experimental Protocols*, Methods in Molecular Biology, vol. 1397, DOI 10.1007/978-1-4939-3353-2_15, © Springer Science+Business Media New York 2016

both molecular and cellular levels [1]. Outstanding in this multifunctional repertoire is the ability of fetuin-A to bind and organize nascent mineral nuclei, forming colloidal high molecular weight complexes, termed calciprotein particles (CPP), in an analogous manner to the solubilization of lipid cargo by apolipoproteins [2]. CPP formation limits further crystal growth, retards the ripening of amorphous mineral to more crystalline phases [3], and is thought to facilitate the rapid, immunologically inert, clearance of mineral debris from the extracellular fluid [4]. Removal of such debris may help to reduce ectopic deposition of precipitates within the tissues and undesirable downstream inflammatory sequelae. Elevated CPP levels, as seen in patients with Chronic Kidney Disease (CKD), may reflect increased genesis of CPP from bone or calcifying tissues and/or reduced clearance due to impairment of renal elimination pathways or inadequate buffering through incorporation into bone.

CPP formation is a multistep process, starting with the aggregation of small clusters of fetuin-A-bound mineral called calciprotein monomers (similar to Posner clusters and other ion-association complexes), forming spherical nanoparticles, called primary CPP (approx. 50–100 nm in diameter), which contain amorphous calcium phosphate. These primary CPP undergo spontaneous rearrangement to more stable, densely packed, prolate needle-shaped particles, called secondary CPP (approx. 100–200 nm in diameter), that contain mineral in a crystalline phase [2, 3, 5, 6]. The primary-to-secondary CPP transition is dependent on time, fetuin-A concentration, the ionic activity of calcium, phosphate and magnesium, pH, temperature [2, 5, 7] in addition to other unidentified small molecule/protein modulators.

Arguably, the first in vivo evidence of CPP and their putative role in mineral trafficking comes from studies by Gersh [8, 9]. In the late 1930s, he reported the appearance of a colloidal mineral phase in the serum of dogs, following IV loading with supraphysiological amounts of calcium phosphate, which was rapidly cleared from the circulation by the reticuloendothelial system. Herrmann and colleagues elucidated the clearance pathway of fetuin-A-containing CPP in more contemporary studies, which appears principally mediated by the class A scavenger receptor [4]. In 2002, Price et al. identified a high molecular weight complex of fetuin-A, matrix Gla protein, and mineral (termed fetuin mineral complexes) in the serum of etidronate-treated rats [10]. The appearance of these particles after administration of various other bone-modulating agents in the same animals led these authors to speculate on a bone origin for CPP [11]. Importantly CPP were not detected in the serum of rats treated with vehicle. Heiss et al. made the first identification of CPP in humans, in the ascitic fluid of a dialysis patient with calcifying peritonitis (with albumin and fetuin-A as major protein components) [5]. Subsequently, Young et al. isolated relatively pleomorphic CPP-like mineraloprotein nanoparticles from 0.2 μm-filtered human and bovine serum [12]. Proteomic

different these particles and related "granulations" isolated from different human body fluids (ascites, urine, cerebrospinal and synovial fluid) revealed a complement of proteins that varied with each fluid, but which consistently included fetuin-A, albumin, complement C3, α1-antitrypsin, prothrombin, and apolipoprotein A1 [13].

With respect to kidney disease, studies by Matsui et al. revealed the appearance of circulating CPP in rats with adenine-induced renal failure [14]. Again, CPP were undetectable in control animals. Importantly, as implied in earlier studies from Price and colleagues [15], elevations in serum CPP *preceded* aortic calcium accrual and histological evidence of arterial mineralization [14]. Treatment with alendronate completely abrogated the development medial calcification and serum CPP became undetectable within 5 weeks. Despite the disappearance of CPP from serum, total serum fetuin-A and hepatic levels (mRNA and protein) were unchanged compared to vehicle treated uremic rats. Intriguingly, further characterization of CPP isolated from these uremic rats demonstrated that virtually all of the fetuin-A circulating in CPP was in the fully phosphorylated state [14]. The functional significance of this protein modification remains obscure, as fetuin-A phosphorylation does not appear to be a requisite for the inhibition of mineral precipitation in solution [16, 17].

Despite the well-defined role of fetuin-A in the control of matrix mineralization, and the strong calcifying phenotype of fetuin-A knockout mice [18, 19], epidemiological studies evaluating the relationship between serum concentrations this protein, declining renal function, vascular calcification, and patient outcome have yielded somewhat conflicting findings (reviewed in [20]). The apparent disconnect between the clinical and preclinical evidence base may reflect the complex molecular heterogeneity of fetuin-A [21], which has largely been overlooked in epidemiological studies to date. Indeed, accumulating evidence indicates that ascertainment of specific subfractions present in serum, such as fetuin-A containing-CPP, may reveal important pathobiological relationships that are not evident with total fetuin-A determinations alone [17, 22–24]. Using fairly nonspecific insensitive methodology, CPP appear undetectable in the serum of healthy adults, but relatively abundant in patients with non-dialysis dependent CKD [17, 23], in whom levels are strongly associated with systemic inflammatory markers, bone resorption markers and aortic stiffness [17]. Circulating CPP load also appears increased in patients with chronic rheumatologic inflammatory disease but normal renal function and markedly so in a number of patients with calcific uremic arteriolopathy (calciphylaxis) [22, 25], where levels were found to track closely with serial changes in inflammatory cytokine concentrations. Other authors have correlated serum CPP levels with coronary calcification scores in CKD patients, and linked reductions in CPP with decrements in PTH following parathyroidectomy or instigation of cinacalcet therapy [23].

Proteomic analysis of the CPP fraction revealed a similar complement of proteins to those previously identified in CPP-like particles fetuin-A, albumin, fibrinogen, fibronectin, β-actin, immunoglobulin κ light chains, and apolipoprotein A1 [23].

In prospective analyses, higher serum CPP levels have emerged as a strong inflammation-associated predictor of an increased risk of all-cause mortality in the pre-dialysis CKD setting, while the transformation time of primary amorphous mineral containing-CPP to secondary crystalline-mineral containing CPP (calcification propensity) was associated with the risk of death, independent of conventional renal and cardiovascular risk factors [24]. More recently, increased calcification propensity has been associated with an increased risk of cardiovascular mortality and graft failure in renal transplant recipients [26, 27]. Thus, although more confirmatory data is awaited, disturbances in this humoral homeostatic system and resultant accumulation of these nanoparticles, in addition to the propensity for their formation in the uremic milieu, appear at least predictive of vascular dysfunction and poor patient outcome.

This chapter describes optimized sample collection and preparative procedures for the isolation of CPP from human biological fluids using centrifugation, gel filtration chromatography, and immuno-enrichment with magnetic beads. Methods for the quantitation of CPP in serum and other fluids are also discussed in detail.

2 Materials

2.1 Sample Collection

1. BD Vacutainer™ SST™ (Becton Dickinson, Franklin Lakes, NJ, USA) or serum 10 mL venous blood collection tubes.
2. TBS: 50 mM Tris–HCl, 140 mM NaCl pH-adjusted with 10 M NaOH to 7.40 at 37 °C (keep separate stocks at RT and 4 °C).
3. Eppendorf Centrifuge 5702 R with A-4-38 swing-bucket rotor and appropriate tube adaptors (Eppendorf, Hamburg, Germany).

2.2 CPP Isolation

2.2.1 Centrifugal Fractionation

1. Eppendorf Centrifuge 5430 R equipped with FA-45-24-11-HS rotor.
2. Eppendorf Safe-lock 2.0 mL tubes.
3. THP depletion buffer: 1 % CHAPS (w/v) in TBS.
4. –80 °C freezer.

2.2.2 Size Exclusion Chromatography

1. HiPrep 16/60 Sephacryl™ S-500 HR SEC column (GE Healthcare Life Sciences, Pittsburgh, PA, USA).
2. ÄKTA™ pure chromatography system (GE Healthcare Life Sciences).

3. Ca-TBS: 50 mM Tris–HCl, 140 mM NaCl, 10 mM $CaCl_2$ pH-adjusted with 10 M NaOH to 7.40 at 37 °C.
4. Centrifugal filter units; Amicon Ultra-15 and 0.5 mL units with 100 kDa MWCO regenerated cellulose membrane (Millipore, Billerica, MA, USA).

2.2.3 CPP Immuno-Enrichment

1. Exosome–Dynabeads® Streptavidin for Isolation/Detection (Life Technologies, Carlsbad, CA, USA).
2. Goat Anti-AHSG polyclonal antibody for biotinylation and bead capture (Abcam, Cambridge, UK).
3. EZ-Link™ Sulfo-NHS-Biotin kit (Thermo Scientific, Waltham, MA, USA).
4. Zeba™ Spin Desalting Columns, 7K MWCO, 0.5 mL (Thermo Scientific).
5. Magnet.
6. Mixer.
7. Flat-bottom microcentrifuge 2.0 mL tubes with screw caps (Neptune Scientific, San Diego, CA, USA).
8. Isolation Buffer: TBS with 0.1 % bovine serum albumin (BSA), filtered through a 0.22 μm Millex-GP filter (Millipore).

2.3 SDS-PAGE/ Western Blotting Analysis

1. Micro BCA™ Protein Assay Kit (Thermo Scientific).
2. Mouse Anti-AHSG monoclonal antibody [2H2] (Abcam) for flow cytometry and Western blotting.
3. Goat Anti-APOA1 polyclonal antibody (LifeSpan Bioscience, Seattle, CA, USA).
4. APEX™ Alexa Fluor® 488/647 Antibody Labeling Kits (Life Technologies).
5. 1× Laemmli Sample Buffer with 50 mM EDTA (Bio-Rad, Hercules, CA, USA).
6. 2-Mercaptoethanol reducing agent (Bio-Rad).
7. Thermal Cycler (Bio-Rad).
8. Mini-PROTEAN™ TGX Stain-Free precast gels 4–15 % (Bio-Rad).
9. Mini-PROTEAN™ Tetra Cell Vertical Electrophoresis System (Bio-Rad).
10. PowerPac Basic (Bio-Rad).
11. 10× Tris–HCl/Glycine/SDS Running Buffer (Bio-Rad).
12. Trans-Blot Turbo Transfer System (Bio-Rad).
13. Trans-Blot Turbo Transfer Pack, mini format 0.2 μm nitrocellulose (Bio-Rad).
14. Precision Plus Protein WesternC Standards (Bio-Rad).

15. Precision Protein StrepTactin-HRP conjugate (Bio-Rad).
16. Mouse TrueBlot™ ULTRA: Anti-Mouse Ig HRP (Rockland, Limerick, PA, USA).
17. Goat TrueBlot™ ULTRA: Anti-Goat Ig HRP (Rockland).
18. SuperSignal West Dura Extended Duration Substrate (Thermo Scientific).
19. ChemiDoc MP Imaging System (Bio-Rad).
20. TBS-T: TBS with 0.05 % Tween 20 (v/v) (Bio-Rad).
21. Blocker™ BLOTTO in TBS (Thermo Scientific).

2.4 CPP Characterization and Quantitation

1. BD FACSVerse™ flow cytometer (Becton Dickinson).
2. Precision™ Automated Microplate Pipetting System (BioTek, Winooski, VT, USA).
3. EDTA-TBS: 100 mM Tris–HCl, 140 mM NaCl, 20 mM EDTA pH-adjusted with 10 M NaOH to 7.40 at 37 °C.
4. Fetuin-A (AHSG) Human ELISA kit (Biovendor, Karasek, Czech Republic).
5. DS2—2-plate ELISA processing system (Dynex Technologies, Chantilly, VA, USA).

3 Methods

3.1 Purification of CPP from Body Fluids

This section gives a simple method for obtaining crude CPP preparations from body fluids, which have been successfully applied to the isolation of CPP from blood (serum), urine and ascetic fluid/dialysate effluent. Pre-analytical conditions can impinge greatly on the quality of the sample obtained, so best sampling practice should be observed throughout. Given the sensitivity of mineralization biochemistry to changes in pH and ionic strength, cell-depleted samples are immediately diluted in buffered saline to maintain neutrality prior to isolation. Dilution also reduces sample viscosity and thus improves pelleting efficiency. A stepped centrifugation protocol is used to remove cells, dead cells, debris, and large protein aggregates/cryo-precipitates and to separate the CPP from the bulk fluid phase, which contains high concentrations of unbound proteins also present as bound components of CPP (e.g., fetuin-A, albumin). The pellet is then washed and resuspended in a small volume of buffered saline to concentrate the CPP present. Alternatively, size exclusion chromatography can be used to separate out CPP from major serum constituents including free monomeric fetuin-A. It is important to appreciate that the CPP-containing fraction obtained from these steps is likely to be contaminated to a variable extent by large extracellular vesicles (variously termed microparticles or microvescicles), apoptotic bodies and exosomes depending on the density of their cargo. Some protein/chromatin complexes/aggregates and, even denser lipoprotein particles, may also co-sediment or co-elute with this fraction. Thus, for some downstream applications, an additional selective purification step (e.g., magnetic bead-based immunoaffinity isolation) is needed to accomplish greater CPP enrichment. In our hands, ultrafiltration-based methods for CPP isolation from biological fluids have proved unreliable due to frequent membrane clogging, loss of buffering (serum retentates can become quite strongly alkaline) and poor recovery characteristics, and are not recommended for the isolation of CPP from biological fluids.

3.2 Sample Collection

3.2.1 Serum

1. Collect blood into 10 mL plastic serum or SST tubes (*see* **Note 1**).
2. Gently invert tubes at least five times to mix the activator with the blood (*see* **Note 2**).
3. Allow samples to clot completely. Place capped tubes upright in a rack at room temperature: 60 min for plain serum and at least 30 min for SST tubes (*see* **Note 3**).
4. Centrifuge blood tube in swinging-bucket rotor for 10 min at 1200 ×*g* at room temperature (*see* **Note 4**).
5. Transfer serum to a clean tube, dilute 1:2 in an appropriate volume of sterile-double filtered TBS and mix carefully with a pipet (*see* **Note 5**).
6. Diluted serum is stable at room temperature for 4 h, up to 48 h at 2–4 °C. For long-term storage, aliquots should be frozen at −80 °C in safe-lock capped tubes. Do not store in the freezer at −20 °C.

3.2.2 Urine and Other Fluids (e.g., Peritoneal Dialysate Effluent, Ascites)

1. Collect mid-stream second morning void or other fluid into a sterile 50 mL container (*see* **Note 6**).
2. Centrifuge in swinging-bucket rotor for 10 min at 1200 ×*g* at 4 °C to remove debris (*see* **Note 7**).
3. Recover supernatant and dilute 1:1 in an appropriate volume of sterile-filtered TBS. Mix carefully.
4. Diluted fluid should be used immediately or aliquots taken and stored frozen at −80 °C.

3.3 CPP Isolation

After the removal of large contaminants (Subheading 3.3.1), crude CPP isolates can be obtained by high-speed centrifugation (Subheading 3.3.2) or size exclusion chromatography (Sub heading 3.3.3). Equivalent recoveries are achieved using either technique, but centrifugal isolation is likely to be more accessible to most users and is used routinely in our laboratory. Crude CPP isolates can be further enriched for CPP by bead-based immunoisolation (Subheading 3.3.4).

3.3.1 Removal of Large Contaminants

1. Stored serum samples should be thawed at 37 °C to avoid the formation of cryoprecipitates. Mix all samples thoroughly after thawing.
2. Centrifuge diluted samples for 30 min at 10,000 × *g* and 4 °C in an appropriate clean tube or bottle to pellet residual cell debris, cryoprecipitates, and other large contaminants (*see* Fig. 1a).

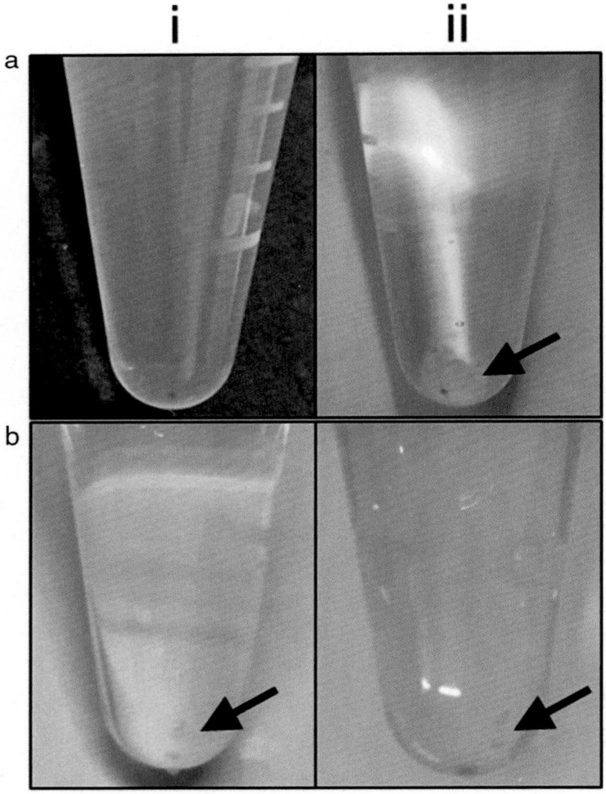

Fig. 1 The appearance of CPP-containing sediments and contaminating debris following centrifugation of human serum. (**a**) Fibrin/red cell debris following centrifugation of serum for 30 min at 10,000 × *g* and 4 °C (*i*) and after TBS wash (*ii*). (**b**) CPP-containing pellets after centrifugation of supernatant in (**a**) for 2 h at 30,000 × *g* and 4 °C (*i*) and TBS wash (*ii*). As noted in text, pellets are small and translucent or slightly off-white in colouration and difficult to visualize. *Arrows* denote position of pellet

3. Carefully transfer the supernatant to a clean safe-lock capped tube, taking care not to contaminate the supernatant with the pellet. If isolating CPP from urine do not discard the pellet at this stage.

For urine samples only:

4. Deplete polymeric THP from urine by resuspending pellets from **step 3** in 1 % (w/v) CHAPS in TBS and re-spinning for 30 min at 10,000 × *g* and 4 °C (*see* **Note 8**).
5. Pool supernatants from **steps 3** and **4** to maximize CPP recovery.

3.3.2 CPP Isolation by Centrifugation

1. Mark the outer side of each safe-lock capped tube with a pen to indicate the orientation of the tube in the rotor and help locate the pellet (*see* **Note 9**).
2. Centrifuge for 2 h at 30,000 × *g* and 4 °C to sediment the CPP-containing fraction (*see* **Note 10** and Fig. 1b).
3. Remove and discard as much supernatant as possible, then carefully wash and resuspend the pellet in 100–200 μL ice-cold TBS using a pipette. If the pellet is not visible, resuspend by flushing the pipette tip up and down the side of the tube where the pellet should be located (bottom of the tube on the side of the mark). Add ice-cold TBS to fill the tube.
4. Centrifuge for 1 h at 30,000 × *g* and 4 °C and resuspend the pellet in 100 μL warmed (37 °C) TBS.
5. Pool aliquots of the same sample and store at −80 °C until analysis.

3.3.3 CPP Isolation by Size Exclusion Chromatography (SEC)

1. Apply 3 mL aliquots of diluted sample to a HiPrep 16/60 Sephacryl™ S-500 HR column equilibrated with calcium-supplemented TBS (Ca-TBS).
2. Elute column isocratically with Ca-TBS at a flow rate of 0.5 mL/min with monitoring at 280 and 400 nm (Fig. 2).
3. Collect the 5 mL fraction eluting immediately after the void volume (*see* **Note 11**).
4. Concentrate CPP-containing fraction, by passing eluate through a centrifugal filter unit (100 kDa MWCO). Collect the retentate.
5. Pool as necessary and store at −80 °C until analysis.

3.3.4 CPP Immuno-Enrichment Using Magnetic Beads

Here, fetuin-A-containing CPP are selectively enriched from crude isolates by labeling particles with high-affinity biotinylated anti-Fetuin-A antibodies, which can then be captured using streptavidin-coated paramagnetic Dynabeads. Separation of this bound solid phase from the unbound phase is achieved with the use of a magnet.

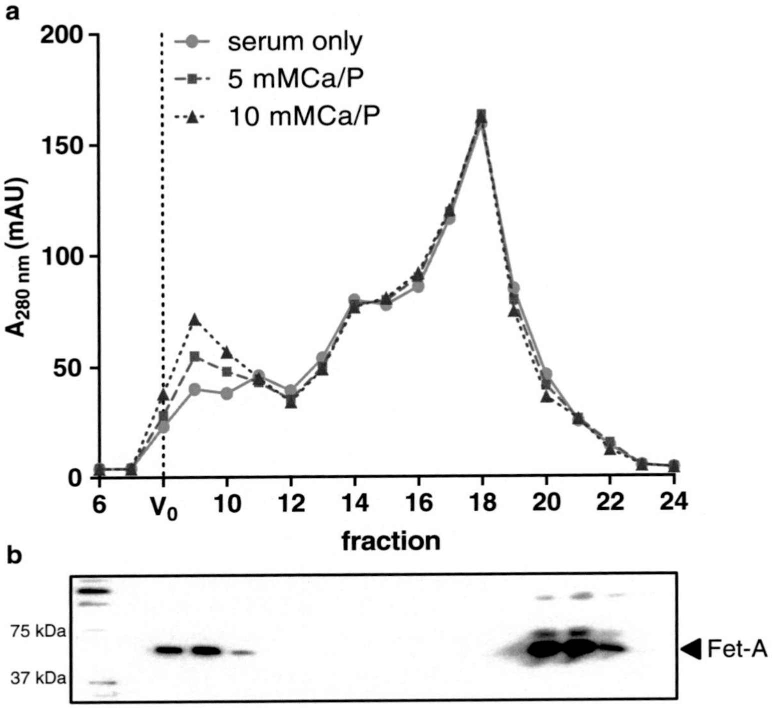

Fig. 2 CPP isolation by size exclusion chromatography. (**a**) Exemplar chromatographs showing the fractionation of fetuin-A-containing CPP from the fetuin-A monomer in human serum and serum spiked with buffered calcium and phosphate solutions (at the stated concentrations) monitored at 280 nm. CPP are eluted directly after the void volume (V_0). (**b**) Immunoblot of fetuin-A in eluted fractions

1. Titrate the volume of the pre-enriched CPP obtained from Subheading 3.3.2 or 3.3.3 to ~20 μg of total protein and make up to a final volume of 1 mL in Isolation Buffer in a clean safe-lock capped tube (*see* **Note 12**).
2. Add 10 μg biotinylated anti-Fetuin-A antibody to the tube and mix.
3. Incubate for 60 min at 2 °C with gentle mixing (*see* **Note 13**).
4. Mark the outer side of each safe-lock capped tube as described in **step 1**, Subheading 3.3.2. Centrifuge for 1 h at 30,000 × *g* and 4 °C to pellet CPP and separate from unbound antibody.
5. Remove and discard as much supernatant as possible then carefully wash and resuspend the pellet in 200 μL ice-cold Isolation Buffer using a pipette. If the pellet is not visible, resuspend by flushing the pipette tip up and down the side of the tube where the pellet should be located (bottom of the tube on the side of the mark).

6. Repeat spin **steps 4** and **5**, and resuspend pellet in 1 mL Isolation Buffer.
7. Transfer 50 μL beads into a flat-bottomed tube and wash beads with 300 μL of Isolation Buffer. Mix well (*see* **Note 14**).
8. Place the tube on the magnet for 1 min and discard the supernatant.
9. Remove the tube from the magnet and add 1 mL of the antibody-bound CPP complex mixture. Mix well and incubate the tube for 20 h at 4 °C with mixing.
10. Centrifuge the tube in a microfuge for 5 s to collect the sample at the bottom of the tube.
11. Wash the bead-bound CPP by adding 500 μL of Isolation Buffer. Mix gently by pipetting up and down.
12. Place the tube on the magnet for 1 min and discard the supernatant.
13. Remove the tube from the magnet and repeat wash with 500 μL of Isolation Buffer. Mix gently by pipetting.
14. Repeat **steps 9** and **10** twice.
15. CPP-bound beads can now be denatured for SDS-PAGE/ Western blotting (Subheading 3.5.1) or stained for flow cytometry (Subheading 3.5.2) to confirm successful isolation.

3.4 Confirming the Isolation of CPP

The techniques described in this section can be applied to bead-bound CPP isolates (Subheading 3.4.1). Conventional Western blotting and flow cytometry can be used to confirm the presence of protein markers ubiquitously present in CPP (e.g., fetuin-A, Apo-AI). The use of bead-bound CPP for staining for flow cytometry offers the advantage of be able to use a magnet to immobilize the bead for washing during separation of bound from unbound antibody, instead of more cumbersome and time-consuming centrifugation steps to pellet CPP after each wash step. It should be stressed that although Western blotting can be used to detect proteins present in purified CPP isolates, it does not confirm that they are necessarily of CPP origin and that contaminants are absent. Nonetheless, in our hands, markers of apoptotic bodies (e.g., calreticulin) and exosomes (e.g., CD63, CD81, CD9) are not usually detected in these preparations. Flow cytometry is more amenable to multiplexing targets and determining the relative abundance of various surface proteins. However, neither Western blotting nor flow cytometry of bead-bound CPP enable the user to discriminate between differences in the number of particles from changes in molecular composition/enrichment and therefore do not allow quantitation.

Analysis of particle size, morphology, crystallinity (e.g., diffraction studies), elemental composition (e.g., electron energy loss spectroscopy) and immunolabeling of free CPP can be made under cryo-TEM as reported previously (Fig. 3) [28]. An advantage of this EM modality is the avoidance of sample dehydration, collapse of the carbon (protein) matrix, and the use of chemical fixatives which can lead to artifactual appearances and compositional changes. Imaging studies can also help to assess sample heterogeneity and the presence of contaminants. Details of these EM methods are beyond the scope of this chapter and the interested reader is directed to other recent texts [29].

3.4.1 SDS-PAGE/Western Blotting Analysis of CPP

1. Place the tube containing bead-bound CPP on the magnet for 1 min and discard the supernatant (*see* **Note 15**).
2. Add 25 μL 2× reducing Laemmli Sample Buffer (containing 40 mM EDTA) and vortex vigorously for 30 s.
3. Incubate at 95 °C for 10 min.
4. Place the tube in the magnet and load supernatant and molecular weight standards onto 4–15 % TGX gels. Run for ~45 min at 200 V.
5. Transfer to 0.2 μm nitrocellulose using Trans-Blot Turbo (3 min, 2.5 A, 25 V).
6. Block the membranes in BLOTTO for 2 h at room temperature.
7. Incubate overnight at 2 °C (with gentle mixing) with primary anti-Fetuin-A or anti-ApoAI antibodies diluted 1 in 1000 in BLOTTO.
8. Wash extensively in TBS-T (6 × 10 min).
9. Incubate with species-matched TrueBlot HRP-conjugated secondary antibody (1 in 2000) plus StrepTactin-HRP conjugate (1 in 10,000) in BLOTTO for 2 h at room temperature.
10. Wash extensively in TBS-T (6 × 10 min).
11. Develop signal using SuperSignal West Dura Extended Duration Substrate and image after 5 min incubation at room temperature protected from light.

3.4.2 Staining CPP for Flow Cytometry

1. Prepare staining antibody mix: combine 1 μL APOAI antibody (AF647), 1 μL Fetuin-A antibody (AF488) in 20 μL Isolation Buffer (*see* **Note 16**). This is sufficient for ten labeling reactions.
2. Add 2 μL staining antibody mix to 100 μL of bead-bound CPP from **step 8**, Subheading 3.4.1 in a flow tube and mix gently by pipetting (*see* **Note 17**).
3. Incubate for 30 min at RT protected from light with mixing.
4. Wash the bead-bound CPP by adding 300 μL of Isolation Buffer. Mix gently by pipetting.
5. Place the tube on the magnet for 1 min and discard the supernatant.

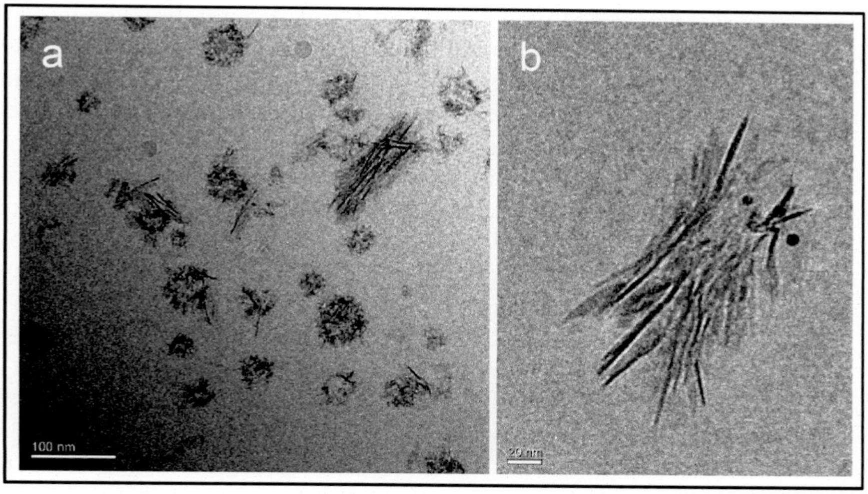

Fig. 3 Cryo-TEM analysis of fetuin-A-containing CPP isolated from human serum. (**a**) Micrograph showing heterogeneous population of mineral-containing nanoparticles isolated from human serum by differential centrifugation and immuno-enrichment (scale bar = 100 nm). (**b**) Immunogold labeling of fetuin-A in mature secondary CPP (scale bar = 20 nm)

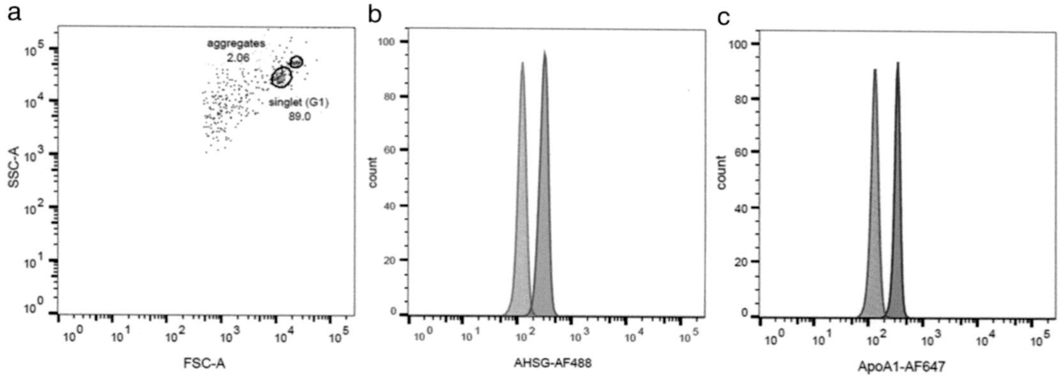

Fig. 4 Flow cytometric detection of CPP in serum from patient with CKD. CPP were pre-enriched by ultracentrifugation, bound via fetuin-A-biotin to streptavidin-coated dynabeads during an overnight incubation at 4 °C. Bound CPP were stained with the AF488/647-conjugated isotype antisera (grey), or specific antibodies to fetuin-A-AF488 (green; **b**) or ApoA1-AF647 (blue; **c**) and analysed on the FACSVerse. Events were gated on singlet bead/CPP complexes (G1) on FSC/SSC plots (panel **a**).

6. Repeat the **steps 4** and **5** once and resuspend in the desired volume of Isolation Buffer for analysis. See Fig. 4 for an example of flow cytometric analysis of CPP bound to dyanbeads.

3.5 CPP Quantitation

Hamano et al. developed an ELISA-based protocol for measuring circulating CPP levels based on the apparent reduction in (total) serum fetuin-A concentrations after high-speed centrifugation of serum [23] (Fig. 5). Using this analytical approach, CPP were detectable in serum from CKD patients but not in serum from healthy individuals (NB, this is likely to reflect the relative insensitivity and imprecision of the method and does *not* indicate their absence in healthy sera) [23]. Moreover, whereas CPP levels were associated with the severity of coronary artery calcification on CT, total serum fetuin-A concentrations showed no significant association with calcification scores [23], underscoring the importance of considering CPP-associated fetuin-A levels rather than total serum fetuin-A. Our group has applied the same methodology (with some modifications) to the measurement of serum CPP in other CKD and non-CKD cohorts, where, consistent with their inflammatory potential in vitro [28], relationships between fetuin-A reduction ratios (RR) and systemic inflammatory status has been reproducibly observed [17, 22, 24].

A protocol for the determination of serum fetuin-A reduction ratios is given below. However, a number of theoretical and technical shortcomings of this procedure should be acknowledged from the outset. Overall, the technique lacks satisfactory analytical specificity or sensitivity. Fetuin-A may bind other high molecular weight species (possibly nonspecifically) that are sedimented in the

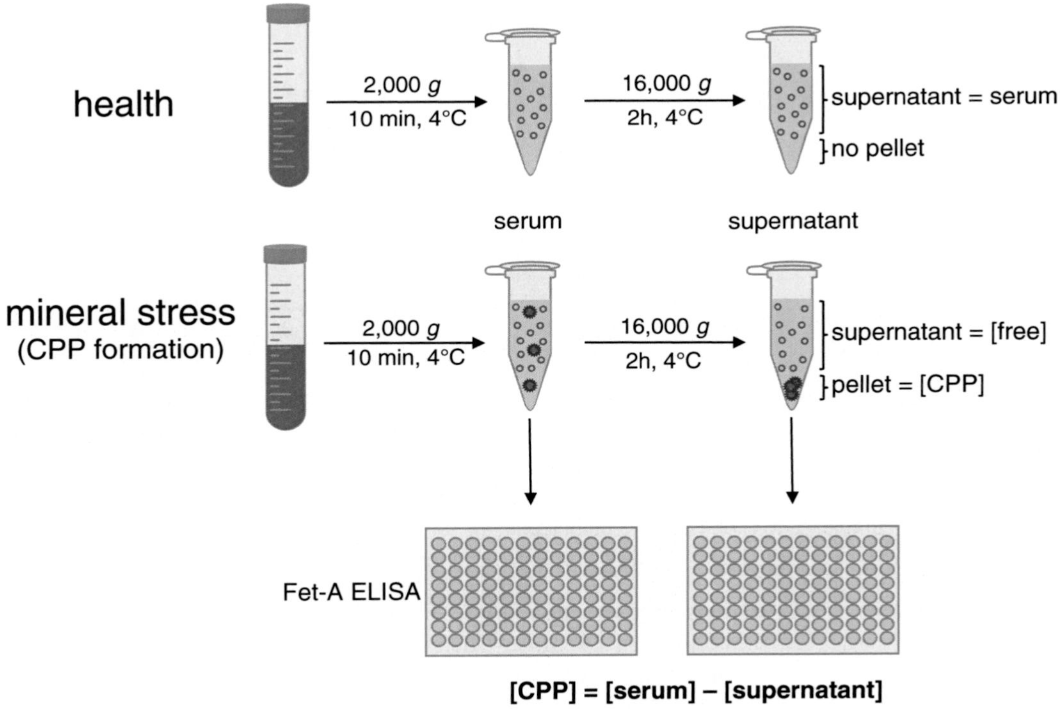

Fig. 5 CPP detection by serum fetuin-A reduction with high-speed centrifugation as originaly described by Hamano et al. [Ref. 23]. In health, CPP are undetectable and free fetuin-A does not sediment after centrifugation [16,000 × *g* 2 h, 4 °C]. In states of mineral stress, CPP are formed in extracellular fluid to inhibit mineral precipitation and are sedimented with centrifugation [16,000 × *g* 2 h, 4 °C], resulting in an apparent reduction in supernatant fetuin-A concentrations

same fraction (e.g., extracellular vesicles/cell debris), which may result in an over-estimation of circulating levels and give a false positive signal. In particular, heavily mineral-laden matrix vesicles (and other exosomal bodies) are likely to be pelleted in this fraction which may, in part, explain the apparent discrepancy of the proportion of fetuin-A present in the CPP-containing fraction in serum and that predicted from studies of simple saturated solutions (<5 %) [6]. Moreover, such a test readout does not provide a measure of particle number (rather of fetuin-A mass/occupancy), nor distinguish between primary and secondary CPP, which potentially have quite disparate biological effects. Indeed, theoretical considerations suggest that denser heavily crystalline mineral-laden secondary CPP may be relatively enriched by ultracentrifugation. The sensitivity of the analysis is also inherently constrained, as estimates are based on the small numerical difference in two values, both measured with error, following substantial (1 in 10,000), multistep dilution. Indeed, the relative abundance of fetuin-A in serum (~0.5 g/L), as well as its evident functional and structural heterogeneity, does not favor measurement of this protein by immunoassay. Furthermore, despite the crude nature of the separation, exceedingly low analytical imprecision (at the limit of attainability for ELISA) is an absolute requirement and necessitates the use of expensive high-precision pipetting systems (e.g., Precision™ Automated Microplate Pipetting Systems, BioTek; Liquidator 96, Mettler Toledo). Even then, adequate analytical control is difficult to achieve even in experienced hands due to variation in the manufacture of commercial ELISA kit.

An alternative analytical strategy is to quantitate the amount of mineral contained within the CPP-fraction. Standard colorimetric calcium assays (e.g., o-cresolphthalein complexone) are not sufficiently sensitive to detect the very small quantities of calcium typically found in this fraction (<1 μg/mL of serum [17, 23]). Acid-dissolved sediments can be analyzed for calcium by ultrasensitive graphite furnace atomic absorption spectrophotometry, as described by us previously [17], but instrumentation for this is not widely available and is not suitable for high-throughput analysis. Fluorometric-based assays using various proprietary calcium-sensitive probes are commercially available (e.g., Rhod Red or calcein) and demonstrate much greater sensitivity than their colorimetric counterparts. The use of fluorophore-conjugated bisphosphonate-based imaging agents (e.g., Osteosense, PerkinElmer) may enhance specificity for the mineral phase (e.g., OCP, or HAP) and permit direct measurement in fluids, (without the need for acid dissolution or separation) but lack validation in such quantitative applications. Indeed, a variety of mineral phases are likely to coexist in extracellular fluid that may not only differ in their affinity for probe binding but also complicates calibration and standardization of the readout. Qualitative and quantitative differences in the protein corona of mineral nanoparticles (e.g., primary vs. secondary CPP) may also significantly impact on probe binding and accessibility to the mineral core. Fundamentally, measurement of the mineral phase alone offers little advantage over protein-directed strategies and arguably even less analytical specificity given the abundance of calcium-containing contaminants that are present in the CPP-fraction. Like fetuin-A reduction ratios by ELISA, mineral detection does not yield information about absolute particle number or type. Furthermore, while significantly correlated, fetuin-A reduction ratios and CPP mineral content do not appear consistently related [17].

3.5.1 Determination of Fetuin-A Reduction Ratios in Serum

1. Collect blood and prepare serum samples according to Subheading 3.2.1.
2. Remove large insoluble debris/contaminants by centrifugation as described in **steps 1–3** in Subheading 3.3.1.
3. Dilute each sample supernatant 1 in 100 with TBS to a total volume of 2 mL (*see* **Note 18**).
4. Combine 1 mL of each diluted supernatant with an equal volume of TBS containing 20 mM EDTA into a safe-lock tube and label "TF"—'total fetuin-A'(*see* **Note 19**).
5. To the remaining 1 mL, mix with an equal volume of TBS and label "PF"—'pellet fetuin-A'.
6. Mix both aliquots for 2 h at 4 °C.
7. For "PF" samples, mark the outer side of tube with a pen to indicate the orientation of the tube in the rotor and help locate the pellet. Centrifuge for 2 h at 30,000 × *g* and 4 °C to sediment the CPP-containing fraction.
8. Carefully transfer the supernatant to another clean tube and label "SF"—'supernatant fetuin-A'.
9. Dilute TF and SF samples 1 in 10 in 1× Dilution Buffer and mix thoroughly.
10. Measure fetuin-A concentration for each sample (in duplicate) using commercial ELISA kit according to the manufacturer's instructions (*see* **Note 20**).
11. Determine the fetuin-A reduction ratio for each sample according to the following formula:

$$\left([\text{fetuin-A}]_{\text{TF}} - [\text{fetuin-A}]_{\text{SF}}\right) / [\text{fetuin-A}]_{\text{TF}} \times 100$$

4 Notes

1. Blood collection tubes containing calcium-chelating anticoagulants (EDTA, citrate) which make result in the dissolution of CPP are not suitable for testing. Use of tubes with heparin-

based anticoagulants is not recommended due to variable effects on CPP recovery and downstream techniques.

2. Serum and SST tubes contain a powdered silica clot activator and a silicone-coated interior and must be inverted to ensure proper mixing and sample clotting. Improper mixing can result in a gelatinous/fibrinous serum sample which may interfere with pelleting efficiency and the recovery of CPP. Do not shake blood tubes vigorously as this can result in hemolysis. Heavily hemolysed samples are not suitable for analysis due to the formation of hemoglobin aggregates and hemoglobin–haptoglobin precipitation complexes which may contaminate the CPP-fraction. The release of large amounts of intracellular phosphate from lysed cells may also result in the ex vivo production of CPP (i.e., false positive) with prolonged storage.

3. Clotting times may have to be extended for patients who are receiving anticoagulant therapy (e.g., heparin or Coumadin). Tubes must remain stoppered as changes in pH may occur with loss of dissolved CO_2. Premature centrifugation may result in a gelatinous/fibrinous serum that is unsuitable for further processing. Re-spinning may result in loss of CPP. Excessive delays (>48 h) in separation from cells may result in further changes in pH (due to anaerobic metabolism of cells) and leakage of cell contents into serum (e.g., phosphate). Relatively short delays in centrifugation (<2 h) may also result in the release of contaminating extracellular vesicles.

4. Always spin blood at room temperature to avoid the formation of cryoprecipitates and other insoluble protein aggregates.

5. HEPES-buffered saline can be used as an alternative to TBS [50 mM HEPES, 140 mM NaCl pH-adjusted with 10 M NaOH to 7.40 at 37 °C]. A minimum volume of 2 mL of diluted serum (400 μL) is needed for reliable downstream analysis. Higher dilutions (1:10) can be employed with more viscous samples.

6. CPP urinary load varies greatly between individuals and over time in the time in the same individual. To maximize recovery, it is recommended that at least 30 mL is collected to provide sufficient material for analysis. The use of "fresh" urine is desirable and samples should be processed within 4 h of collection to avoid bacterial growth/contamination. Ideally, urine samples should be stored on ice or at 4 °C until the preparative steps are complete.

7. Pellets can be difficult to see and may be frankly invisible to the eye: to avoid potential contamination leave the bottom 10 mL of supernatant so not to disturb the pellet.

8. Tamm-Horsfall protein (THP) is abundant in urine and can polymerize to form dense aggregates that trap particulates like CPP and which sediment at relatively low speeds.

9. Safe-lock tubes should be used to prevent evaporative losses during high-speed centrifugation (even at low temperature). This is imperative if performing quantitative volume-based measurements.
10. The CPP-containing pellet is typically very small and translucent or slightly off-white in colouration. In some instances, the pellet may not be visible.
11. The void volume (V_0) can be determined using 1 μm diameter polystyrene sizing beads (Spherotech, Lake Forest, IL, USA).
12. Estimation of total protein can be performed by Micro BCA™ Protein Assay Kit (Thermo Scientific).
13. Adequate mixing is essential for successful CPP isolation. Use a mixer that provides titling and rotation of tubes to ensure beads do not settle during incubation. End-over-end mixing is suboptimal.
14. This protocol has been optimized to achieve high CPP depletion efficiency from pre-enriched serum samples suitable for quantitation and morphological/elemental analysis by cryo-TEM or proteomic workflows. Fewer beads (20 μL per 1 mL) may be needed for some flow cytometry applications to maximize the coupling of particles per bead. The protocol can be scaled proportionately if larger or smaller input volumes are used.
15. If analyzing unbound CPP from **step 8**, Subheading 3.4.2 for Western blotting, samples will first need to be concentrated using a microcentrifugal filter unit (e.g., Amicon 0.5 mL Ultracel 100K). Mix with an appropriate volume of 2× reducing Laemmli Sample Buffer (containing 40 mM EDTA), heat to 95 °C for 5 min and then process sample as for bead-bound CPP.
16. Remove unreacted dye and protein aggregates from labelled antibody by centrifugation (10,000 g, 4 °C, 30 min) before use in staining.
17. If staining unbound CPP from **step 8**, Subheading 3.4.2 for flow cytometry, labeled particles will need to be separated from unbound labeled antibody by centrifugation. Follow **steps 4** and **5** in Subheading 3.4.1, but resuspend pellet in 100 μL Isolation Buffer after final wash.
18. Note requirement for high-precision pipetting for dilution steps (<1.0 % for 10 μL).
19. Direct detection of CPP-bound fetuin-A on the solid phase of the ELISA plate is unreliable, is nonstoichiometric and tends to underestimate total fetuin-A present due to the masking of epitopes. Estimates of total fetuin-A (TF) should therefore be based on samples following dissolution with calcium chelation (EDTA).
20. Note requirement for very low analytical coefficients of variation (<2.5 %) to permit accurate derivation of fetuin-A reduction ratios.

Acknowledgements

Thanks to Dr. Eric Hansen (Bio21 Electron Microscopy Unit, University of Melbourne) for technical assistance with the EM analyses. This work was supported by an unrestricted investigator-initiated grant from Amgen Australia Pty Ltd, Baxter Pty Ltd, the Kincaid-Smith Research Foundation, and a RMH Home Lottery Research Grant-in-aid.

References

1. Jahnen-Dechent W, Heiss A, Schafer C et al (2011) Fetuin-A regulation of calcified matrix metabolism. Circ Res 108(12):1494–1509
2. Heiss A, DuChesne A, Denecke B et al (2003) Structural basis of calcification inhibition by alpha 2-HS glycoprotein/fetuin-A. Formation of colloidal calciprotein particles. J Biol Chem 278(15):13333–13341
3. Heiss A, Jahnen-Dechent W, Endo H et al (2007) Structural dynamics of a colloidal protein-mineral complex bestowing on calcium phosphate a high solubility in biological fluids. Biointerphases 2(1):16–20
4. Herrmann M, Schafer C, Heiss A et al (2012) Clearance of fetuin-A-containing calciprotein particles is mediated by scavenger receptor-A. Circ Res 111(5):575–584
5. Heiss A, Eckert T, Aretz A et al (2008) Hierarchical role of fetuin-A and acidic serum proteins in the formation and stabilization of calcium phosphate particles. J Biol Chem 283(21):14815–14825
6. Heiss A, Pipich V, Jahnen-Dechent W et al (2010) Fetuin-A is a mineral carrier protein: small angle neutron scattering provides new insight on Fetuin-A controlled calcification inhibition. Biophys J 99(12):3986–3995
7. Rochette CN, Rosenfeldt S, Heiss A et al (2009) A shielding topology stabilizes the early stage protein-mineral complexes of fetuin-A and calcium phosphate: a time-resolved small-angle X-ray study. Chembiochem 10(4):735–740
8. Gersh I (1938) The fate of colloidal calcium phosphate in the dog. Am J Physiol 121(3): 589–594
9. Gersh I (1938) Histochemical studies on the fate of colloidal calcium phosphate in the rat. Anat Rec 70(3):331–349
10. Price PA, Thomas GR, Pardini AW et al (2002) Discovery of a high molecular weight complex of calcium, phosphate, fetuin, and matrix gamma-carboxyglutamic acid protein in the serum of etidronate-treated rats. J Biol Chem 277(6):3926–3934
11. Price PA, Caputo JM, Williamson MK (2002) Bone origin of the serum complex of calcium, phosphate, fetuin, and matrix Gla protein: biochemical evidence for the cancellous bone-remodeling compartment. J Bone Miner Res 17(7):1171–1179
12. Young JD, Martel J, Young D et al (2009) Characterization of granulations of calcium and apatite in serum as pleomorphic mineralo-protein complexes and as precursors of putative nanobacteria. PLoS One 4(5):e5421
13. Martel J, Young D, Young A et al (2011) Comprehensive proteomic analysis of mineral nanoparticles derived from human body fluids and analyzed by liquid chromatography-tandem mass spectrometry. Anal Biochem 418(1):111–125
14. Matsui I, Hamano T, Mikami S et al (2009) Fully phosphorylated fetuin-A forms a mineral complex in the serum of rats with adenine-induced renal failure. Kidney Int 75(9):915–928
15. Price PA, Roublick AM, Williamson MK (2006) Artery calcification in uremic rats is increased by a low protein diet and prevented by treatment with ibandronate. Kidney Int 70(9):1577–1583
16. Schinke T, Amendt C, Trindl A et al (1996) The serum protein alpha2-HS glycoprotein/fetuin inhibits apatite formation in vitro and in mineralizing calvaria cells. A possible role in mineralization and calcium homeostasis. J Biol Chem 271(34):20789–20796
17. Smith ER, Ford ML, Tomlinson LA et al (2012) Phosphorylated fetuin-A-containing calciprotein particles are associated with aortic stiffness and a procalcific milieu in patients with pre-dialysis CKD. Nephrol Dial Transplant 27(5):1957–1966
18. Merx MW, Schafer C, Westenfeld R et al (2005) Myocardial stiffness, cardiac remodeling, and diastolic dysfunction in calcification-prone fetuin-A-deficient mice. J Am Soc Nephrol 16(11):3357–3364
19. Schafer C, Heiss A, Schwarz A et al (2003) The serum protein alpha 2-Heremans-Schmid

glycoprotein/fetuin-A is a systemically acting inhibitor of ectopic calcification. J Clin Invest 112(3):357–366

20. Cai MM, Smith ER, Holt SG (2015) The role of fetuin-A in mineral trafficking and deposition. Bonekey Rep 4:672
21. Smith ER, Holt SG (2010) Important differences in measurement of fetuin-A. Ann Intern Med 153(6):419, author reply 419–420
22. Smith ER, Cai MM, McMahon LP et al (2013) Serum fetuin-A concentration and fetuin-A-containing calciprotein particles in patients with chronic inflammatory disease and renal failure. Nephrology (Carlton) 18(3):215–221
23. Hamano T, Matsui I, Mikami S et al (2010) Fetuin-mineral complex reflects extraosseous calcification stress in CKD. J Am Soc Nephrol 21(11):1998–2007
24. Smith ER, Ford ML, Tomlinson LA et al (2014) Serum calcification propensity predicts all-cause mortality in predialysis CKD. J Am Soc Nephrol 25(2):339–348
25. Cai MM, Smith ER, Brumby C et al (2013) Fetuin-A-containing calciprotein particle levels can be reduced by dialysis, sodium thiosulphate and plasma exchange. Potential therapeutic implications for calciphylaxis? Nephrology (Carlton) 18(11):724–727
26. Pasch A, Bachtler M, Smith ER et al (2014) Blood calcification propensity and cardiovascular events in hemodialysis patients in the EVOLVE trial. ASN Kidney Week
27. Keyzer CA, de Borst MH, van den Berg E et al (2015) Calcification propensity and survival among renal transplant recipients. J Am Soc Nephrol
28. Smith ER, Hanssen E, McMahon LP et al (2013) Fetuin-A-containing calciprotein particles reduce mineral stress in the macrophage. PLoS One 8(4):e60904
29. Aronova MA, Leapman RD (2013) Elemental mapping by electron energy loss spectroscopy in biology. Methods Mol Biol 950:209–226

Chapter 16

Combining Near Infrared Fluorescent Imaging for Calcification and Inflammation in Vascular Tissue Samples Ex Vivo

Matthias Derwall

Abstract

Quantification of vascular lesions in models of vascular calcification is crucial for testing novel treatments, but remains a challenging endeavor. Conventional methods include Oil-Red-O staining of whole tissue samples, calcium quantification in incinerated samples, or stereoisometric histologic processing. While most techniques offer fairly high levels of reliability, all of them share the fact that samples are not available for other assays following the analysis, as tissue is altered or destroyed in the course of the procedure. Furthermore, none is capable of measuring both calcification and inflammation at the same time. Here we present a novel technique for the simultaneous quantification of vascular inflammation and calcification, after which samples are still available for further histologic processing.

Key words Atherosclerosis, Inflammation, Vascular calcification, Fluorescence imaging

1 Introduction

The use of experimental imaging using fluorescence has seen a huge boost since the 1980s, when Chance and others described the capability of near-infrared photons (650–900 nm) to penetrate certain tissues [1, 2]. In fluorescing molecules, these photons are emitted when light at a particular wavelength is absorbed. These findings enabled the development of experimental near-infrared fluorescence (NIRF) imaging [3], finally leading to the development of targeted fluorescence imaging probes [4–6]. By coupling NIRF-molecules with substrates, or creating enzyme-activated probes, targeted NIRF imaging has evolved to become a valuable tool in models of cancer [7] atherosclerosis [8], thrombosis [9], osteoporosis [10], and myocardial infarction [11].

NIRF probes can be prepared for activation by specific proteases that are abundant in certain cell types or pathologies [12]. Hence, an increase of the signal intensity at the particular wavelength

Tim D. Hewitson et al. (eds.), *Kidney Research: Experimental Protocols*, Methods in Molecular Biology, vol. 1397,
DOI 10.1007/978-1-4939-3353-2_16, © Springer Science+Business Media New York 2016

signifies an accumulation of cellular mediators of inflammation such as macrophages, when the probe is cleaved by cathepsin (Prosense™). When coupled to bisphosphonates, accumulation of the probe signifies pro-osteogenic activity in osteoblast-like cells (Osteosense™).

2 Materials

2.1 Consumables

1. Sterile Phosphate buffered saline (PBS). Sterilization is recommended prior to intravenous injection.
2. Osteosense™ and Prosense™ probes. Probes are stable in aqueous solution for 6 months at 6 °C. Keep away from light to avoid loss of activity. Both reagents should be reconstituted using PBS and stored in a light-proof container at 6 °C.
   (a) Osteosense™ 680 EX (NEV10020EX, Perkin Elmer, Waltham, MA, USA) is available in several wavelengths. When combining with other near-infrared probes, make sure to choose wavelengths that are a) available on the NIRF-reader you use, and b) that all probes used differ in the wavelength they emit. Prepare a stock solution with 1.2 mL of 1× PBS to obtain the recommended dilution of 2 nmol/100 μL Osteosense™. One batch provides sufficient volume for 10 animals with an injection volume of 100 μL per mouse. Note that the injection volume should not exceed 5 mL/kg of bodyweight (hence, 150 μL max in a 30 g mouse). Therefore consider choosing a higher concentration depending on the number of probes used simultaneously in one mouse (*see* **Note 1**).
   (b) Prosense™ 750 EX (NEV10001EX, Perkin Elmer is available at different wavelengths. The probe is supplied as lyophilized solid, and must be reconstituted with 1.2 mL of 1× PBS, resulting in 2 nmol/100 μL Prosense™. One batch is sufficient volume for 10 animals with an injection volume of 100 μL per mouse (weighing about 25 g each).
3. 30G hypodermic needles taken from insulin syringes to prepare catheters for intravenous injection of the NIRF probes.
4. 20G cannulas to be used as Luer-slip connector with the catheter for intravenous injections of the NIRF probes.
5. PE20 Polyethylene tubing (Braintree Scientific, Braintree, MA, USA) to prepare catheters for intravenous injection of NIRF probes.
6. 1 mL Tuberculin syringes with μL scale and Luer Slip for intravenous injection of NIRF-probes.
7. Anesthetics for intraperitoneal injection (e.g., ketamine/xylazine) or inhalation (Isoflurane). Follow your institutions recommendations for inducing anesthesia or euthanizing mice.

8. Mouse restrainer.
9. Incandescent heat lamp.
10. Polypropylene foil. Can be cut from a transparent sheet protector.
11. Saline.
12. Petri dish.
13. Fixative or cryogenic embedding compound and liquid nitrogen.

2.2 Instruments and Software

1. Microsurgical instruments: Straight, pointed tweezers (e.g., Moria Ultra Fine Tipped Forceps, Fine Science Tools, Foster City, CA, USA); small pointed spring scissors (Fine Science Tools).
2. Odyssey infrared imaging system (LI-COR, Lincoln, NE, USA) with the appropriate software to acquire signal intensities of defined regions of interest. Other laser scanners capable of emitting and detecting light at the desired wavelengths may be suitable.
3. ImageJ software (NIH, Bethesda, Maryland, USA). This software is free to download and can be used to create heat maps from the acquired greyscale pictures to visualize signal intensities in different regions of interest.

3 Methods

3.1 Preparing the Intravenous Catheter

1. Use forceps to gently detach the 30G needle from the plastic hub of the cannula by twisting the syringe while holding the needle in place.
2. Take roughly 6 cm of PE20 tubing and insert about half of the blunt end of the 30G needle into the tubing.
3. Fix needle by adding a single drop of superglue to the joint of needle and tubing. Make sure not to occlude tubing with superglue by fixing glued catheters in an upright position until glue is solid.
4. Finally, use microscope and forceps to gently introduce a 20G Cannula into the other end of PE20 tubing.
5. Test for leakage or occlusion of the final assembly by gently flushing the catheter with sterile saline or PBS. Sterilize assembly according to your institutions protocols before using it for intravenous injections.

3.2 Intravenous Injection of NIRF Probes (See Note 2)

1. 24 h prior to imaging, prepare 1 mL syringes with the appropriate volume to inject, depending on the animal's bodyweight (not exceeding 5 mL/kg for i.v. injections). Wrap drawn syringes with aluminum foil to protect the agent from light while letting the fluid reach room temperature.

2. Position animal in a tube restrainer to gain access to one of the tail veins on each side of the tail. Consider use of an incandescent lamp or warm water to increase filling and visibility of the veins.
3. Swab the point of injection with alcohol and insert the needle at a slight angle without a syringe attached to the catheter. You may notice a flush of blood in the catheter when entering the vessel.
4. When injecting the probe, the vein should become clear, and the fluid should be injectable without significant pressure. When a bump occurs at the site of injection, stop the injection and try again and puncture the vessel proximal to the first site of injection.

3.3 Dissection

Keep tissue moist throughout the procedure using saline or PBS.

1. Euthanize the animal according to your institutional protocol.
2. Dissect the animal using a microscope and microsurgical instruments: Perform a thoracic and abdominal midline incision and remove thymoid tissue, the lungs, and abdominal organs.
3. Finally, remove connective tissue surrounding the aorta and cut out the heart with a circular incision, leaving the aortic valve in situ (*see* Fig. 1 for details) (*see* **Notes 3** and **4**).
4. Flush aorta with saline to reduce noise.
5. Before removing the aorta from the body, make sure that the connective tissue and smaller vascular branches are fully removed from the aorta. Then detach aorta starting with the femoral-, then renal-, subclavian-, and finally the carotid arteries.
6. Put vessel in a small petri dish. Flush again with saline or similar solutions.

3.4 Imaging

1. After dissection, put vessels on polypropylene-foil in the desired orientation. Consider imaging vessels from treatment- and control-group side by side to account for serial differences in exposure conditions over time (*see* **Note 5**).
2. Move prepared specimens immediately to laser scanner (Odyssey Imaging System, LI-COR Biotechnology, Lincoln, NE). As the laser scanner emits and detects light from below, put prepared foil upside down on the scanner, allowing for direct contact of the aorta to the scanner.
3. Use a saline-filled syringe with a small needle to fill the capillary gap between foil and scanner with fluid. This way, specimens are pushed down in position, while being kept moist during the process of the scan. Make sure to remove any air-bubbles before starting the scan.

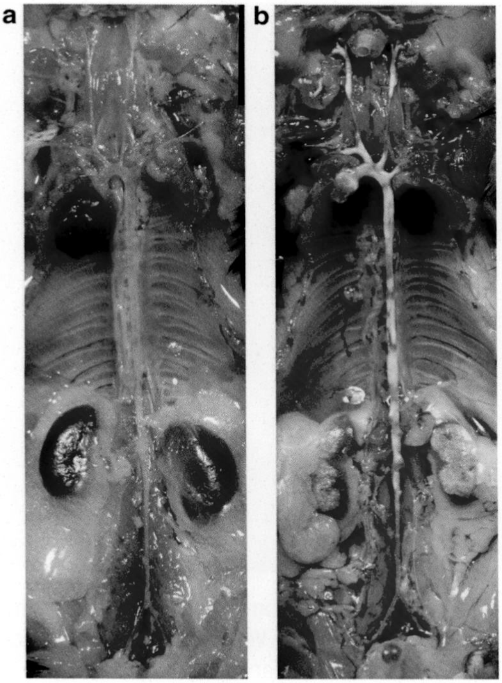

Fig. 1 Anatomic structures in dissected animals. (**a**) Exposed aorta in 26-week-old wild type mouse on high-fat diet. Note that the aorta appears transparent following an Intraventricular flush with saline. Heart and kidneys still in situ to allow for anatomic orientation. (**b**) Exposed aorta in 26-week-old LDL receptor-deficient mice ($LDLR^{-/-}$) on high-fat diet for 20 weeks. Note visible white atherosclerotic plaques in all mayor vessels including abdominal aorta and carotid arteries

4. Adjust position and size of the scanned area according to your samples. Set exposure and resolution according to your needs for detail and signal. Note that higher resolution results in longer scans. Make sure that signal intensities do not exceed the chosen sensitivity. Try to use the same set of settings throughout all of your experiment to make serially acquired samples comparable.
5. Following your scan, samples should be immediately fixed or frozen in a cryogenic compound to allow for histologic processing later on.

3.5 Analysis and Post-processing

1. Define regions of interest within the scanned pictures and measure signal intensity of each channel using the scanner software.
2. Greyscale pictures generated by the scanner software can be further processed in Image J. Here, adjust the display range before applying lookup tables to visualize signal intensities in color (The lookup table used in Fig. 2 is 'Thal').

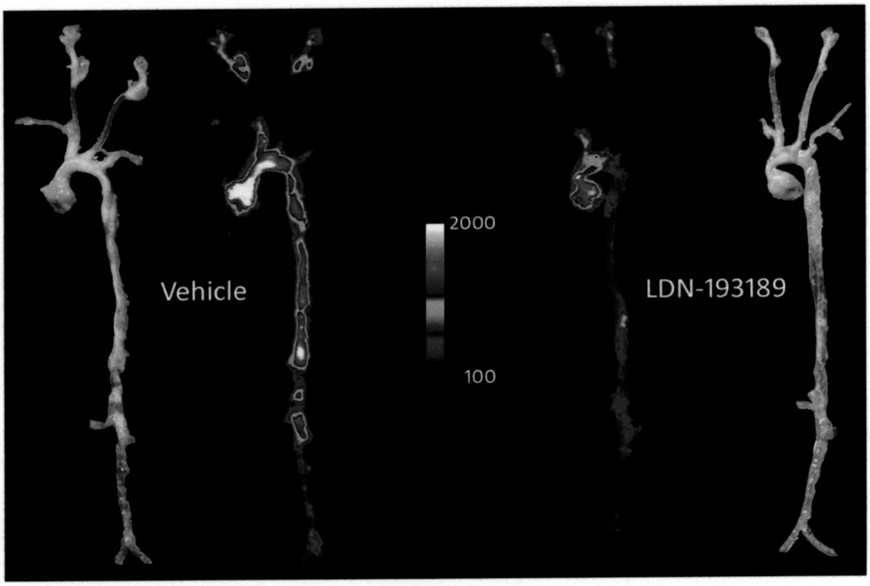

Fig. 2 Heat map of two aortas taken from 26-week-old LDLR$^{-/-}$ mice on high-fat diet for 20 weeks and treated with either 2.5 mg/kg LDN-193189 or Vehicle. Heat maps are accompanied by the respective daylight photomicrographs of the isolated vessels (far *left* and *right*). Signals imply signal intensity of Osteosense™ 680, measuring osteogenic activity in each vessel

4 Notes

1. Recommended injection-volumes per kg bodyweight may be exceeded when combining NIRF probes at the dilution recommended by the manufacturer. Consider choosing a lower dilution of the stock-solution, depending on your animal's bodyweight.
2. The recommended procedure for in vivo imaging with OsteoSense™ and Prosense™ is administration via tail vein injection and imaging 24 h post tail vein injection. However, similar results may be obtained with i.p. injections of both imaging agents and following shorter incubation.
3. Removing surrounding tissue is key to obtain clear readings. This includes removing the heart.
4. It is difficult, if not impossible, to remove adhering connective tissue from the aorta as soon as the large vessels are cut. Due to its elasticity, the tissue will collapse and make it difficult to tell apart vascular wall and connective tissue.
5. Imaging of the vasculature in whole animals is not recommended, as high intensity uptake in other tissues (bones) will overwhelm details from the vessels.

Acknowledgements

This work was supported by a research fellowship award of the German Research Foundation (Deutsche Forschungsgemeinschaft, DE-1685-1/1). The author wishes to thank Drs. K. D. Bloch and P.B. Yu for experimental advice. I am gratefully indebted for the passionate work and skillful technical assistance of Y. Beppu.

References

1. Chance B (1991) Optical method. Ann Rev Biophys Chem 20:1–28
2. Jobsis FF (1977) Noninvasive, infrared monitoring of cerebral and myocardial oxygen sufficiency and circulatory parameters. Science 198(4323):1264–1267
3. Weissleder R, Tung CH, Mahmood U, Bogdanov A Jr (1999) In vivo imaging of tumors with protease-activated near-infrared fluorescent probes. Nat Biotechnol 17(4): 375–378
4. Tatsuta M, Iishi H, Ichii M, Baba M, Yamamoto R, Okuda S, Kikuchi K (1989) Diagnosis of gastric cancers with fluorescein-labeled monoclonal antibodies to carcinoembryonic antigen. Lasers Surg Med 9(4): 422–426
5. Folli S, Wagnieres G, Pelegrin A et al (1992) Immunophotodiagnosis of colon carcinomas in patients injected with fluoresceinated chimeric antibodies against carcinoembryonic antigen. Proc Natl Acad Sci U S A 89(17): 7973–7977
6. Pelegrin A, Folli S, Buchegger F, Mach JP, Wagnieres G, van den Bergh H (1991) Antibody-fluorescein conjugates for photoimmunodiagnosis of human colon carcinoma in nude mice. Cancer 67(10):2529–2537
7. Bremer C, Tung CH, Bogdanov A Jr, Weissleder R (2002) Imaging of differential protease expression in breast cancers for detection of aggressive tumor phenotypes. Radiology 222(3):814–818
8. Deguchi JO, Aikawa M, Tung CH et al (2006) Inflammation in atherosclerosis: visualizing matrix metalloproteinase action in macrophages in vivo. Circulation 114(1):55–62
9. Tung CH, Gerszten RE, Jaffer FA, Weissleder R (2002) A novel near-infrared fluorescence sensor for detection of thrombin activation in blood. Chembiochem 3(2-3):207–211
10. Hjortnaes J, Butcher J, Figueiredo JL et al (2010) Arterial and aortic valve calcification inversely correlates with osteoporotic bone remodelling: a role for inflammation. Eur Heart J 31(16):1975–1984
11. Chen J, Tung CH, Allport JR, Chen S, Weissleder R, Huang PL (2005) Near-infrared fluorescent imaging of matrix metalloproteinase activity after myocardial infarction. Circulation 111(14):1800–1805
12. Tung CH, Bredow S, Mahmood U, Weissleder R (1999) Preparation of a cathepsin D sensitive near-infrared fluorescence probe for imaging. Bioconjug Chem 10(5): 892–896

Part V

Analytical Techniques

Chapter 17

Laser Capture Microdissection of Archival Kidney Tissue for qRT-PCR

Tim D. Hewitson, Michael Christie, and Edward R. Smith

Abstract

Whole-organ molecular analysis of the kidney potentially misses important factors involved in the pathogenesis of disease in glomeruli and tubules. Organ wide analysis can however be augmented by using laser capture microdissection (LCM) to isolate morphologically similar cells and nephron structures from a heterogeneous tissue section via direct visualization of the cells. The protocol here provides a practical approach utilizing LCM in combination with RNA isolation techniques for downstream analysis. This technique is readily applicable to study mRNA expression in isolated glomeruli and tubules in both experimental animal models and human kidney biopsy material.

Key words Laser capture microdissection, Kidney, Glomeruli, Tubules, RNA, PCR

1 Introduction

Studies of whole kidney tissue have for many years provided invaluable insights into both pathophysiology and mechanisms of renal disease. These have been supplemented by histological and histochemical analysis, the underlying basis of diagnosis in most forms of kidney disease. Notwithstanding the value of these approaches, molecular insights into these mechanisms are hampered by the interrelated processes involved, heterogeneity in gene expression across renal compartments [1], and the existence of more than 20 different cell types in the kidney [2]. While global screening has identified candidate genes, these techniques therefore often fall short of identifying compartmental or cellular culprits of altered gene expression.

A number of different approaches have been used to overcome these issues. In situ hybridization offers spatial resolution through annealing a labeled nucleic acid probe to complementary sequences in a tissue section. However, the method lacks sensitivity for all but the most abundantly transcribed genes, is only semiquantitative, and is in practical terms limited to those targets for which a cloned cDNA/cRNA probe can be made [3].

Tim D. Hewitson et al. (eds.), *Kidney Research: Experimental Protocols*, Methods in Molecular Biology, vol. 1397, DOI 10.1007/978-1-4939-3353-2_17,

A major technical innovation has been the development of laser capture microdissection (LCM) [4]. LCM allows detailed molecular interrogation of individual structures by isolating pure structures and cells from a heterogeneous tissue section. It is by design ideally suited to the kidney's anatomically distinct compartments, and can be combined with quantitative PCR analysis, making it a useful adjunct to other analyses.

Several proprietary LCM platforms currently exist. First developed at the National Institute of Health (NIH) laboratories, both manual (e.g., Arcturus PixCell™, Bio-Rad Clonis™) and automated robotic systems (e.g., Veritas™) are now available commercially. The available platforms use a variety of mechanisms to isolate and capture the cells or tissues of interest. In several platforms, including that described in this chapter, a laser is used to capture individual structures by melting a plastic film to cells of interest. Peeling away the film then removes the targeted cells from the tissue section.

Although originally used and validated with frozen tissue [5], LCM has now also been applied to paraffin-embedded fixed tissue. Fixation confers chemical stability on tissue, hardens tissue for sectioning and most importantly, halts autolysis and degradation. However, changes to the molecular form mean that fixation is often therefore a compromise between retention and preservation. Comparative studies suggest RNA stability is better with denaturing fixatives (methanol, acetone, acetic acid) than cross linking fixatives (formalin, paraformaldehyde) [5]. Recovery from paraffin-embedded tissue is possible with the short processing schedules routinely used in most pathology laboratories [6, 7]. Validating LCM with this tissue has both improved resolution of the technique, and made available an abundance of archived material [7].

In this chapter we describe a method for preparing material for the quantitative analysis of gene expression in different compartments of the kidney. The technique uses LCM to isolate RNA from glomeruli and tubules for subsequent analysis of gene expression with quantitative PCR (qRT-PCR).

2 Materials

2.1 Tissue Collection and Processing

2.1.1 Fixative

Methyl Carnoy's fixative: 60 % methanol, 30 % chloroform, 10 % glacial acetic acid v/v/v.

2.1.2 Paraffin Embedding and Sectioning

1. Double deionized H_2O (ddH_2O).
2. Labeled glass vials.
3. Graded ethanol: 50, 70, 95, and 100 % v/v mixture of ethanol and ddH_2O.

4. Chloroform.
5. Paraplast™ paraffin embedding medium (McCormick Scientific, St. Louis, MO, USA), or equivalent low temperature embedding wax.
6. Embedding molds and cassettes.
7. Scalpel and No. 22 blade.
8. Artist's paint brush.
9. Shallow bowl filled with a 20 % v/v mixture of ethanol and water.
10. Microtome.
11. RNase free LCM plastic slides (Arcturus Bioscience, Mountain View, CA, USA).
12. Histological water bath.
13. Slide racks.
14. Microslide box.
15. Silica desiccant.
16. Kimwipes™ (Kimberly-Clark, Roswell, GA, USA).
17. RNaseZAP™ (Sigma-Aldrich, St Louis, MO, USA).

2.1.3 Section Staining

1. HistoGene™ Frozen Section Staining Kit (Arcturus).
2. Plastic slide jars or 50 mL centrifuge tubes.
3. Slide forceps.

2.2 Laser Capture Microscopy

1. CapSure™ Macro LCM Caps (Arcturus).
2. Laser capture microdissection instrument (Veritas™ microdissector, Arcturus).

2.3 RNA Isolation and Amplification

1. Paradise PLUS™ Reagent System (Arcturus) including reagents for RNA isolation, extraction, Whole Transcript RT, and quality assessment.
2. SuperScript™ III Reverse Transcriptase, 200 U/μL (Invitrogen, Thermo Fisher, Waltham, MA, USA).
3. Incubator (e.g., hybridization oven).
4. Temperature-controlled microcentrifuge.
5. Cold block/ice bath.
6. Snap-top nuclease-free microcentrifuge tubes (0.2, 0.5 mL).
7. Sterile nuclease-free filter pipette tips.
8. UltraPURE™ DNase, RNase-free water (Invitrogen).
9. C1000 thermal cycler (Bio-Rad, Hercules, CA, USA).
10. Freezer (−80 °C) for storage.

2.4 qRT-PCR

1. CFX96 Real-time PCR detection system (Bio-Rad).
2. SYBR Select™ Master Mix (Applied Biosystems).
3. UltraPURE™ DNase, RNase-free water (Invitrogen).
4. 96-well Hard-Shell optical PCR plate (Bio-Rad).
5. Optical adhesive film (Bio-Rad).
6. Primers for gene of interest (Sigma).
7. Universal Mouse Reference RNA (Agilent Technologies).
8. High Pure PCR Clean-Up Kit (Roche Applied Sciences).

3 Methods

The following protocols detail the necessary steps for tissue processing, LCM, RNA extraction, amplification, and analysis. The techniques described are based on the proprietary Arcturus Bioscience work flows.

3.1 Processing and Collection of Kidney Tissue

In summary, kidney tissue fixed in Methyl Carnoy's and embedded in paraffin wax is sectioned at ~10 μm. Sections are attached to RNase free plastic slides (Arcturus). The slides are de-waxed using an RNase-free technique (*see* **Note 1**), and stained using the HistoGene™ kit (Arcturus). The HistoGene™ stain has been especially developed to stain tissue sections that are going to be used as a source of RNA. It is a fast penetrating stain that provides good contrast by differential staining of nuclei (purple) and cytoplasm (light pink) while preserving RNA integrity. By providing contrast, it enables the operator to subsequently identify areas of interest for excision.

3.1.1 Fixation and Processing

1. Immerse in cold freshly prepared fixative, 10× the volume of tissue.
2. Store at −4 °C for 2–18 h, depending on size of tissue.
3. Transfer tissue to labeled glass vials.
4. Dehydrate through graded alcohols (50, 70, 95, 100 %) and clear in two changes of chloroform (1 h each). Rotate tissue throughout processing to ensure infiltration (*see* **Note 2**).
5. Infiltrate tissue with two changes of molten paraffin wax (approximately 56 °C) for a total of 4–6 h (*see* **Note 3**).
6. Orientate and embed tissue in fresh wax using warm embedding molds and cassettes.
7. Place tissue and molds in a −20 °C freezer for a minimum of 2 h before separating the block from the mold.

3.1.2 Sectioning of Paraffin-Embedded Tissue

Wear disposable clean gloves throughout the slide preparation procedure.

1. Wipe all accessible surfaces with RNaseZAP™.
2. Fill histological water bath with ddH_2O water and heat to 56 °C.
3. Prepare a transfer bath by filling a shallow bowl with 20 % v/v ethanol in water.
4. Cut 2–5 μm ribbon sections of paraffin-embedded tissue with a microtome, in accordance with the manufacturer's instructions.
5. Use a scalpel to divide ribbon in two to three section pieces. Using the edge of the blade transfer to the 20 % ethanol bath. The section should flatten.
6. Transfer to the heated histological water bath using an uncoated slide.
7. Collect sections on labeled slide, and stand to drain. After 5–10 min, transfer slides to a dry sealed container with desiccant until use.

3.1.3 Staining and Dehydration

Change all solutions in the plastic slide jars between each batch of slides to avoid cross contamination. Do not reuse solutions, and do not transfer solutions back into their original bottles. If you plan to reuse jars, discard all solutions upon completion of the staining and clean them.

1. Label seven plastic slide jars or clean 50 mL Falcon tubes as follows: (a) 75 % ethanol, (b) distilled water, (c) distilled water, (d) 75 % ethanol, (e) 95 % ethanol, (f) 100 % ethanol and (g) xylene.
2. Using the LCM Certified RNase free solutions provided with the HistoGene™ Section Staining Kit (Arcturus), fill the labeled plastic slide jars with 25–40 mL of the appropriate solution.
3. Remove slides from the slide box.
4. Place the slides in plastic slide jar containing 75 % ethanol for 30 s. Use slide forceps (first cleaned with RNaseZAP™) to transfer slides from jar to jar.
5. Transfer the slides to plastic slide jar containing distilled water for 30 s.
6. Place the slides on a Kimwipes™ or a horizontal staining rack.
7. Using an RNase-free pipette tip, apply approximately 100 μL of the HistoGene™ Staining Solution so that it covers the section. Stain for 20 s.
8. Place the slides in plastic slide jar containing distilled water for 30 s.
9. Consecutively transfer slides to plastic slide jars containing 75, 95, and 100 % ethanol for 30 s each.

10. Transfer the slides into plastic slide jar containing xylene for 5 min.
11. Place the slides on a Kimwipes™ to dry for 5 min.
12. Place all slides in a slide box containing fresh desiccant.
13. Remove one slide and perform LCM, keeping the remainder in the desiccator until ready.
14. Discard all used staining and dehydration solutions according to standard procedures.

3.2 Laser Capture Microdissection

The protocol here is based on the use of CapSure™ caps coated with a thermoplastic film. After these are placed over the section, the Veritas™ microdissection instrument uses a narrow UV cutting laser to demarcate the area of interest. The selected area is then pulsed with a larger spot size IR capture laser to fuse the cap polymer with the tissue. Cells adhere to the cap surface when it is lifted from the tissue section, leaving behind the surrounding tissue. Using multiple caps loaded at the same time, it is possible to cut out and retrieve different cell populations from the same section.

1. Load sample slides.
2. Load LCM Caps (CapSure™; Arcturus).
3. Locate target area.
4. Use cursor to demarcate area for cutting and capturing.
5. Repeat to obtain as many cells as possible (*see* **Note 4**).
6. Adjust laser spot size to produce overlapping laser pulses covering as much of the area of interest as you can (*see* **Note 5**; Fig. 1).
7. Select cut and capture.
8. Create additional capture groups.
9. Cut and capture again as above.
10. Offload caps and slides.

3.3 Extraction, Isolation, and Purification of Total RNA from LCM of Archival Tissue Samples

Molecular analysis of archival fixed paraffin-embedded LCM tissue presents a significant technical challenge to the researcher. As already noted, RNA is highly susceptible to fragmentation (strand breaks) and chemical modification during the fixation and paraffin embedding preservation process. The quality of the specimen, duration of preservation prior to LCM and the expediency of microdissection and subsequent processing also all impinge greatly upon the quantity and quality of the final RNA yield. The following procedures are based around the use of proprietary Arcturus kit reagents and standardized protocols that are designed to maximize the recovery of RNA of sufficient quality from LCM samples to permit downstream gene expression analysis.

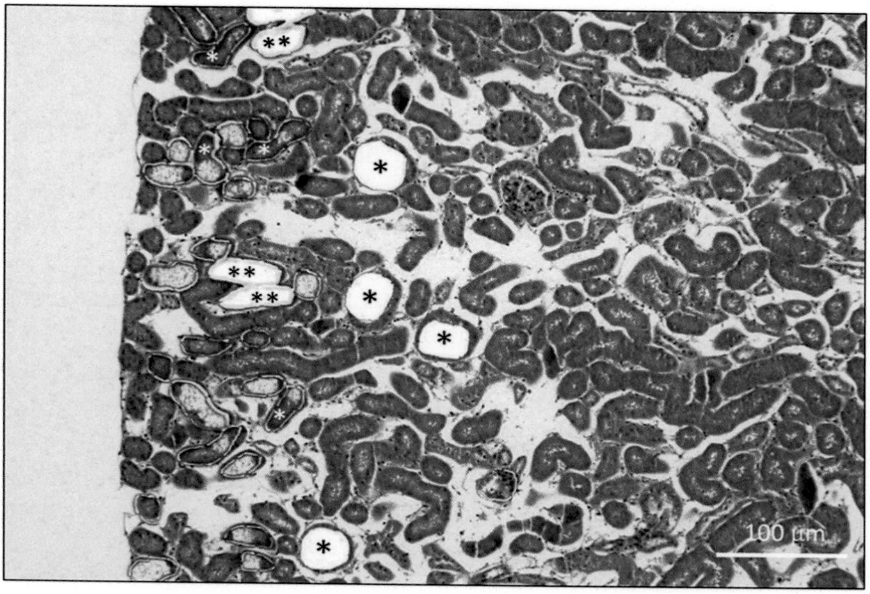

Fig. 1 Methyl Carnoy's-fixed paraffin-embedded tissue section after LCM of glomeruli (*black asterisk*) and proximal tubules (*double black asterisk*). In each case demarcation of individual glomeruli and tubules with the cutting laser can clearly be seen. The *white asterisk* shows examples where outlined tubules have not been fully captured. This indicates the importance of using multiple overlapping laser spots to fully fuse the section to the polymer cap. Scale bar = 100 μm

3.3.1 Extracting Captured Tissue from LCM Caps

1. Prepare extraction tubes prior to commencing LCM. Reconstitute Pro K Mix with 300 μL of Reconstitution Buffer (*see* **Note 6**). This provides sufficient mix for 12 extractions. For each cap, pipette 40 μL of the reconstituted Pro K Mix into a 0.5 mL microcentrifuge tube. Keep on ice until use.
2. Once regions of interest are captured, insert CapSure™ Macro LCM Cap into the microcentrifuge tube using the LCM Cap Insertion Tool, taking care not to touch the transfer film surface (*see* **Note 7**). Tap the microcentrifuge tube to ensure the extraction buffer is covering the CapSure™ Macro LCM Cap. Seal the assembly using Parafilm.
3. Incubate assembly for 16–18 h at 50 °C (*see* **Note 8**).
4. Centrifuge assembly at 1000 × *g* for 2 min to collect cell extract and proceed directly with the RNA isolation protocol as detailed below.

3.3.2 Isolation of Total RNA from Microdissected Tissue

RNA (e.g., glomerular RNA) may be extracted from the digested tissue using the Paradise PLUS Reagent System RNA Extraction/ Isolation (Arcturus), which incorporates an on-column DNase digestion step to remove genomic DNA contamination. Briefly, the steps are as follows:

1. Pre-condition the RNA Purification Column:

   (a) Pipette 250 μL Conditioning Buffer (CB) onto the purification column filter membrane and incubate for 5 min at room temperature.
   (b) Centrifuge the purification column in a collection tube at 16,000 × *g* for 1 min.
2. Pipette 50 μL of 70 % Ethanol into the cell extract obtained from **3.3.1**. Mix well by pipetting up and down. Do not centrifuge.
3. Pipette the cell extract and ethanol mixture (total volume ~80 μL) into the preconditioned column.
4. Centrifuge for 2 min at 100 × *g* to bind RNA to Column membrane filter, immediately followed by a centrifugation at 16,000 × *g* for 30 s. Discard flow-through.
5. Pipette 100 μL Wash Buffer (W1) into the purification column and centrifuge for 1 min at 8000 × *g*. Discard flow-through (*see* **Note 9**).
6. Prepare DNase incubation mix by adding 2 μL DNase I Mix to 20 μL DNase Buffer. Mix by gently inverting—do not vortex.
7. Pipette the Complete DNase incubation mix directly into the center of the column membrane. Incubate at room temperature for 25 min (*see* **Note 10**).
8. Pipette 40 μL RNA Kit Wash Buffer 1 (W1) into the purification column membrane. Centrifuge at 8000 × *g* for 15 s.
9. Pipette 100 μL Wash Buffer 2 (W2) into the purification column and centrifuge for 1 min at 8000 × *g*.
10. Pipette another 100 μL Wash Buffer (W2) into the purification column and centrifuge for 2 min at 16,000 × *g*. Check the purification column for any residual wash buffer. If wash buffer remains recentrifuge at 16,000 × *g* for 1 min and discard any flow-though.
11. Transfer the purification column to a new 0.5 mL microcentrifuge tube.
12. Pipette 12 μL Elution Buffer (EB) directly onto the membrane of the purification column. Gently touch the tip of the pipette to the surface of the membrane while dispensing the elution buffer to ensure maximum absorption of EB into the membrane filter.
13. Incubate the purification column for 1 min at room temperature.
14. Centrifuge the column for 1 min at 1000 × *g* to distribute EB in the column, then for 1 min at 16,000 × *g* to elute total RNA. Optional: Retain 1 μL of sample to determine RNA yield (*see* **Note 11**). The remaining sample may be used immediately or stored at −80 °C until use.

3.4 cDNA Synthesis, Purification, and Amplification

The RNA yield from LCM is typically very low (<5–200 ng). Since several micrograms of RNA are needed for array-based applications, the RNA must be amplified. Conventional in vitro transcription (IVT) methods rely on the presence of an intact 3′ polyA tail for amplification by T7 RNA polymerase. However, RNA derived from paraffin-embedded fixed tissue is often degraded, which may lead to preferential amplification of transcripts with intact 3′ ends and thus biased expression analyses. As described here, Whole Transcriptome Amplification (WTA) reagents, such as those employed by the Paradise® PLUS Reagent (Arcturus) system, circumvent the issue of 3′ end bias by the use of unique chimeric DNA–RNA primers. The DNA portion of these oligonucleotides hybridizes randomly across the entire transcriptome, while the RNA portion encodes a sequence that serves as a priming site for linear amplification by T7 RNA polymerase. Reverse transcription proceeds from the 3′ DNA end of each primer generating first strand cDNA. Second strand synthesis is then directed by the addition of a second exogenous primer, which generates dsDNA. Column-based purification of this cDNA followed by IVT with T7 RNA polymerase yields antisense RNA (aRNA) before isolation and a further round of amplification (optional for some applications). Detailed protocol notes can be found in the manual for the Paradise PLUS reagent system but the key steps and suggested modifications are given below.

3.4.1 Round One: First Strand cDNA Synthesis

1. Add 1.0 μL of Primer 1 (Gray-1) to 10–11 μL of purified RNA (input >20 ng) and incubate mixture for 1 h at 70 °C in a thermal cycler.
2. Cool samples to 4 °C for at least 1 min.
3. Prepare First Strand Synthesis Reaction Mix by combining 2 μL of Enhancer (Yellow), 5 μL of first Strand Master Mix (Red-1), 1 μL of first Strand Enzyme Mix (Red-2) and 1 μL of SuperScript™ III RT Enzyme (Invitrogen). Scale up with 10 % overage for multiple samples.
4. Add 9 μL of this complete First Strand Synthesis Reaction Mix to each sample, mix thoroughly and incubate at 42 °C for 1.5 h.
5. Cool samples to 4 °C for at least 1 min.
6. Add 2.0 μL of First Strand Nuclease Mix (Gold) to each sample, mix thoroughly and incubate in a thermal cycler for 30 min at 37 °C, followed by 95 °C for 5 min.
7. Cool samples to 4 °C.
8. Proceed directly to Subheading 3.4.2 (second strand cDNA synthesis) or store sample at −20 °C overnight.

3.4.2 Round One: Second Strand cDNA Synthesis

1. Add 1.0 μL of Primer 2 (Gray-2) to sample and incubate mixture for 2 min at 95 °C in a thermal cycler.
2. Cool samples to 4 °C for at least 1 min.

3. Prepare Second Strand Synthesis Reaction Mix by combining 29 μL of second Strand Master Mix (White-1), 1 μL of second Strand Enzyme Mix (White-2). Scale up with 10 % overage for multiple samples.
4. Add 30 μL of this complete Second Strand Synthesis reaction mix to each sample, mix thoroughly, and incubate in a thermal cycler as follows:

   25 °C 10 min

   37 °C 30 min

   70 °C 5 min

   4 °C hold until ready to proceed with purification (max. 30 min)

3.4.3 Round One: cDNA Purification

1. Precondition the DNA binding column by adding 250 μL of DNA Binding Buffer to each column and incubate for 5 min at room temperature before centrifugation into a collection tube at 16,000 × *g* for 1 min.
2. Pipette 200 μL of DNA Binding Buffer into the sample, mix thoroughly, and transfer entire volume into the purification column.
3. Centrifuge for 2 min at 100 × *g* to bind DNA to column membrane filter, immediately followed by a centrifugation at 16,000 × *g* for 1 min. Discard flow-through.
4. Pipette 250 μL of DNA Wash Buffer into the purification column and centrifuge for 2 min at 16,000 × *g*. Discard flow-through and collection tube.
5. Transfer column into 0.5 mL microcentrifuge tube and pipette 12 μL of DNA Elution Buffer onto the center of the purification membrane filter and incubate for 1 min at room temperature.
6. Centrifuge at 1000 × *g* for 1 min and then at 16,000 × *g* for 1 min. Discard the column and retain the cDNA-containing eluate for further processing.
7. Proceed directly with Subheading 3.4.4 (In Vitro Transcription) or store purified cDNA at −20 °C overnight.

3.4.4 Round One: In Vitro Transcription

1. Prepare the Complete IVT Reaction Mix by combining 2 μL of IVT Buffer (Blue-1), 6 μL of IVT Master Mix (Blue-2), 2 μL of IVT Enzyme Mix (Blue-3), and 2 μL of Enhancer (Yellow). Scale up with 10 % overage for multiple sample.
2. Add 12 μL of this complete IVT Reaction Mix to each samples, mix thoroughly and incubate for 8 h at 42 °C.
3. Cool samples to 4 °C.

4. Add 1 μL DNase Mix (Blue-4), mix thoroughly, and incubate for 20 min at 37 °C.
5. Cool samples to 4 °C and proceed immediately with aRNA purification.

3.4.5 Round One: aRNA Purification

1. Precondition the RNA binding column by adding 250 μL of RNA Binding Buffer to each column and incubate for 5 min at room temperature before centrifugation into a collection tube at 16,000 × *g* for 1 min.
2. Pipette 120 μL of RNA Binding Buffer into the sample, mix thoroughly, and transfer entire volume into the purification column.
3. Centrifuge for 2 min at 100 × *g* to bind DNA to column membrane filter, immediately followed by a centrifugation at 16,000 × *g* for 1 min. Discard flow-through.
4. Pipette 200 μL of RNA Wash Buffer into the purification column and centrifuge for 1 min at 10,000 × *g*. Discard flow-through.
5. Repeat the previous wash step, and discard flow-through and collection tube.
6. Transfer column into a 0.5 mL microcentrifuge tube and pipette 12 μL of RNA Elution Buffer onto the center of the purification membrane filter and incubate for 1 min at room temperature.
7. Centrifuge at 1000 × *g* for 1 min and then at 16,000 × *g* for 1 min. Discard the column and retain the aRNA-containing eluate for further processing.
8. Proceed directly with Subheading 3.4.6 (second round amplification) or store purified aRNA at −80 °C overnight.

3.4.6 Round Two: cDNA Synthesis, Purification and Amplification

1. For first strand cDNA synthesis: add 1.0 μL of Primer 2 (Gray-2) to 11 μL of purified aRNA product and incubate mixture for 5 min at 70 °C in a thermal cycler. Cool samples to 4 °C for at least 1 min and then add 9.0 μL of complete first Strand Synthesis Mix to each sample (as detailed under **step 3**, Subheading 3.4.1) and incubate for 10 min at 25 °C followed by 1.5 h at 37 °C. Cool samples to 4 °C then either proceed with second strand cDNA synthesis and a further round of linear amplification or store overnight al −20 °C. If performing qRT-PCR proceed to Subheading 3.5.
2. For second strand cDNA synthesis: repeat steps under Subheading 3.4.1, but substitute Primer 3 (Gray-3) for Primer 2.
3. For further cDNA purification, IVT and aRNA purification: repeat steps under Subheadings 3.4.3, 3.4.4 and 3.4.5 as described above. The yield and purity of aRNA can now be determined by one of several methods (*see* **Note 12**). The

purified amplified aRNA generated by these reactions is then amenable to direct labeling with biotin or fluorescent moieties (e.g., Cy3 or Cy5) or labeling of cDNA following reverse transcription. Alternative IVT kits are also available to generate labeled aRNA following Round Two cDNA purification. Refer to the relevant manufacturer's protocol for specific details.

3.5 qRT-PCR

1. Gene-specific primers can be designed using shareware software like Primer3 (NIH, Bethesda, Maryland, USA). To reduce the possibility of generating artifacts from contaminating genomic DNA, we routinely design our primers to span exon–exon boundaries of <200 bp within 400 bp of the 3′ terminus of the gene of interest. In our lab, we perform RT-PCR on a CFX96 (Bio-Rad) using SYBR Select Master Mix reagents (Applied Biosystems) according to the manufacturer's protocol.
2. Set up qRT-PCR reactions in duplicate in a Hard-Shell 96-well PCR plate (Bio-Rad). For each reaction, add 3 μL cDNA (*see* **Note 13**), 10 μL SYBR Select Master Mix (2×), and a final concentration of primers of 300 nM in a total volume of 20 μL made up with RNase-free water. Seal the plate with an optical adhesive cover. Carry out thermocycling with denaturation at 95 °C for 15 s, annealing/extension at 60 °C for 1 min for 50 cycles followed by a dissociation temperature gradient to obtain a melt curve. Use built-in software for data analysis (CFX Manager): background should be set close to exponential rise in signal. Threshold cycle (Ct) of each reaction should be set to 10 standard deviations above this background level.
3. During primer validation it is important to ensure that the melting temperatures of individual cycles are matching by comparing the dissociation curves of the amplified products; melting temperatures should also be identical (Fig. 2).
4. In order to compare the expression of different genes it is necessary to demonstrate that the relative amplification efficiencies of each primer pair are >90 % (RSq > 0.99) and close to those of the reference gene: gradient of log[cDNA] vs. ΔCt ($Ct_{target} - Ct_{ref}$) <0.1. If equivalent, the comparative CT method ($2^{-\Delta\Delta Ct}$) can be used for relative quantitation. If relative efficiencies are not comparable then reactions must be run separately and each compared to a calibration curve constructed from cDNA of known concentration (*see* **step 5** below).
5. The relationship between threshold cycle and input amount as well as the relative amounts of different transcripts can be determined by including a standard curve with each run constructed from dilutions of cDNA (100, 10, 1, 0.1 ng) generated from a universal reference RNA standard (Agilent Technologies) with primers that amplify the 3′ end of the β-actin gene.

4 Notes

1. Maintain nuclease-free technique throughout: wear disposable gloves and change frequently; use certified nuclease-free reagents and avoid kit substitutions; use new plasticware, tips and tubes that are certified nuclease-free; clean all work surfaces thoroughly with RNaseZap™ or other decontamination solution prior to commencing work.
2. Nucleic acids are not adversely affected by exposure of tissues to chloroform and xylene for short periods of time.
3. Infiltrate tissue with wax at the lowest possible temperature to keep it molten.
4. Aim to collect approximately 50 glomeruli or tubule cross sections.
5. If laser spots do not cover enough of the area of interest, cut and capture may only remove a portion of the field (Fig. 1).
6. Mix and/or reconstitute enzyme solutions carefully by inverting several times, flicking the tube followed by a brief spin in a microcentrifuge. Do not vortex and avoid excessive foaming. Always keep on ice or in a 4 °C block until use.
7. Process caps as quickly as possible after capture (do not batch). Immersion in extraction buffer helps to stabilize the nucleic acid.
8. Proteinase K digestion is essential for all formalin-fixed specimens. The duration of digestion may need to be optimized depending on the age of the sample, the specimen and fixation type and the size of the dissected tissue. A 16–18 h digestion is suggested as a starting point but shorter digestion times may suffice but need to be determined empirically.
9. Ensure column does not come into contact with any ethanol-containing flow-through: recentrifuge at 10,000–16,000 $\times g$ to remove any traces of flow-through if this occurs.
10. DNase treatment is essential as substantial quantities of contaminating genomic DNA may amplify along with RNA.
11. The amount of RNA extracted from LCM tissues is too small to quantitate by standard spectrophotometry or assess quality (degradation and purity) by agarose gel electrophoresis (>200 ng). More sensitive, sample-sparing (>50 pg/μL) methods such as using a PicoChip™ on an Agilent Bioanalyser (Agilent Technologies) are available but in practice are rarely performed because of the paucity of material for subsequent amplification. The quality of the extracted RNA may also be assessed using the Paradise Quality Assessment kit. Here, two sets of β-actin-specific primers, one directed against the 3′ end,

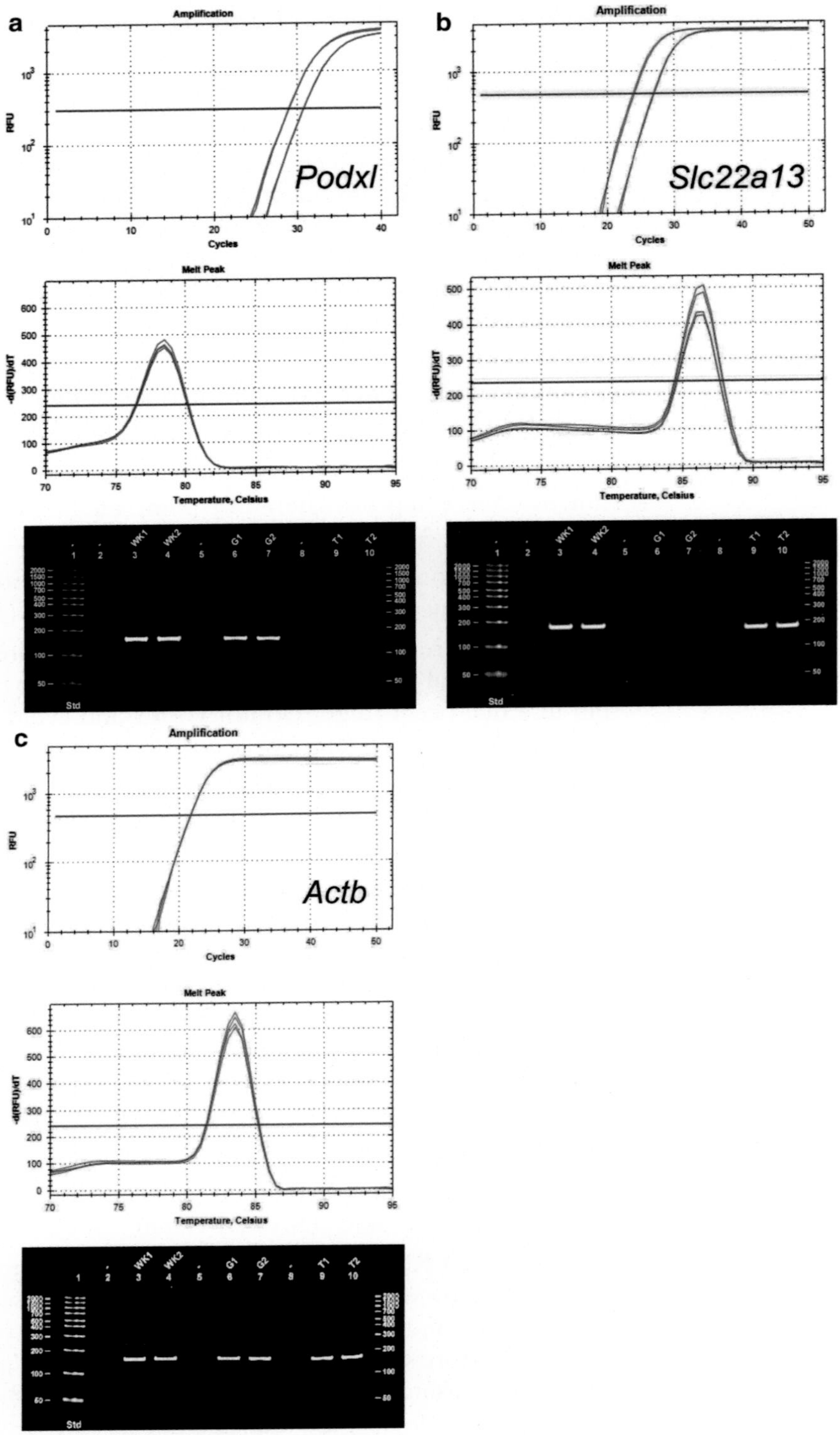

Fig. 2 Amplification plots, melt curve, and electrophoretic analysis of cDNA products of (**a**) glomerular-specific transcript *Podxl*, (**b**) tubule-specific transcript *Slc22a13*, and (**c**) *Actb*, of RNA extracted from laser capture microdissected glomerular (G), tubular (T) segments or from whole kidney (WK) ($n = 2$ in duplicate). PCR analysis was performed using CFX manager software (Bio-Rad). Amplification products were resolved in 10 % TBE precast gels (Bio-Rad), visualized with SYBR Gold nucleic acid stain (Invitrogen) and sized using the AmpliSize™ Molecular Ruler (Bio-Rad)

and the other set to the 5′ end, are employed to amplify cDNA which can be quantitated against cDNA generated from a universal reference RNA standard (Agilent Technologies). Partial loss of the 3′ polyA tail in degraded specimens results in shorter transcripts (and hence greater yield) and thus 3′:5′ ratios >1. Ratios <15 are deemed satisfactory.

12. Following amplification, the quantity and quality of RNA can be assessed using standard methodology. For visualization of RNA after electrophoresis we recommend staining with SYBR Gold or SYBR Green II (Invitrogen).

13. An additional cleanup of the pre-amplified cDNA may help minimize PCR inhibition (e.g., using the High Pure™ PCR Clean-Up Kit from Roche Applied Science).

References

1. Higgins JP, Wang L, Kambham N et al (2004) Gene expression in the normal adult human kidney assessed by complementary DNA microarray. Mol Biol Cell 15:649–656
2. Henger A, Schmid H, Kretzler M (2004) Gene expression analysis of human renal biopsies: recent developments towards molecular diagnosis of kidney disease. Curr Opin Nephrol Hypertens 13:313–318
3. Darby IA, Hewitson TD (2008) In situ hybridization, Molecular biomethods handbook. Humana, Totowa, NJ
4. Espina V, Wulfkuhle JD, Calvert VS et al (2006) Laser-capture microdissection. Nat Protoc 1:586–603
5. Kohda Y, Murakami H, Moe OW et al (2000) Analysis of segmental renal gene expression by laser capture microdissection. Kidney Int 57:321–331
6. Bojmar L, Karlsson E, Ellegard S et al (2013) The role of microRNA-200 in progression of human colorectal and breast cancer. PLoS One 8:e84815
7. Tanji N, Ross MD, Cara A et al (2001) Effect of tissue processing on the ability to recover nucleic acid from specific renal tissue compartments by laser capture microdissection. Exp Nephrol 9:229–234

Chapter 18

Using Phos-Tag in Western Blotting Analysis to Evaluate Protein Phosphorylation

Takahiro Horinouchi, Koji Terada, Tsunehito Higashi, and Soichi Miwa

Abstract

Protein phosphorylation has traditionally been detected by radioisotope phosphate labeling of proteins with radioactive ATP. Several nonradioactive assays with phosphorylation site-specific antibodies are now available for the analysis of phosphorylation status at target sites. However, due to their high specificity, these antibodies they cannot be used to detect unidentified phosphorylation sites. Recently, Phos-tag technology has been developed to overcome the disadvantages and limitations of phosphospecific antibodies. Phos-tag and its derivatives conjugated to biotin, acrylamide, or agarose, form alkoxide-bridged dinuclear metal complexes, which can capture phosphate monoester dianions bound to serine, threonine, and tyrosine residues, in an amino acid sequence-independent manner. Here, we describe our method, which is based on in vitro kinase assay and Western blotting analysis using biotinylated Phos-tag and horseradish peroxidase-conjugated streptavidin, to determine the sites of TRPC6 (transient receptor potential canonical 6) channel phosphorylated by protein kinase A.

Key words Protein phosphorylation, Phos-tag, Western blotting, In vitro kinase assay, Biotin, Horseradish peroxidase-conjugated streptavidin, Protein kinase A

1 Introduction

Protein phosphorylation plays an important role in regulating diverse signaling pathways in eukaryotes [1–3]. Phosphorylation involves the reversible covalent attachment of a negatively charged phosphate group to serine (Ser), threonine (Thr), or tyrosine (Tyr) residues on proteins [4]. In mammalian proteins, the predominant site of phosphorylation is reported to be on Ser (~86 %), as compared to Thr (~12 %) or Tyr (~2 %) [4]. Such posttranslational covalent modifications are by protein kinases, and result in conformational changes in the target proteins [5]. Phosphorylation-dependent conformational changes modulate the activity of various proteins, including enzymes and ion channels [4].

Tim D. Hewitson et al. (eds.), *Kidney Research: Experimental Protocols*, Methods in Molecular Biology, vol. 1397,
DOI 10.1007/978-1-4939-3353-2_18, © Springer Science+Business Media New York 2016

Protein phosphorylation can be detected by several types of analytical methods. For example, traditional approaches include radioisotope phosphate labeling of proteins with radioactive ATP and probing of phosphorylated amino acids with phosphorylation site-specific antibodies, while emerging technologies include protein microarrays, quantitative mass spectrometry, and genetically targetable fluorescent biosensors [4]. Each method has some disadvantages and limitations. The method with radiolabeled ATP (e.g., [γ^{32}P]-ATP) is limited to specimens applicable to metabolic labeling (e.g., the method is not suitable for in vivo studies of human tissues) and poses certain safety and disposal problems. Detection methods with phosphospecific antibodies are frequently employed, since a wide variety of antibodies is now commercially available. However, a phosphospecific antibody specifically recognizes a particular phosphosite so that it is not suitable for the estimation of unidentified phosphorylation sites. Furthermore, the specificity of antibodies varies depending on the quality of the antibody. The emerging technologies require special instrument and expertise to analyze protein phosphorylation.

Recently, Phos-tag™ technology has been developed to overcome the disadvantages and limitations of conventional and current methods [6–8]. Phos-tag™ and its derivatives conjugated to biotin, acrylamide, or agarose, act as novel phosphate binding tags [6–8]. The Phos-tag™ has a vacancy on two metal ions that is suitable for the access of a phosphomonoester dianion (R-OPO_3^{2-}) as a bridging ligand [6–8]. The total charge of resulting 1:1 phosphate-binding complex, R-OPO_3^{2-}-(Phos-tag™)$^{3+}$, is +1 [6–8]. A dinuclear zinc (II) complex (Zn^{2+}-Phos-tag™) can capture phosphate monoester dianions at a neutral pH [6, 7]. On the other hand, a manganese (II) homologue (Mn^{2+}-Phos-tag™) can also bind to the R-OPO_3^{2-} anion at an alkaline pH [8]. Thus, the Phos-tag™ technology has made it possible to detect phosphate monoester dianions bound to serine, threonine, and tyrosine residues, in an amino acid sequence-independent manner. Advantages of the approach with the Phos-tag™ over traditional and emerging methods are as follows: (a) the Phos-tag™ can simultaneously detect multiple phosphorylation sites in an amino acid sequence-independent manner, (b) safety and low cost as compared to radioisotope labeling, (c) methods with Phos-tag™ are simple and do not require any special instrument, and (d) the Phos-tag™ enables to identify new phosphorylated proteins.

TRPC6 (transient receptor potential canonical 6), a voltage-independent, Ca^{2+}-permeable nonselective cation channel, plays a pathophysiological role in the development of cardiovascular diseases such as pulmonary arterial hypertension associated with excessive production of endothelin-1 [9]. The activity of TRPC6 is regulated by various protein kinases, including

Ca^{2+}/calmodulin-dependent protein kinase II [10], the Src tyrosine kinase family [11], protein kinase C [12], protein kinase G [13], and protein kinase A (PKA) [14, 15]. PKA is an arginine-directed serine/threonine protein kinase and arginine residue is preferred at positions -4 to -1 amino-terminal to the phosphorylation site [16]. Searching for the primary sequence of TRPC6, we have found that potential sequences for PKA-mediated phosphorylation are present within TRPC6 sequence, namely R-R-G-G-S at Ser^{14}, R-R-N-E-S at Ser^{28}, R-R-Q-T at Thr^{69}, and R-K-L-S at Ser^{321} [15].

Here, we describe the stepwise approach to determine the phosphorylation sites for PKA on TRPC6. First, we have carried out immunoprecipitation to collect proteins of TRPC6 wild-type and mutants with Ser^{14}, Ser^{28}, Thr^{69}, and Ser^{321} replaced to alanine and an in vitro kinase assay to phosphorylate them by PKA. Then, we have applied the biotin-pendant Zn^{2+}-Phos-tag™ to Western blotting analysis to probe TRPC6 proteins phosphorylated in vitro by PKA. Finally, we have detected phosphorylated TRPC6 proteins bound to Zn^{2+}-Phos-tag™ on PVDF (polyvinylidene difluoride) membrane using ECL (enhanced chemiluminescence) system based on a method of biotin/avidin-peroxidase detection. Using these protocols, we have found that PKA phosphorylates TRPC6 on Ser^{28} and Thr^{69} [15].

2 Materials

Prepare all solutions using ultrapure water (resistivity > 18.0 MΩ cm at 25 °C) and reagents of the highest grade in purity. Prepare and store all reagents at room temperature unless indicated otherwise.

2.1 Cells

1. Human embryonic kidney 293 (HEK293) cells stably expressing wild-type and mutant TRPC6 fused with FLAG peptide at the C terminus (TRPC6-FLAG). The HEK293 cells are cultured in 10-cm dish in Dulbecco's modified Eagle's medium supplemented with 10 % heat-inactivated fetal calf serum (v/v), penicillin (100 U/mL), and streptomycin (100 μg/mL) at 37 °C in humidified air with 5 % CO_2.

2.2 Immunoprecipitation

1. Stock solutions: 2 M NaCl in water; 10 mM $MgCl_2$ in water; 500 mM Tris–HCl (pH 6.8) in water; 10 % Nonidet P-40 (NP-40) in water; 10 % sodium dodecyl sulfate (SDS) in water; 200 mM phenylmethylsulfonyl fluoride (PMSF) in ethanol (store at −25 °C); 1 M Na_3VO_4 in water (store at −25 °C); 1 M NaF in water (store at −25 °C); 1 mg/mL leupeptin in water (store at −25 °C); 1 mg/mL aprotinin in 10 mM Tris–HCl (pH 8.0) (store at −25 °C); and 1 mg/mL pepstatin in ethanol (store at −25 °C).

2. Cell lysis buffer: Transfer 75 μL of 2 M NaCl, 150 μL of 10 mM $MgCl_2$, 100 μL of 500 mM Tris–HCl (pH 6.8), 100 μL of 10 % NP-40, 10 μL of 10 % SDS, 5 μL of 200 mM PMSF, 1 μL of 1 M Na_3VO_4, 20 μL of 1 M NaF, 10 μL of 1 mg/mL leupeptin, 10 μL of 1 mg/mL aprotinin, 10 μL of 1 mg/mL pepstatin, and 10 μL of EDTA-free, protease inhibitor cocktail (Thermo Fisher Scientific Inc., Waltham, MA, USA) to a tube containing 500 μL of water. Add 5 mg sodium deoxycholate, mix well and keep on ice. Prepare just before use.
3. Phosphate-buffered saline (PBS) (pH 7.4): Weigh 8.0 g NaCl, 0.2 g KCl, 2.9 g $Na_2HPO_4 \cdot H_2O$, and 0.2 g KH_2PO_4 and transfer to the bottle containing 980 mL of water. Adjust to pH 7.4 with NaOH at 25 °C. Make up to 1 L with water.
4. Protein G Dynabeads™ (Thermo Fisher Scientific Inc.). Store at 4 °C.
5. Binding buffer: PBS (pH 7.4) containing 0.02 % Tween 20 (v/v).
6. Washing buffer: Transfer 750 μL of 2 M NaCl, 1.5 mL of 10 mM $MgCl_2$, 1 mL of 500 mM Tris–HCl (pH 6.8), 1 mL of 10 % NP-40, and 100 μL of 10 % SDS to a tube containing 5.65 mL of water. Add 50 mg sodium deoxycholate and mix well.
7. Ultrasonic disruptor suitable for small volumes.
8. Rotator.
9. DynaMag-2™ magnet (Thermo Fisher Scientific Inc.).
10. Anti-FLAG M2 antibody (1 mg/mL) (Sigma-Aldrich, St. Louis, MO, USA).

2.3 In Vitro Kinase Assay

1. Stock solutions: Prepare 2 M NaCl in water; 10 mM $MgCl_2$ in water; 500 mM Tris–HCl (pH 6.8) in water; 10 % NP-40 in water; 10 % SDS in water; 1 M Tris–HCl (pH 8.0) in water; and 0.5 mM ATP in water (store at −25 °C).
2. Alkaline phosphatase (1 U/μL) (rAPid Alkaline Phosphatase™, Roche Applied Science, Mannheim, Germany). Store at −25 °C.
3. Dephosphorylation buffer: Transfer 750 μL of 2 M NaCl, 1.5 mL of 10 mM $MgCl_2$, 1 mL of 500 mM Tris–HCl (pH 6.8), 1 mL of 10 % NP-40, and 100 μL of 10 % SDS to the tube containing 5.65 mL of water. Add 50 mg sodium deoxycholate and mix well. Adjust to pH 8.5 with NaOH at 25 °C. Warm buffer to 37 °C prior to use.
4. Dephosphorylation reaction mixture: Mix 10 μL of alkaline phosphatase (10 U) and 40 μL of dephosphorylation buffer (37 °C) in a 1.5 mL tube per reaction. Prepare just before use. The mixture needs to be warmed to 37 °C prior to use (*see* **Note 1**).

5. cAMP-dependent protein kinase catalytic subunit (Promega Corp., Madison, WI, USA). Store at −70 °C.
6. Phosphorylation buffer: Transfer 2 μL of 1 M Tris–HCl (pH 8.0), 50 μL of 10 mM $MgCl_2$, and 20 μL of 0.5 mM ATP to a tube containing 28 μL of water. The buffer needs to be warmed to 37 °C prior to use.
7. Phosphorylation reaction mixture: Mix 0.5 μL of cAMP-dependent protein kinase catalytic subunit (40 U) and 24.5 μL of phosphorylation buffer (37 °C) in a 1.5 mL tube per reaction (*see* **Note 2**). Prepare just before use.
8. SDS sample buffer (×5): Combine 5 g SDS, 0.04 g bromophenol blue (BPB), 6.25 mL of 1 M Tris–HCl (pH 6.8), and 20 mL of glycerol. Warm to 70 °C and mix well with rotation. The mixture needs to be warmed prior to dilution. Dilute three volumes of the mixture with one volumes of with 2-mercaptoethanol (2-ME) just before use.

2.4 Western Blot Analysis

1. Stock solutions: 1 M Tris–HCl (pH 8.0) in water; 2 M NaCl in water; 10 % Tween-20 in water.
2. 5–20 % polyacrylamide gel (e.g., pre-cast SuperSep Ace™, 5–20 %, 13-well, Wako Pure Chemical Industries, Ltd., Osaka, Japan). Store at 4 °C.
3. PVDF membrane (e.g., Immobilon-P™, pore size 0.45 μm, Millipore Corp., Bedford, MA, USA). Treat with methanol just before use.
4. Filter paper (e.g., ADVANTEC™, 0.93 mm in thickness, Toyo Roshi Co., Ltd., Tokyo, Japan).
5. Standard mini-slab polyacrylamide gel electrophoresis (PAGE) apparatus.
6. Semidry electroblotter apparatus (e.g., Mini-Protean™ 3 Cell, Bio-Rad Laboratories, Hercules, CA, USA).
7. High voltage power supply.
8. SDS-PAGE buffer: Weigh 14.5 g Tris–HCl, 72.0 g glycine, and 5.0 g SDS and transfer to the flask containing 4.5 L of water. Make up to 5 L with water.
9. Transfer buffer: Weigh 15.0 g Tris–HCl and 72.0 g glycine and transfer to a flask containing about 3.5 L of water. Make up to 4 L with water. Add 1 L of methanol to the flask.
10. Tris-buffered saline–Tween 20 (TBS-T): Transfer 10 mL of 1 M Tris–HCl (pH 8.0), 50 mL of 2 M NaCl, and 10 mL of 10 % Tween 20 to the bottle containing 900 mL of water. Make up to 1 L with water.
11. Blocking buffer: 2 % bovine serum albumin (BSA) in TBS-T.

12. ECL Western blotting analysis system (e.g., ECL Western Blotting Detection Reagents, GE Healthcare Limited, Little Chalfont, Buckinghamshire, UK). Store at 4 °C.
13. ECL film (e.g., Hyperfilm ECL™, GE Healthcare Limited).
14. anti-DDDDK-tag HRP-DirecT™ (Medical and Biological Laboratories Co., LTD., Aichi, Japan).

2.5 Phos-Tag

1. 100 mM Phos-tag™ biotin methanol solution: Add 130 μL of methanol to a tube containing 10 mg Phos-tag™ biotin (Wako Pure Chemical Industries, Ltd.).
2. 10 mM Phos-tag™ biotin solution: Add 130 μL of 100 mM Phos-tag™ biotin methanol solution to 1.17 mL of TBS-T.
3. 10 mM $Zn(NO_3)_2$ solution: 0.15 g $Zn(NO_3)_2 \cdot 6H_2O$ in 50 mL of water.
4. Zn^{2+}-Phos-tag™ biotin-bound horseradish peroxidase conjugated streptavidin (HRP-SA) solution: Mix 469 μL of TBS-T, 10 μL of 10 mM Phos-tag™ biotin solution, 20 μL of 10 mM $Zn(NO_3)_2$ solution, and 1 μL of 1 mg/mL HRP-SA (Thermo Fisher Scientific Inc.) and allow to stand for 30 min at room temperature. Transfer the mixed solution to an Amicon Ultra 0.5 mL centrifugal filter (Millipore Corp.). Centrifuge at 14,000 × *g* for 10 min at 25 °C. Dilute the remaining solution (<10 μL) with 30 mL of TBS-T. Store at 4 °C.

3 Methods

Carry out all procedures at room temperature unless otherwise detailed.

3.1 Preparation of Cell Lysates

1. Wash cells twice with 3 mL of ice-cold PBS.
2. Lyse the cells with 1 mL of ice-cold cell lysis buffer.
3. Sonicate the cell lysates for 10 s.
4. Centrifuge at 20,000 × *g* for 20 min at 4 °C.
5. Transfer supernatant to a clean tube and keep on ice.
6. Measure protein content of supernatant according to the method of Bradford [17] using BSA as a standard.

3.2 Immunopreci pitation and In Vitro Kinase Assay

1. Resuspend Dynabeads completely by vortexing for 30 s.
2. Transfer 25 μL of Dynabeads™ to a tube, place on magnet and remove supernatant.
3. Remove tube from magnet and resuspend the Dynabeads™ in 100 μL of binding buffer.
4. Add 1 μL of monoclonal anti-FLAG M2 antibody (1 mg/mL).

5. Incubate for 60 min with rotation at room temperature.
6. Place tube on magnet and remove supernatant.
7. Remove tube from magnet and wash the Dynabeads™-antibody complex by resuspending in 100 μL of binding buffer.
8. Place tube on magnet and remove supernatant.
9. Add the supernatant of cell lysate (equal protein amount: typically 1000 μg) to the Dynabeads™-antibody complex (*see* **Note 3**) and resuspend by vortexing briefly.
10. Incubate for 60 min at room temperature with rotation.
11. Place tube on magnet and remove supernatant.
12. Wash the Dynabeads™-antibody-TRPC6-FLAG complex with 500 μL of the washing buffer by gently vortexing.
13. Transfer the suspension to a clean tube.
14. Place tube on magnet and remove supernatant.
15. Wash the Dynabeads™-antibody-TRPC6-FLAG complex three times with 200 μL of the washing buffer by gently vortexing for each wash.
16. Transfer the suspension to a clean tube.
17. Place tube on magnet and remove supernatant.
18. Add 50 μL of dephosphorylation reaction mixture and incubate for 60 min at 37 °C with shaking at 100 rpm (*see* Fig. 1).
19. Place tube on magnet and remove supernatant.
20. Wash the Dynabeads™-antibody-TRPC6-FLAG complex three times with 200 μL of the washing buffer by gently vortexing for each wash.
21. Transfer the suspension to a clean tube.
22. Place tube on magnet and remove supernatant.
23. Add 25 μL of phosphorylation reaction mixture and incubate for 30 min at 37 °C with shaking at 200 rpm (*see* Fig. 2).
24. Add 6 μL of SDS sample buffer (×5) and incubate for 30 min at 37 °C with shaking at 200 rpm (*see* **Note 4**).
25. Place tube on magnet and transfer supernatant to a clean tube.
26. Store at −80 °C before use.

3.3 Western Blot Analysis

1. Carry out SDS-PAGE in 5–20 %, 13-well polyacrylamide gels. Apply 5 or 15 μL of immunoprecipitated sample to each well. Run the gel at 80 volts for 30 min, and then at 130 volts untill the BPB dye front has reached the bottom of the gel.
2. Immediately following SDS-PAGE, transfer the proteins separated on the gel to a PVDF membrane at 100 volts for 60 min in a refrigerator kept at 4 °C.

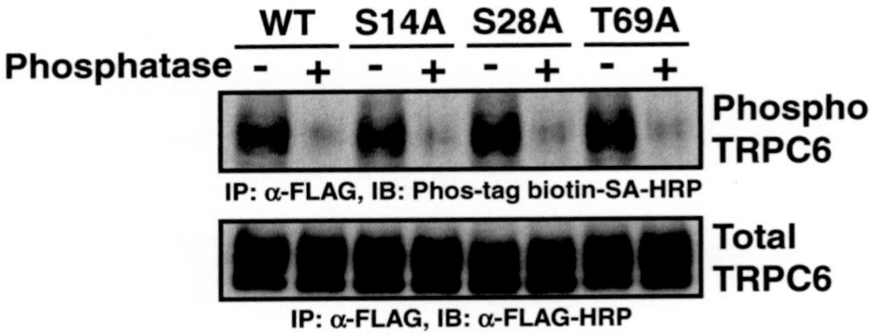

Fig. 1 Effects of phosphatase treatment on the basal phosphorylation of FLAG-tagged TRPC6 proteins immunoprecipitated with anti-FLAG antibody. Immunoprecipitated TRPC6 proteins were incubated in the dephosphorylation reaction mixture for 1 h. *Upper panel* is a representative immunoblot with Phos-tag biotin-bound HRP-SA complex detecting phosphorylated TRPC6 proteins (indicated as phospho TRPC6). *Lower panel* is a representative immunoblot with anti-FLAG-HRP antibody to determine the quantity of TRPC6 in the immunoprecipitate (indicated as total TRPC6). Reproduced with permission from Ref. [15]

3. Wash the membrane three times with TBS-T, 5 min for each wash.
4. Block the membrane with blocking buffer for 60 min (*see* **Note 5**).
5. Wash the membrane two times with TBS-T, 3 min for each wash.
6. Incubate the membrane with Zn^{2+}-Phos-tag™ biotin-bound HRP-SA solution at room temperature for 6 h (for 15 μL of immunoprecipitated sample per lane) or with anti-FLAG-HRP antibody (1:6000 dilution, diluted in blocking buffer, anti-DDDDK-tag HRP-DirecT) overnight at 4 °C (for 5 μL of immunoprecipitated sample per lane).
7. Wash the membrane three times with TBS-T, 5 min for each wash.
8. Detect Zn^{2+}-Phos-tag™ biotin-bound HRP-SA and anti-FLAG-HRP antibody with ECL Western blotting analysis system and ECL film (*see* Figs. 1, 2, and 3).

4 Notes

1. Warm dephosphorylation reaction mixture to 37 °C before use, because addition of cold alkaline phosphatase (−25 °C) to dephosphorylation buffer lowers the temperature of the mixture.
2. Dilute cAMP-dependent protein kinase catalytic subunit with phosphorylation buffer to make a solution of 80 U/μL just before use.

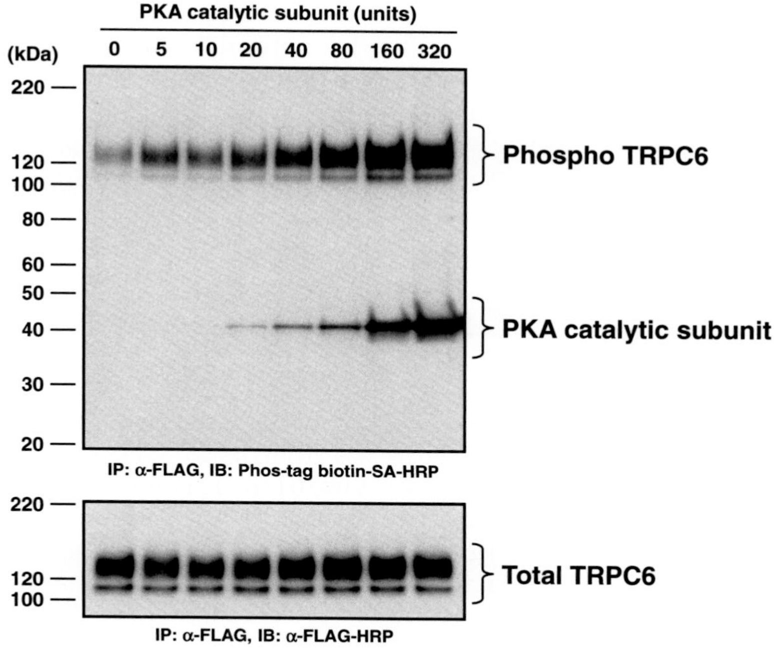

Fig. 2 Dose–response relationship for PKA-mediated phosphorylation of wild-type TRPC6 protein. Immunoprecipitated TRPC6 protein was incubated in the dephosphorylation reaction mixture for 1 h and then was phosphorylated with PKA catalytic subunit at indicated doses for 30 min. *Upper panel* is a representative immunoblot with Phos-tag biotin-bound HRP-SA complex to detect phosphorylated TRPC6 protein (indicated as phospho TRPC6). *Lower panel* is a representative immunoblot with anti-FLAG-HRP antibody to determine the quantity of TRPC6 in the immunoprecipitate (indicated as total TRPC6). The locations of the bands for TRPC6-FLAG and PKA catalytic subunit are indicated to the *right*. The positions of protein size markers are indicated to the *left* in kDa. Reproduced with permission from Ref. [15]

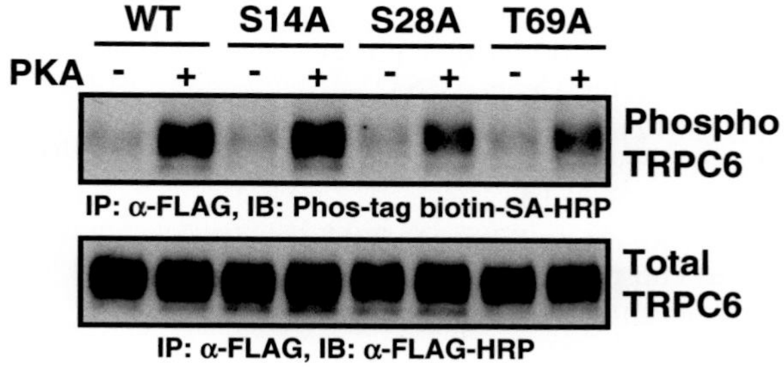

Fig. 3 Phosphorylation of TRPC6 on Ser^{28} and Thr^{69} by PKA. Immunoprecipitated TRPC6 proteins were incubated in the PKA phosphorylation reaction mixture for 30 min. *Upper panel* is a representative immunoblot with Phos-tag biotin-bound HRP-SA complex detecting phosphorylated TRPC6 proteins (indicated as phospho TRPC6). *Lower panel* is a representative immunoblot with anti-FLAG-HRP antibody to determine the quantity of TRPC6 in the immunoprecipitate (indicated as total TRPC6). Reproduced with permission from Ref. [15]

3. Adjust to equal volume (typically 700–1000 μL) with cell lysis buffer. The optimal protein amount needed will depend on the individual proteins tested.
4. The TRPC6 proteins bound to the Dynabeads™ are eluted in the phosphorylation buffer by adding SDS sample buffer.
5. Skim milk is not suitable as a blocking agent since it contains various phosphoproteins which are detected by Zn^{2+}-Phos-tag™ biotin-bound HRP-SA.

Acknowledgement

This work was supported in part by Grants-in-Aid for Young Scientists (B) from Japan Society for the Promotion of Science [grant 21790236] (to T. Horinouchi); Grants-in-Aid for Scientific Research (B) from Japan Society for the Promotion of Science [grant 21390068] (to S.M.); and grants from Smoking Research Foundation of Japan (to S.M.), Mitsubishi Pharma Research Foundation (to T. Horinouchi), the Pharmacological Research Foundation, Tokyo (to T. Horinouchi), the Shimabara Science Promotion Foundation (to T. Horinouchi), and Actelion Pharmaceuticals Japan Ltd. (to T. Horinouchi).

References

1. Hunter T (2000) Signaling—2000 and beyond. Cell 100:113–127
2. Cohen P (2002) Protein kinases— the major drug targets of the twenty-first century? Nat Rev Drug Discov 1:309–315
3. Noble ME, Endicott JA, Johnson LN (2004) Protein kinase inhibitors: insights into drug design from structure. Science 303:1800–1805
4. Newman RH, Zhang J, Zhu H (2014) Toward a systems-level view of dynamic phosphorylation networks. Front Genet 5:263
5. Hunter T (2012) Why nature chose phosphate to modify proteins. Philos Trans R Soc Lond B Biol Sci 367:2513–2516
6. Kinoshita E, Kinoshita-Kikuta E, Takiyama K et al (2006) Phosphate-binding tag, a new tool to visualize phosphorylated proteins. Mol Cell Proteomics 5:749–757
7. Kinoshita E, Kinoshita-Kikuta E, Sugiyama Y et al (2012) Highly sensitive detection of protein phosphorylation by using improved Phos-tag Biotin. Proteomics 12:932–937
8. Kinoshita E, Kinoshita-Kikuta E, Koike T (2009) Separation and detection of large phosphoproteins using Phos-tag SDS-PAGE. Nat Protoc 4:1513–1521
9. Abramowitz J, Birnbaumer L (2009) Physiology and pathophysiology of canonical transient receptor potential channels. FASEB J 23:297–328
10. Shi J, Mori E, Mori Y et al (2004) Multiple regulation by calcium of murine homologues of transient receptor potential proteins TRPC6 and TRPC7 expressed in HEK293 cells. J Physiol 561:415–432
11. Hisatsune C, Kuroda Y, Nakamura K et al (2004) Regulation of TRPC6 channel activity by tyrosine phosphorylation. J Biol Chem 279:18887–18894
12. Kim JY, Saffen D (2005) Activation of M1 muscarinic acetylcholine receptors stimulates the formation of a multiprotein complex centered on TRPC6 channels. J Biol Chem 280:32035–32047
13. Kinoshita H, Kuwahara K, Nishida M et al (2010) Inhibition of TRPC6 channel activity contributes to the antihypertrophic effects of natriuretic peptides-guanylyl cyclase-A signaling in the heart. Circ Res 106:1849–1860
14. Nishioka K, Nishida M, Ariyoshi M et al (2011) Cilostazol suppresses angiotensin II-induced vasoconstriction via protein kinase

A-mediated phosphorylation of the transient receptor potential canonical 6 channel. Arterioscler Thromb Vasc Biol 31:2278–2286

15. Horinouchi T, Higa T, Aoyagi H et al (2012) Adenylate cyclase/cAMP/protein kinase A signaling pathway inhibits endothelin type A receptor-operated Ca^{2+} entry mediated via transient receptor potential canonical 6 channels. J Pharmacol Exp Ther 340:143–151
16. Songyang Z, Blechner S, Hoagland N et al (1994) Use of an oriented peptide library to determine the optimal substrates of protein kinases. Curr Biol 4:973–982
17. Bradford MM (1976) A rapid and sensitive method for the quantitation of microgram quantities of protein utilizing the principle of protein-dye binding. Anal Biochem 72: 248–254

Chapter 19

Qualitative and Quantitative Analysis of Histone Deacetylases in Kidney Tissue Sections

Katherine Ververis, Selly Marzully, Chrishan S. Samuel, Tim D. Hewitson, and Tom C. Karagiannis

Abstract

Fluorescent microscope imaging technologies are increasing in their applications and are being used on a wide scale. However methods used to quantify the level of fluorescence intensity are often not utilized—perhaps given the result may be immediately seen, quantification of the data may not seem necessary. However there are a number of reasons given to quantify fluorescent images including the importance of removing potential bias in the data upon observation as well as quantification of large numbers of images gives statistical power to detect subtle changes in experiments. In addition discreet localization of a protein could be detected without selection bias that may not be detectable by eye. Such data will be deemed useful when detecting the levels of HDAC enzymes within cells in order to develop more effective HDAC inhibitor compounds for use against multiple diseased states. Hence, we discuss a methodology devised to analyze fluorescent images using Image J to detect the mean fluorescence intensity of the 11 metal-dependent HDAC enzymes using murine kidney tissue sections as an example.

Key words Fluorescence microscopy, Histone deacetylases, Image J analysis, Immunofluorescence staining, Kidney tissue

1 Introduction

In eukaryotes, DNA is packaged into chromatin, which, apart from DNA, consists of histone and non-histone proteins. Epigenetic modulation of gene expression reflects environment–gene interactions such as exposure to cytokines and growth factors. It leads to changes in protein expression, many of which are inherited in disease, through cell division. Epigenetic modifications have been described in diverse forms of renal disease including renal carcinoma, transplantation, autoimmune disease, and diabetes (reviewed in ref. 1). Under normal conditions these modifications are balanced and reversible, but they may be altered in disease states.

Epigenetic regulation of gene expression in the kidney includes posttranslational histone modifications [2], postreplicative DNA

Tim D. Hewitson et al. (eds.), *Kidney Research: Experimental Protocols*, Methods in Molecular Biology, vol. 1397, DOI 10.1007/978-1-4939-3353-2_19, © Springer Science+Business Media New York 2016

methylation [3], and RNA interference [4]. In the first case, histone residues can be modified by mono-, di-, and tri-methylation and acetylation, phosphorylation, sumoylation, or ubiquitination, with the first two widely implicated in disease progression.

The combination of site and type of modification attached to the histones gives rise to the 'histone code' that regulates activation or repression of transcription [5]. Although there are many histone sites that can be modified, lysine (K) modifications on H3 and H4 are common [6]. In the simplest form, histone acetylation results in a more open chromatin structure which is more accessible for transcription. However, we now appreciate that this only occurs in about 50 % of target genes, and that acetylation can be both expression permissive and repressive depending on individual modifications [7, 8]. The differential mono-, di-, or tri-methylation provides further functional diversity. Global histone methylation and acetylation are consistent features in chronic kidney disease.

Numerous enzymes regulate histone methylation and acetylation and it has now been shown that expression of both these epigenetic regulators as well as their target modifications change during disease. Histone acetylation marks exist in a steady state balance between histone acetylation (HAT) and histone deacetylation (HDAC) (Fig. 1) [9]. Broad spectrum HDAc inhibitors (HDAC*i*) [2, 10] and more recently a class I specific HDAC*i* [11] ameliorate progression of chronic kidney disease by altering the steady state balance of acetylation and deacetylation. Some 11 different HDACs are relevant, which can be subdivided into Class I, IIa, IIb, and IV based on specificity (Table 1).

Systematic organ and disease specific analysis of histone deacetylases therefore offers us a valuable insight into the regulation of kidney disease, and the mechanisms of progression. In this chapter we describe a histochemical based technique to detect and analyze HDAC expression in routinely collected tissue sections.

2 Materials

All aqueous solutions are made in laboratory grade deionized water (dH_2O).

2.1 Fixative

1. 0.01 M phosphate buffered saline (PBS): Prepare a 10× stock by dissolving 80 g of NaCl, 2.0 g of KCl, 14.4 g of Na_2HPO_4, and 2.4 g of KH_2PO_4 in 900 mL of dH_2O. Adjust to pH 7.4 and make up to 1 L. Dilute 1:10 for a 1× working concentration.
2. 4 % PFA: Add 4 g paraformaldehyde (BDH) to 100 mL 0.01 M phosphate buffered saline (PBS) pH 7.4, heat to a maximum of 60 °C with stirring and add 5–10 drops of 1 M NaOH to clear.

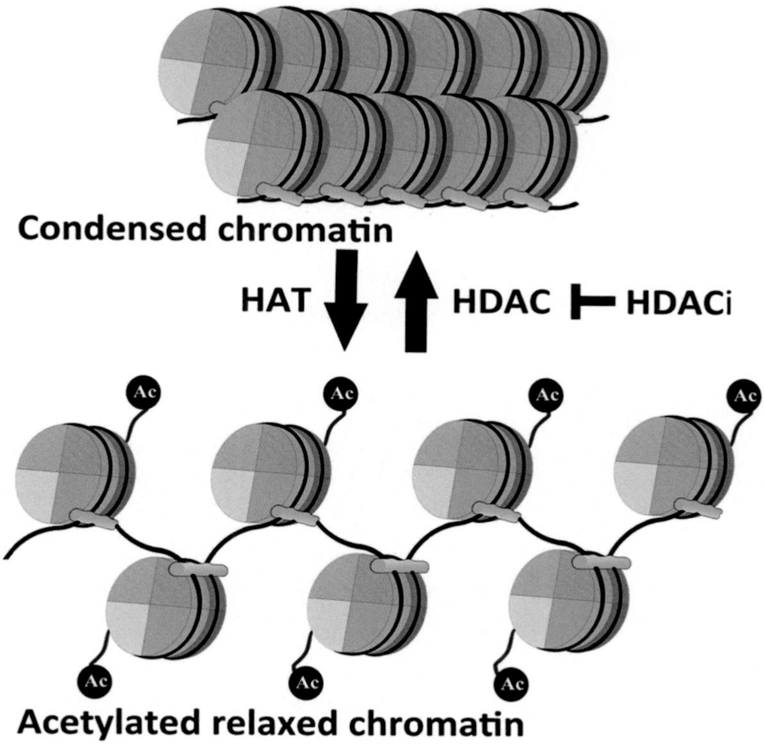

Fig. 1 Effects of histone acetylation and deacetylation through the actions of histone acetyl transferase (HAT) and histone deacetylase (HDAC) respectively. HDAC inhibitors (HDACi) increase acetylation by inhibiting the action of HDAC

Table 1
The 11 metal-dependent HDAC enzyme and their subdivisions

Class	Enzyme	Location	Expression
Class I (Rpd3)	HDAC1 HDAC2 HDAC3 HDAC8	Nucleus	Ubiquitous
Class IIa (Hda1)	HDAC4 HDAC5 HDAC7 HDAC9	Nucleus/cytoplasm	Tissue specific
Class IIb (Hda1)	HDAC6 HDAC10	Cytoplasm	Tissue specific
Class IV (Rpd3/Hda1)	HDAC11	Nucleus/cytoplasm	Tissue specific

2.2 Histological Processing and Tissue Sectioning

1. Labeled glass vials.
2. Graded ethanol: 50, 70, 95, and 100 % v/v mixture of ethanol and dH_2O.
3. Chloroform.
4. Paraplast™ paraffin embedding medium (McCormick Scientific, St. Louis, MO, USA), or equivalent low temperature embedding wax.
5. Embedding molds and cassettes.
6. Scalpel and No. 22 blade.
7. Artist's paint brush.
8. Shallow bowl filled with a 20 % v/v mixture of ethanol in dH_2O.
9. Microtome.
10. Histological water bath.
11. Slide racks.
12. Microscope slides with frosted edges (*see* **Note 1**).
13. Chloroform.

2.3 HDAC Staining

1. Xylene (or less toxic equivalent such as Histolene™).
2. Ethanol (100, 90, 70 % v/v in dH_2O).
3. PBST: 0.1 % Tween 20 in PBS (without Ca^{2+} and Mg^{2+}).
4. PBST with 1 % bovine serum albumin (BSA) (v/v).
5. Dako Target Retrieval Solution (Dako, Glostrup, Denmark) (*see* **Note 2**).
6. 0.1 % Triton™ X-100 (Sigma, St. Louis, MO, USA).
7. 1 % w/v BSA in PBS (without Ca^{2+} and Mg^{2+}).
8. Polyclonal rabbit anti-HDAC1-11 antibody panel (Biovision, Milpitas, CA, USA).
9. Polyclonal goat anti-β-actin (Abcam, Cambridge, UK).
10. Alexa Fluor™ goat 546 anti-rabbit IgG (H + L) (Invitrogen; Life Technologies, Grand Island, NY, USA).
11. Alexa Fluor™ donkey 488 anti-goat IgG (H + L) (Invitrogen).
12. ProLong™ Gold Antifade Solution with DAPI (Invitrogen)
13. Water-repelling immunostaining pen (e.g., Pap Pen™).
14. Coverslips.
15. Nail polish.
16. Rotating platform.
17. Slide racks.
18. Staining jars.

19. Humidified incubation chamber.
20. Microwave oven.

2.4 Image Acquisition

1. Fluorescence microscope, with digital camera (e.g., Olympus BX61 automated upright microscope with FVII camera).

2.5 Heat Map Analysis

1. Image J analysis software version (1.146d or later) for the appropriate platform (Downloadable as freeware from Fiji Is Just Image J) [12].
2. Microsoft Excel (Microsoft, Seattle, CA, USA).

3 Methods

The procedures below describe techniques to label paraffin-embedded histological material. The methods outline preparation of sections, the use of immunofluorescent staining to localize intracellular HDAC expression in situ, and finally, a quantitative analysis of immunofluorescent intensity to examine changes in HDAC expression.

3.1 Tissue Fixation, Embedding, and Sectioning

1. Immerse in cold freshly prepared 4 % PFA, 10× the volume of tissue.
2. Store at −4 °C for 2–18 h, depending on size of tissue.
3. Wash in 0.01 M PBS for 10 min.
4. Dehydrate tissue through graded alcohols (50, 70, 95, 100 %) and clear in two changes of chloroform (45 min to 1 h each).
5. Infiltrate tissue with two changes of paraffin wax (approx 56 °C) for a total of 4–6 h. Excess heat should be avoided as it may potentially affect antigenicity.
6. Orientate and embed tissue in fresh wax using molds and embedding cassettes.
7. Place tissue and molds in a −20 °C freezer for a minimum of 2 h before separating the block from the mold.
8. Heat water bath to 56 °C.
9. Prepare a flotation bath by filling a small bowl with 20 % v/v ethanol in water.
10. Cut 2–5 μm sections of paraffin-embedded tissue with a microtome, in accordance with the manufacturer's instructions.
11. Float sections on 20 % ethanol to flatten and transfer to a water bath using a glass slide.
12. Collect sections on slides and stand upright to drain. Transfer to a slide rack and allow to dry overnight in an oven at 40 °C.

3.2 Immunofluo rescent Staining of HDAC

1. Use 2–5 μm paraffin sections.
2. Deparaffinize paraffin sections by immersing tissue sections in two changes of xylene (5 min each) and rehydrate tissue through graded alcohols (100, 90, 70 %) for 1 min each (*see* **Note 3**).
3. Dilute Dako Target Retrieval solution 1:10 and fill staining jars.
4. Microwave on high for 1 min and then on low for 10.0 min (*see* **Note 4**).
5. Cool slides for 20 min to room temperature before decanting the target retrieval solution.
6. Circle the tissue using a pap (wax) pen.
7. Permeabilize the tissue with 100 μL of 0.1 % Triton X-100 for 10 min.
8. Equilibrate slides in two changes of PBST for 15 min each.
9. Block the tissue using 100 μL of 1 % BSA for 1 h at room temperature or overnight at 4 °C.
10. Incubate with 100 μL primary antibody diluted in 1 % BSA antibody diluent overnight at 4 °C on a rotating platform (*see* **Note 5**), e.g., Polyclonal rabbit anti-HDAC1 (Biovision) (10 μg/mL) + donkey anti-β-actin (5 μg/mL). Separate slide sections are required for each antibody HDAC1 to HDAC11.
11. Wash three times in PBST with 1 % BSA for 10 min each.
12. Incubate with 100 μL secondary antibody diluted in 1 % BSA antibody diluent for 1 h at room temperature on a rotating platform, e.g., Goat anti-rabbit Alexa 594 + goat anti-donkey Alexa 488
13. Wash three times in PBST with 1 % BSA for 10 min each.
14. Permanently mount the slides using Prolong Gold Antifade™ mounting media with DAPI (*see* **Note 6**).
15. Coverslip slides and blot off excess mounting media then nail-polish all sides to seal (*see* **Note 7**).

3.3 Image Aquisition

1. Acquire images using fluorescence microscope with digital camera (*see* **Note 8**).
2. Select an exposure time for all slides which does not show saturated pixels and adjust the acquisition settings to image each filters on every slide at the same exposure time (*see* **Note 9**).
3. Image a minimum of five frames per sample/slide and save the images in the standard format as TIF files.

3.4 Heat Map Analysis of Staining Using Image J Software

1. Open the TIF file in Image J and you will see a black and white image with three slices across the slide bar—each image corresponding to a separate fluorophore.

2. To color the image slices with the corresponding color go to: Image > Color > Make Composite.
   (a) Select Color in the dropdown bar.
   (b) Tick Chanel 1 > select more ≫ and choose the appropriate color (i.e., blue).
3. Repeat for Chanel 2 and 3.
4. Split the image into its separate colors channels—go to Image > Color > Split channels.
5. Convert each image separately to an 8-bit image by first clicking the image then go to Image > Type > 8-bit.
6. To analyze the average HDAC expression within the nucleus only, adjust the threshold according to the nuclei by first selecting the nuclei image channel and then go to Image > adjust > Threshold.
7. Tick dark background and adjust the threshold using the horizontal toggle bar so that the background is minimized and only the strong nuclear staining is evident. Clicks apply.
8. Generate a binary image of this threshold blue channel by going to Process > Math > Divide. Divide by 255 to generate a binary image corresponding to a distribution of 0 and 1 values. This image is the mask.
9. Go to Process > Image Calculator and multiply the channel of interest (red—HDAC image) by the blue nuclear channel.
10. Ensure create new window is selected. The final product is an image that encompasses the staining of interest restricted specifically to within the nuclear regions.
11. To measure mean fluorescence intensity in the nuclear regions only, go to Image > adjust > Threshold. Adjust threshold to 1, this selects only the nuclear region.
12. Go to Set measurements > tick (1) Area; (2) Mean gray value; (3) Integrated density; (4) Limit to threshold; (5) Display label.
13. Go to Analyse > Measure and the results will be given in a new results window.
14. To measure the average HDAC expression within the cytoplasm only. Repeat **steps 7–9** according to the cytoplasm (green) channel.
15. To make a mask of the cytoplasm without the nuclear staining subtract the nuclear mask from the cytoplasm mask—go to Process > Image Calculator and subtract the blue nuclei mask from the green cytoplasm mask.
16. To create regions on the red HDAC channel that encompasses the staining of interest restricted specifically to within the cytoplasm regions. Go to > Image Calculator and multiply the channel of interest by the new mask.

17. Repeat **steps 12–14** to measure the mean fluorescence intensity in the cytoplasm regions only.
18. Record data in Microsoft Excel spreadsheets.
19. Data obtained from the color controls should be subtracted from the test results.

3.5 Interpretation

There are 11 metal-dependent enzymes which are known as the classical HDAC family [13]. The classical HDACs represent class I, II, and IV of the HDAC superfamily and although they share a highly conserved catalytic deacetylase domain they differ in their N- and C- terminal tails, cellular localization and biological roles. Class I involves HDAC1, 2, 3, and 8 and shares homology to the yeast reduced dependency 3 (RDP3). These HDACs are primarily found in the nucleus and play a primary role in proliferation and cell survival. Class II HDACs share homology to the yeast histone deacetylase-1 (Hda1) and includes the class IIa HDAC4, 5, 7, and 9 and class IIb HDAC6 and 10. Class II HDACs are primarily found in the cytoplasm and can act on histone and non-histone substracts. HDAC6 is unique as it contains an ubiquitin binding site and is a specific and exclusive deacetylase of numerous non-histone substrates including: cell motility mediators, α-tubulin and contractin; chaperones, HSP90 and HSP70; DNA repair proteins, Ku70 and signaling mediators such as β-catenin. HDAC11 shares homology to both class RDP3 and Hda1 and represent class IV and found in both the nucleus and cytoplasm [14, 15].

In this protocol, we investigate the expression and localization of the HDACs by co-localizing the HDAC staining (red) with either nuclear staining (blue) and cytoplasmic staining (green). In the example shown, we found HDAC2 to be expressed mainly in the nucleus and HDAC9 to be expressed primarily in the cytoplasm as previously described [14]. Following post-unilateral ureteric obstruction in the kidney, the expression of HDAC2 elevated significantly over 10 days, where opposingly, HDAC9 was found to decrease. Previous studies have shown that using a specific class I inhibitor, MS-275, all fibrotic responses were suppressed in a UUO murine model after 7 days [11]. Therefore, it is necessary to understand the role of HDACs and their increased or decreased activities in not only kidney fibrosis but other diseases states. Using class specific or HDAC specific inhibitors may be more beneficial in targeting these responses (Fig. 2).

4 Notes

1. We routinely use slides coated with 3-Aminopropyltriethoxysilane (APES). This bonds tissue sections to the slides and prevents the loss of tissue during the staining procedure.

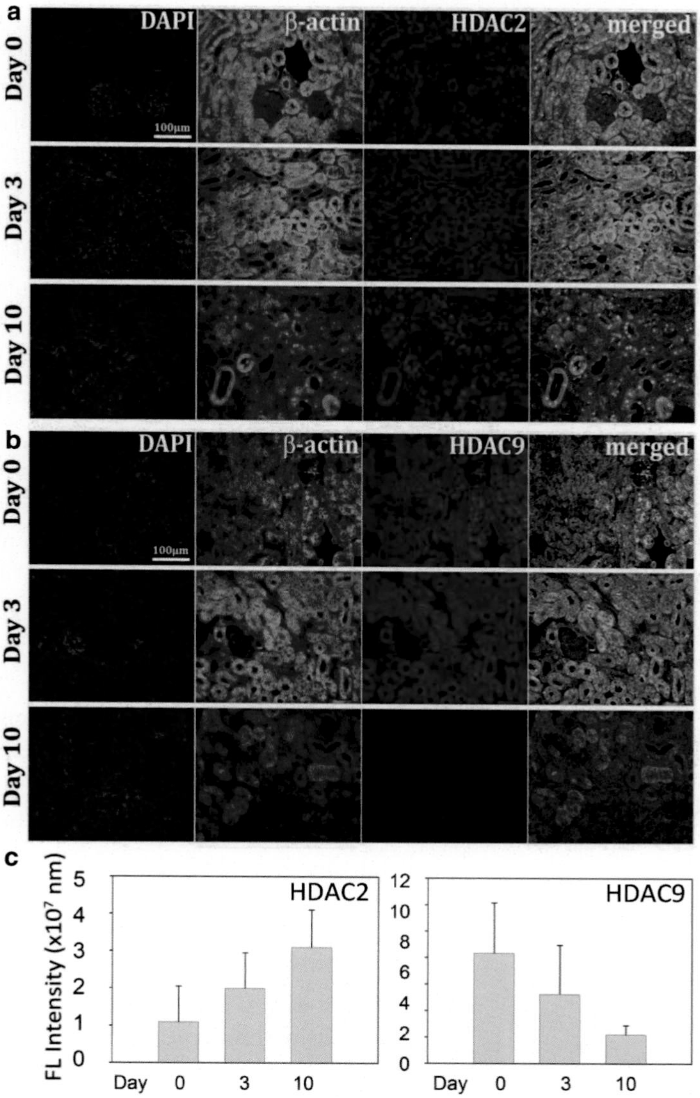

Fig. 2 Immunofluorescence post-unilateral ureteric obstruction (UUO) showing an increase in nuclear HDAC2 (**a**) and decrease in cytoplasmic HDAC9 (**b**) staining post-UUO. Results are quantitatively represented as a fluorescence (FL) intensity (mean ± SD) (**c**)

2. Sodium citrate buffer may be used as a substitute to Target Retrieval buffer. Sodium Citrate Buffer (10 mM sodium citrate, 0.05 % Tween 20, pH 6.0): Add 2.94 g tri-sodium citrate (dehydrate) to 1000 mL distilled water. Adjust pH to 6.0 with 1 N HCl and then add 0.5 mL Tween 20 and mix well. This solution can be stored at 4 °C.

3. Following rehydration of tissue, prevent drying out the tissue at any point in time throughout the staining procedure and use a dark humidified chamber for incubations throughout.
4. When you have more slides than can fit into one Coplin jar, microwave each jar separately for 1 min, then place them all into the microwave for the second heating, and cooling period. Ensures temperature stability when trying to heat more than one, at a time, on the high setting.
5. It is recommended to set up color control slides to test for nonspecificity of the primary and secondary antibodies. Additional slides should be set up as follows:
   (a) no primary or secondary antibodies—this test for autofluorescence of the tissue.
   (b) no primary antibodies, with both secondary antibodies—this detects nonspecific binding of secondary antibodies.
   (c) both primary antibodies with no secondary antibodies—this detects any autofluorescence of the primary antibodies.
   (d) using rabbit primary (red) antibody with the secondary anti-goat Alexa 488 (green)—this detects cross-reactivity of the fluorophores.
   (e) using the goat primary (green) antibody with the secondary anti-rabbit Alexa 546 (red)—this detects cross-reactivity of the fluorophores.
6. Do not dry the cells completely before adding Antifade™.
7. Seal the coverslip with two coats of nail polish and incubate the slides at 4 °C overnight before imaging. Be sure to remove any air bubbles in the mounting medium before sealing with nail polish.
8. Select a 40× Uplan FL objective. Depending on the tissue size and cells within in you can image using 20× objective or above. It is not recommended to image the cells using less than a 20× objective.
9. It is recommended to image any comparative images on the same day to prevent any changes to the fluorescence intensity over time on the microscope and the slides.

Acknowledgements

TDH, CSS, and TK are supported by National Health and Medical Research Council of Australia (NHMRC) Project Grant (APP1078694). TK is an Australian Research Council Future Fellow and CCS a NHMRC Senior Research Fellow (APP1041766).

References

1. Chmielewski M, Lindholm B, Stenvinkel P, Ekstrom JT (2011) The role of epigenetics in kidney diseases. Prilozi 32:45–54
2. Pang M, Kothapally J, Mao H, Tolbert E, Ponnusamy M, Chin YE, Zhuang S (2009) Inhibition of histone deacetylase activity attenuates renal fibroblast activation and interstitial fibrosis in obstructive nephropathy. Am J Physiol Renal Physiol 297:F996–F1005
3. Bechtel W, McGoohan S, Zeisberg EM et al (2010) Methylation determines fibroblast activation and fibrogenesis in the kidney. Nat Med 16:544–550
4. Kantharidis P, Wang B, Carew RM, Lan HY (2011) Diabetes complications: the microRNA perspective. Diabetes 60:1832–1837
5. Mann J, Mann DA (2013) Epigenetic regulation of wound healing and fibrosis. Curr Opin Rheumatol 25:101–107
6. Perugorria MJ, Wilson CL, Zeybel M, Walsh M, Amin S, Robinson S et al (2012) Histone methyltransferase ASH1 orchestrates fibrogenic gene transcription during myofibroblast transdifferentiation. Hepatology 56:1129–1139
7. Reddy MA, Natarajan R (2011) Epigenetics in diabetic kidney disease. J Am Soc Nephrol 22:2182–2185
8. Ververis K, Hiong A, Karagiannis TC, Licciardi PV (2013) Histone deacetylase inhibitors (HDACIs): multitargeted anticancer agents. Biologics 7:47–60
9. Royce SG, Karagiannis TC (2014) Histone deacetylases and their inhibitors: new implications for asthma and chronic respiratory conditions. Curr Opin Allergy Clin Immunol 14:44–48
10. Marumo T, Hishikawa K, Yoshikawa M, Hirahashi J, Kawachi S, Fujita T (2010) Histone deacetylase modulates the proinflammatory and -fibrotic changes in tubulointerstitial injury. Am J Physiol Renal Physiol 298:F133–F141
11. Liu N, He S, Ma L et al (2013) Blocking the class I histone deacetylase ameliorates renal fibrosis and inhibits renal fibroblast activation via modulating TGF-beta and EGFR signaling. PLoS One 8:e54001
12. Fiji is Just Image J. http://fiji.sc/wiki/index.php/Fiji. Accessed 26 Feb 2015
13. Gregoretti IV, Lee YM, Goodson HV (2004) Molecular evolution of the histone deacetylase family: functional implications of phylogenetic analysis. J Mol Biol 338(1):17–31
14. Marks PA, Dokmanovic M (2005) Histone deacetylase inhibitors: discovery and development as anticancer agents. Expert Opin Investig Drugs 14(12):1497–1511
15. Ververis K, Karagiannis TC (2011) Potential non-oncological applications of histone deacetylase inhibitors. Am J Transl Res 3(5):454–467

Index

Tim D. Hewitson et al. (eds.), *Kidney Research: Experimental Protocols*, Methods in Molecular Biology, vol. 1397, DOI 10.1007/978-1-4939-3353-2,